Quantum Optics of Confined Systems

NATO ASI Series

Advanced Science Institutes Series

A Series presenting the results of activities sponsored by the NATO Science Committee, which aims at the dissemination of advanced scientific and technological knowledge, with a view to strengthening links between scientific communities.

The Series is published by an international board of publishers in conjunction with the NATO Scientific Affairs Division

A Life Sciences	Plenum Publishing Corporation
B Physics	London and New York
C Mathematical and Physical Sciences	Kluwer Academic Publishers
D Behavioural and Social Sciences	Dordrecht, Boston and London
E Applied Sciences	
F Computer and Systems Sciences	Springer-Verlag
G Ecological Sciences	Berlin, Heidelberg, New York, London,
H Cell Biology	Paris and Tokyo
I Global Environmental Change	

PARTNERSHIP SUB-SERIES

1. **Disarmament Technologies**	Kluwer Academic Publishers
2. **Environment**	Springer-Verlag / Kluwer Academic Publishers
3. **High Technology**	Kluwer Academic Publishers
4. **Science and Technology Policy**	Kluwer Academic Publishers
5. **Computer Networking**	Kluwer Academic Publishers

The Partnership Sub-Series incorporates activities undertaken in collaboration with NATO's Cooperation Partners, the countries of the CIS and Central and Eastern Europe, in Priority Areas of concern to those countries.

NATO-PCO-DATA BASE

The electronic index to the NATO ASI Series provides full bibliographical references (with keywords and/or abstracts) to more than 50000 contributions from international scientists published in all sections of the NATO ASI Series.
Access to the NATO-PCO-DATA BASE is possible in two ways:

– via online FILE 128 (NATO-PCO-DATA BASE) hosted by ESRIN,
Via Galileo Galilei, I-00044 Frascati, Italy.

– via CD-ROM "NATO-PCO-DATA BASE" with user-friendly retrieval software in English, French and German (© WTV GmbH and DATAWARE Technologies Inc. 1989).

The CD-ROM can be ordered through any member of the Board of Publishers or through NATO-PCO, Overijse, Belgium.

Quantum Optics of Confined Systems

edited by

Martial Ducloy

and

Daniel Bloch

Laboratoire de Physique des Lasers,
Institut Galilée,
Université Paris-Nord,
Villetaneuse, France

Kluwer Academic Publishers

Dordrecht / Boston / London

Published in cooperation with NATO Scientific Affairs Division

Proceedings of the NATO Advanced Study Institute on
Quantum Optics of Confined Systems
Les Houches, France
May 23–June 2, 1995

A C.I.P. Catalogue record for this book is available from the Library of Congress.

ISBN 0-7923-3974-6

Published by Kluwer Academic Publishers,
P.O. Box 17, 3300 AA Dordrecht, The Netherlands.

Kluwer Academic Publishers incorporates the publishing programmes of
D. Reidel, Martinus Nijhoff, Dr W. Junk and MTP Press.

Sold and distributed in the U.S.A. and Canada
by Kluwer Academic Publishers,
101 Philip Drive, Norwell, MA 02061, U.S.A.

In all other countries, sold and distributed
by Kluwer Academic Publishers Group,
P.O. Box 322, 3300 AH Dordrecht, The Netherlands.

Printed on acid-free paper

Printed in the Netherlands

This book contains the proceedings of a NATO Advanced Study Institute held within the programme of activities of the NATO Special Programme on Nanoscale Science as part of the activities of the NATO Science Committee.

Other books previously published as a result of the activities of the Special Programme are:

NASTASI, M., PARKING, D.M. and GLEITER, H. (eds.), *Mechanical Properties and Deformation Behavior of Materials Having Ultra-Fine Microstructures.* (E233) 1993 ISBN 0-7923-2195-2

VU THIEN BINH, GARCIA, N. and DRANSFELD, K. (eds.), *Nanosources and Manipulation of Atoms under High Fields and Temperatures: Applications.* (E235) 1993 ISBN 0-7923-2266-5

LEBURTON, J.-P., PASCUAL, J. and SOTOMAYOR TORRES, C. (eds.), *Phonons in Semiconductor Nanostructures.* (E236) 1993 ISBN 0-7923-2277-0

AVOURIS, P. (ed.), *Atomic and Nanometer-Scale Modification of Materials: Fundamentals and Applications.* (E239) 1993 ISBN 0-7923-2334-3

BLÖCHL, P. E., JOACHIM, C. and FISHER, A. J. (eds.), *Computations for the Nano-Scale.* (E240) 1993 ISBN 0-7923-2360-2

POHL, D. W. and COURJON, D. (eds.), *Near Field Optics.* (E242) 1993 ISBN 0-7923-2394-7

SALEMINK, H. W. M. and PASHLEY, M. D. (eds.), *Semiconductor Interfaces at the Sub-Nanometer Scale.* (E243) 1993 ISBN 0-7923-2397-1

BENSAHEL, D. C., CANHAM, L. T. and OSSICINI, S. (eds.), *Optical Properties of Low Dimensional Silicon Structures.* (E244) 1993 ISBN 0-7923-2446-3

HERNANDO, A. (ed.), *Nanomagnetism* (E247) 1993. ISBN 0-7923-2485-4

LOCKWOOD, D.J. and PINCZUK, A. (eds.), *Optical Phenomena in Semiconductor Structures of Reduced Dimensions* (E248) 1993. ISBN 0-7923-2512-5

GENTILI, M., GIOVANNELLA, C. and SELCI, S. (eds.), *Nanolithography: A Borderland Between STM, EB, IB, and X-Ray Lithographies* (E264) 1994. ISBN 0-7923-2794-2

GÜNTHERODT, H.-J., ANSELMETTI, D. and MEYER, E. (eds.), *Forces in Scanning Probe Methods* (E286) 1995. ISBN 0-7923-3406-X

GEWIRTH, A.A. and SIEGENTHALER, H. (eds.), *Nanoscale Probes of the Solid/Liquid Interface* (E288) 1995. ISBN 0-7923-3454-X

CERDEIRA, H.A., KRAMER, B. and SCHÖN, G. (eds.), *Quantum Dynamics of Submicron Structures* (E291) 1995. ISBN 0-7923-3469-8

WELLAND, M.E. and GIMZEWSKI, J.K. (eds.), *Ultimate Limits of Fabrication and Measurement* (E292) 1995. ISBN 0-7923-3504-X

EBERL, K., PETROFF, P.M. and DEMEESTER, P. (eds.), *Low Dimensional Structures Prepared by Epitaxial Growth or Regrowth on Patterned Substrates* (E298) 1995. ISBN 0-7923-3679-8

MARTI, O. and MÖLLER, R. (eds.), *Photons and Local Probes* (E300) 1995. ISBN 0-7923-3709-3

GUNTER, L. and BARBERA, B. (eds.), *Quantum Tunneling of Magnetization - QTM '94* (E301) 1995. ISBN 0-7923-3775-1

MARTIN, T.P. (ed.), *Large Clusters of Atoms and Molecules* (E313) 1996. ISBN 0-7923-3937-1

PERSSON, B.N.J. and TOSATTI, E. (eds.), *Physics of Sliding Friction* (E314) 1996. ISBN 0-7923-3935-5

CONTENTS

FOREWORD

In the last few years it was seen the emergence of various new quantum phenomena specifically related with electronic or optical confinement on a sub-wavelength-size. Fast developments simultaneously occurred in the field of Atomic Physics, notably through various regimes of Cavity Quantum Electrodynamics, and in Solid State Physics, with advances in Quantum Well technology and Nanooptoelectronics. Simultaneously, breakthroughs in Near-Field-Optics provided new tools which should be widely applicable to these domains. However, the key concepts used to describe these new and partly related effects are often very different and specific of the Community involved in a given development.

It has been the ambition of the Meeting held at "Centre de Physique des Houches" to give an opportunity to specialists of different Communities to deepen their understanding of advances more or less intimately related to their own field, while presenting the basic concepts of these different fields through pedagogical Introductions. The audience comprised advanced students, postdocs and senior scientists, with a balanced participation of Atomic Physicists and Solid State Physicists, and had a truly international character. The considerable efforts of the lecturers, in order to present exciting new results in a language accessible to the whole audience, were the essential ingredients to achieve successfully what was the main goal of this School.

In these proceedings, which were amended or written by the authors after confrontation with the mixed audience of the School, it is expected that the reader will find the same flavor of pedagogical effort than the one present during the whole School. The ordering of the chapters in the present book is somewhat arbitrary and cannot retrace the deliberate mixture expected from a School. The present contributions are restricted to the main lectures delivered at Les Houches and may be grouped in three parts. One is related to Cavity Quantum Electrodynamics with a general Introduction by J.M. Raimond, presentation of squeezing (J. Rarity), and optical applications on microspheres and photonic bandgaps (R.K. Chang and J.W. Haus). Another one deals with the Optical Properties of Solids, as introduced by A. Quattropani and examplified by C. Delalande, while a large section devoted to the Semiconductor Laser Physics is given by Y. Yamamoto and completed in more technological aspects by K. Ebeling. At last, near-field microscopy is presented (U. Fischer) in view of spectroscopic applications (V. Letokhov), and methods of spectroscopy in a confined space are presented by G. Nienhuis and by L. Moi in the case of the vicinity to a surface, while A. Weis and S. Kanorsky tackles the problem of confinement in a nanocavity of dense He environment.

Finally, it is our pleasure to acknowledge the various supports which made possible the organization of the School, notably the Nanoscale Program Commitee of NATO, the European Commission - DGXII - and the C.N.R.S. We would like to thank also M. Leduc because she suggested the organization of the School, M. Voos for his constant support, and naturally all the members of the Scientific Commitee who contributed to the elaboration of the program. Technical and secretarial supports were kindly provided by A. Wilk and E. Lakrout in Villetaneuse (Laboratoire de Physique des Lasers) and G. Chioso and B. Rousset at Les Houches.

M. DUCLOY   D. BLOCH

PRÉFACE

Ces dernières années ont été marquées par l'émergence de divers effets nouveaux liés au confinement électronique ou optique sur des dimensions plus petites que la longueur d'onde. Des développements rapides sont apparus tant en Physique Atomique qu'en Physique des Solides, notamment par l'observation de régimes variés prédits par l'Electrodynamique Quantique en cavité ou en relation avec des avancées dans la technologie des puits quantiques et en nanooptoélectronique. Parallèlement, l'éclosion de la microscopie en champ proche a fourni des outils nouveaux qui devraient trouver de larges applications dans l'étude des systèmes confinés. Cependant, les concepts de base utilisés pour décrire ces différents effets nouveaux restent très divers et gardent l'empreinte des champs disciplinaires qui les ont forgés spécifiquement.

L'ambition de l'Ecole qui s'est tenue au centre de Physique des Houches était de fournir l'occasion à des spécialistes d'origines diverses d'approfondir leur compréhension d'avancées liées de façon plus ou moins intime à leur propres centres d'intérêts. Il s'agissait donc pour l'essentiel que soient présentés, dans des cours à vocation fortement pédagogique, les concepts de base des différentes disciplines impliquées . Le public, d'une très large représentativité internationale, incluait aussi bien doctorants ou postdocs que chercheurs confirmés et respectait l'équilibre entre Physiciens Atomistes et Physiciens du Solide. Les conférenciers enseignants ont su faire des efforts considérables pour présenter d'une façon accessible à tout le public les résultats les plus saillants dans leurs domaines et ils ont donc été les principaux artisans de la réussite de cette rencontre.

Nous espérons que dans les Actes de cette Ecole, dont les textes ont été écrits ou remaniés par leurs auteurs après les réactions du public, le lecteur saura retrouver le même souci pédagogique que celui qui anima toute l'Ecole. L'ordonnancement des chapitres est certes ici quelque peu arbitraire et ne peut retracer l'imbriquement plus étroit que permet la tenue d'une Ecole. Les textes présentés ici, qui n'incluent pas des séminaires plus spécialisés ayant également animé l'École, peuvent être regroupés en trois parties. L'aspect Electrodynamique Quantique en Cavité est présenté dans une large introduction par

J.M. Raimond, complétée par des considérations sur les états comprimés (J.Rarity) et illustré par des applications aux microsphères et aux bandes photoniques (R.K. Chang et J.W. Haus respectivement). Les propriétés optiques des solides sont passées en revue dans la contribution de A. Quattropani et approfondies dans le cas des excitons par C. Delalande, cependant que ces concepts éclairent la vaste contribution de Y. Yamamoto sur la Physique des Lasers à semiconducteurs, dont les progrès technologiques sont présentés par K. Ebeling. En dernier lieu, les concepts de la microscopie en champ proche sont introduits par U. Fischer en vue d'applications spectroscopiques (V. Letokhov), cependant que les diverses méthodes de spectroscopie en espace confiné sont présentées par G. Nienhuis et L. Moi, et le cas spécifique d'un environnement en nanocavité d'hélium dense est abordé avec la contribution de S. Kanorsky et A. Weis.

Enfin, nous avons le plaisir d'exprimer notre reconnaissance aux différents organismes qui ont soutenu l'organisation de cette Ecole, notamment le Comité du programme "Nanoscale" de l'O.T.A.N., la Commission européenne (DG XII), la Formation Permanente du C.N.R.S, le Ministère des Affaires Etrangères et l'Université Paris-Nord. Nous tenons aussi à remercier M. Leduc pour nous avoir incités à l'organisation de cette Ecole, M. Voos pour son encouragement constant, et naturellement tous les membres du Comité Scientifique qui ont participé à l'élaboration du programme. Enfin, la logistique technique et le secrétariat a largement bénéficié de la participation de A. Wilk et E. Lakrout au Laboratoire de Physique des Lasers et de G. Chioso et B. Rousset au Centre des Houches.

M. DUCLOY   D. BLOCH

SCIENTIFIC COMMITEE

M. DUCLOY (Université Paris-Nord)

T.W. HÄNSCH (Max-Planck-Institut and Munich University)

S. HAROCHE (E.N.S Paris and Institut Universitaire de France)

V. LETOKHOV (Institute of Spectroscopy, Moscow
and Université Paris-Nord)

Y.Z. WANG (Institute of Fine Mechanics and Optics, Shanghaï)

C. WEISBUCH (DRET, Paris)

Y. YAMAMOTO (Stanford University)

LIST OF CONTRIBUTORS

R.K. CHANG, Department of Applied Physics and Center for Laser Diagnostics,
Yale University, New Haven, Connecticut 06520-8284, USA

G. CHEN, Department of Applied Physics and Center for Laser Diagnostics,
Yale University, New Haven, Connecticut 06520-8284, USA

C. DELALANDE, Laboratoire de Physique de la Matière Condensée
de l'Ecole Normale Supérieure, 24 rue Lhomond, 75005 Paris, France

K.J. EBELING, Department of Optoelectronics, University of Ulm,
89069 Ulm, Germany

U.C. FISCHER, Physikalisches Institut, Westfälische Wilhelms-Universität,
Wilhelm Klemmstr. 10, 48149 Münster, Germany

H. FUCHS, Physikalisches Institut, Westfälische Wilhelms-Universität,
Wilhelm Klemmstr. 10, 48149 Münster, Germany

J.W. HAUS, Rensselaer Polytechnic Institute, Troy, NY 12180-3590, USA

S.I. KANORSKY, Max-Planck-Institut für Quantenoptik, Hans Kopfermann Str. 1,
85748 Garching, Germany

J. KOGLIN, Physikalisches Institut, Westfälische Wilhelms-Universität,
Wilhelm Klemmstr. 10, 48149 Münster, Germany

V.S. LETOKHOV, Institute of Spectroscopy, Russian Academy of Sciences, Troitzk,
Moscow region, 142092, Russia and
Université Paris-Nord, Institut Galilée, 93430 Villetaneuse, France

E. MARIOTTI, INFM-Dipartimento di Fisica, Universita di Siena,
via Banchi di Sotto 55, 53100 Siena-Italy

Md. M. MAZUMDER, Department of Applied Physics and Center for
Laser Diagnostics, Yale University, New Haven, Connecticut 06520-8284, USA

L. MOI, INFM-Dipartimento di Fisica, Universita di Siena,
via Banchi di Sotto 55, 53100 Siena Italy

A. NABER, Physikalisches Institut, Westfälische Wilhelms-Universität,
Wilhelm Klemmstr. 10, 48149 Münster, Germany

G. NIENHUIS, Huygens Laboratorium, Rijksuniversiteit Leiden
Postbus 9504, 2300 RA Leiden, The Netherlands

A. QUATTROPANI, Institut de Physique Théorique, Ecole Polytechnique Fédérale,
1015 Lausanne, Switzerland

J.M. RAIMOND, Laboratoire Kastler Brossel, Ecole Normale Supérieure,
Département de Physique, 24 rue Lhomond 75005 Paris, France

J.G. RARITY, Defence Research Agency, St Andrews Rd, Malvern,
Worcestershire WR14 3PS, United Kingdom

A. RASCHEWSKI, Physikalisches Institut, Westfälische Wilhelms-Universität,
Wilhelm Klemmstr.10, 48149 Münster, Germany

V. SAVONA, Institut de Physique Théorique, Ecole Polytechnique Fédérale,
1015 Lausanne, Switzerland

P. SCHWENDIMANN, Defense Procurement and Technology Agency,
System Analysis Division, 3003 Bern, Switzerland

P.R. TAPSTER, Defence Research Agency, St Andrews Rd, Malvern,
Worcestershire, WR14 3PS United Kingdom

F. TASSONE, Institut de Physique Théorique, Ecole Polytechnique Fédérale,
1015 Lausanne, Switzerland

R. TIEMANN, Physikalisches Institut, Westfälische Wilhelms-Universität,
Wilhelm Klemmstr.10, 48149 Münster, Germany

A. WEIS, Max-Planck-Institut für Quantenoptik, Hans Kopfermann Str. 1,
85748 Garching, Germany

Y. YAMAMOTO, ERATO Quantum Fluctuation Project, E.L. Ginzton Laboratory,
Stanford University, Stanford CA 94305, U.S.A.

BASICS OF CAVITY QUANTUM ELECTRODYNAMICS

J.M. RAIMOND
Laboratoire Kastler Brossel[†]
Ecole Normale Supérieure, Département de Physique
24 Rue Lhomond 75005 Paris France

The mode structure of a confined electromagnetic field is drastically modified by the boundary conditions. The radiative properties of an "atom" in confined space can thus be very different from the free space ones. For instance, spontaneous emission rates can be altered, the energy levels may be shifted. In extreme situations, the spontaneous emission may even become a reversible, oscillatory process. The study of these effects constitutes the growing field of Cavity Quantum Electrodynamics (CQED).

The history of CQED opened on a short note by Purcell [1] mentioning the possible modification of spontaneous lifetime for a magnetic spin transition coupled to a resonant circuit. Intensive theoretical work on the quantum electrodynamics in confined space soon followed [2]. Most of the effects predicted then dealt with what is called now the "weak coupling regime" of CQED. In this regime, the radiative properties of the atoms are quantitatively modified, but the qualitative behaviour remains the same as in free space. The first experiments by Drexhage [3] on the modification of spontaneous emission for dye molecules near mirrors were also in this weak coupling regime. The modern history of CQED began when experimental techniques made it possible to prepare and study Rydberg states. These levels, with very high principal quantum numbers, combine a very strong coupling to the field in the millimeter-wave domain and a relatively long lifetime. Coupled to very high quality superconducting cavities, they made it possible to observe many CQED effects, such as large modifications of spontaneous emission rates (either enhancement [4] or inhibition [5]), the associated energy levels shifts [6], and, in the strong coupling regime, transient masers [7, 8], one-atom or micromasers [9, 10, 11], Rabi precession in quantized fields [12], "vacuum Rabi splitting" [13] and finally the manifestations of the dispersive atom-field interaction [14]. The field of CQED with

[†]Laboratoire de l'Ecole Normale Supérieure et de l'Université Pierre et Marie Curie, Associé au CNRS (URA 18)

M. Ducloy and D. Bloch (eds.), Quantum Optics of Confined Systems, 1-46.
© *1996 Kluwer Academic Publishers. Printed in the Netherlands.*

Rydberg atoms is now evolving towards the realization of very fundamental tests of our understanding of quantum theory. The very simple system of a single atom radiating in an almost undamped field mode can be directly analyzed in terms of the basic quantum postulates. Any experiment performed with this system is indeed a study of the quantum measurement theory. Experiments on non-local quantum entangled states, reminiscent of the EPR paradox [15], on quantum superpositions of states differing through their macroscopic attributes (sometimes called "Schrödinger cats" [16]) are planned or in progress [17, 18, 19].

More recently, the field of CQED in atomic physics expanded to the optical domain. Spontaneous emission alteration for atoms or dye molecules in micron-sized structures [20, 21, 22, 23], cavity induced frequency shifts [24], vacuum Rabi splittings for atoms in very high Q optical cavities [25] have been observed. In this domain, the promising development of low volume optical resonators made of small dielectric spheres is extremely promising [26, 27, 28]. At the same time, the concepts of CQED emerged in the field of semiconductor physics, with the first experiments on integrated microcavities. Most "classical" CQED effects, such as spontaneous emission alteration [29, 30, 31], vacuum Rabi splitting [32], have been observed in this context. This domain expanded in a quite spectacular way, this school being a clear illustration of the convergence between the quantum optics and solid state physics communities. Of course, many applications are expected from this field ranging from thresholdless lasers, to lasers emitting non-classical light [33] or even photon number states (see the lectures notes by Y. Yamamoto in this book).

The purpose of these lecture notes is twofold. First, they are a very basic introduction to the main concepts of cavity QED. The great simplicity of the system under study makes cavity QED an excellent primer to quantum optics. These notes will focus mainly on the strong atom-single mode cavity coupling regime (the weak coupling one being retrieved as a limiting case). Second, they show how the cavity QED situations can be used to perform very subtle tests of the quantum measurement theory.

In the first section, the basic notations for the atom-cavity system, and the atom or cavity mode relaxation, are introduced. The dressed states picture, which is extremely useful to analyze most situations will also be discussed. In the second section, the spontaneous emission inside a cavity, both for the case of a single atom and for the one of collective emission, is analyzed, as well as the micromaser. In the third section, a brief introduction to present experiments, performed in the dispersive – or non-resonant – atom-field coupling regime and aiming at tests of the quantum measurement theory is given. Finally, the last section will be devoted to the description of very high quality optical resonators made of dielectric microspheres, and of possible applications.

The limited size of this paper allows only a very brief overview of each subject. The interested reader should refer, for more detailed information on the standard techniques of quantum optics, to standard textbooks [34, 35]. For CQED itself, a number of review papers provide a very detailed overview of the field and extensive reference to more specialized papers [36, 37, 38, 39].

1. The Atom–Field System

1.1. QUANTUM DESCRIPTION OF THE CAVITY MODE

In all the situations considered in this chapter, the atom is nearly resonant with one of the modes of a small-size high-Q cavity. The frequency difference between adjacent modes being much larger than the atom-cavity detuning and mode width, only this mode plays an important role.

A single mode of the field (angular frequency ω) is formally equivalent to an harmonic oscillator [35]. The energy spectrum is made of equally spaced, non degenerate, mutually orthogonal, energy levels $|n\rangle$. The energy of the nth state is

$$E_n = (n + \frac{1}{2})\hbar\omega, \tag{1}$$

with $n \geq 0$. n is the number of elementary excitations in the mode, called photons. $|n\rangle$ is therefore called a photon-number-state, or Fock-state of the radiation field. $|0\rangle$ represents the vacuum field in the cavity.

Fock states are connected to each other by the photon annihilation or creation operators, a and $a^\dagger$ ($a|n\rangle = \sqrt{n}|n-1\rangle$ and $a^\dagger|n\rangle = \sqrt{n+1}|n+1\rangle$). a and $a^\dagger$ obey bosonic commutation rules: $[a, a^\dagger] = 1$. In terms of these operators, the field hamiltonian writes:

$$H_F = \hbar\omega(a^\dagger a + \frac{1}{2}) = \frac{\hbar\omega}{2}(a^\dagger a + aa^\dagger). \tag{2}$$

To describe the atom-field coupling on a dipole transition, the relevant operator is the electric field $\boldsymbol{E}(\boldsymbol{r})$ at the atomic position $\boldsymbol{r}$ in the cavity mode:

$$\boldsymbol{E}(\boldsymbol{r}) = \mathcal{E}_0 \left[a\boldsymbol{f}^*(\boldsymbol{r}) + a^\dagger\boldsymbol{f}(\boldsymbol{r}) \right], \tag{3}$$

where $\mathcal{E}_0$ is a normalization factor. $\boldsymbol{f}(\boldsymbol{r})$ is a complex vector function describing the field polarization and relative amplitude at position $\boldsymbol{r}$ in the cavity mode. It obeys both the wave equations inside the mode and the boundary conditions at the cavity walls. It is normalized so that its maximum value (assumed, for sake of simplicity, to be at $\boldsymbol{r} = 0$) is unity. The normalization $\mathcal{E}_0$ can be obtained by expressing that a single photon field energy is $\hbar\omega$ above the vacuum level. This yields:

$$\mathcal{E}_0 = \sqrt{\frac{\hbar\omega}{2\varepsilon_0 \mathcal{V}}}, \tag{4}$$

where the effective cavity mode volume, $\mathcal{V}$, is defined by:

$$\mathcal{V} = \int |\boldsymbol{f}(\boldsymbol{r})|^2 \, d^3\boldsymbol{r}. \tag{5}$$

Fock states of the radiation field have no classical counterpart. The expectation value of the electric field, $\langle n|\boldsymbol{E}|n\rangle$, is zero which means that the phase of the field is random, while the energy is perfectly defined. This results from the "energy-phase uncertainty relations". Though there is no properly defined phase operator (a state with a perfectly defined phase has an infinite energy), all happens as if the photon number and phase variance were linked by an Heisenberg uncertainty relation $\Delta n \Delta \Phi \geq 1$ (a less qualitative discussion can be undertaken in terms of the Wigner distributions for the field quadrature components, analog of the harmonic oscillator position and momentum).

The fields emitted by classical sources, such as electronic oscillators or lasers, have a well defined phase, and are described by the Glauber coherent states [40], eigenstates of the annihiliation operator:

$$a|\alpha\rangle = \alpha|\alpha\rangle; \quad |\alpha\rangle = \sum_n c_n|n\rangle; \quad c_n = e^{-|\alpha|^2/2} \frac{\alpha^n}{\sqrt{n!}}, \tag{6}$$

where α is an arbitrary complex number (a is non hermitian, and admits therefore complex eigenvalues).

The photon number distribution is a Poisson one, well localized for large $|\alpha|$. The mean photon number is $\overline{n} = |\alpha|^2$ and the relative photon number variance is $\sigma = \Delta n^2/\overline{n} = 1$. The phase is well defined (the electric field average in such a field is non zero), with a variance $\Delta\Phi = 1/\sqrt{\overline{n}}$. These states are minimal uncertainty ones, since the product $\Delta n \Delta\Phi = 1$ has the minimum value authorized by the "intensity-phase uncertainty relations".

A classical current coupled to the mode produces in the initially empty cavity a coherent field whose amplitude α is proportional to the current and to the cavity-source coupling time. When the source acts on a cavity already containing a coherent field, the coherent amplitude fed by the source merely adds up with the previous one. It is therefore possible to add different coherent amplitudes, exactly as in classical electrodynamics. Let us mention finally that, besides a global phase factor on the wavefunction, the time evolution of a free coherent state amplitude is the same as the one of the classical electric field in the cavity. The field amplitude circles in the complex plane at the mode frequency around the origin: $\alpha(t) = \alpha(0)e^{-i\omega t}$.

For most applications, the cavity relaxation cannot be neglected. The more widely used model is a weak coupling of the cavity harmonic oscillator to a bath of harmonic oscillators in thermal equilibrium at temperature T, spanning a frequency range much larger than the cavity mode spectral width. Eliminating the bath of oscillators, one can write a rate equation for the field density matrix ρ_F [36, 38] (as soon as relaxation is taken into account, the state of the system can no longer be described by a wave function). The evolution equation of ρ_F reads:

$$\frac{d\rho_F}{dt} = \frac{1}{i\hbar}[H_F, \rho_F] + \Lambda_F \rho_F. \tag{7}$$

The first term describes the free hamiltonian evolution, the second the relaxation. The action of the Liouvillian operator Λ_F on ρ_F is given by:

$$\begin{aligned}
\Lambda_F \rho_F &= -\frac{\omega}{2Q}[a^\dagger a, \rho_F]_+ + \frac{\omega}{Q}a\rho_F a^\dagger \\
&\quad -\frac{\omega}{2Q}\frac{n_{\mathrm{B}}}{n_{\mathrm{B}}+1}[a^\dagger a, \rho_F]_+ + \frac{\omega}{Q}\frac{n_{\mathrm{B}}}{n_{\mathrm{B}}+1}a\rho_F a^\dagger,
\end{aligned} \tag{8}$$

where $[\ ]_+$ is an anti-commutator. The phenomenological parameter Q is the cavity mode quality factor, $Q/\omega = t_{\mathrm{cav}}$ being the cavity field energy damping time. $n_{\mathrm{B}} = 1/(\exp(\hbar\omega/k_BT)-1)$ is the mean number of blackbody photons per mode at temperature T (k_B is the Boltzmann constant).

From Eq.(8), one deduces easily coupled rate equations for the Fock states occupation probabilities (or photon number distribution) $p(n) = \langle n|\rho_F|n\rangle$:

$$\begin{aligned}
\dot{p}(n) &= \frac{\omega}{Q}[-n_{\mathrm{B}}(n+1)p(n) + n_{\mathrm{B}}np(n-1) \\
&\quad -(n_{\mathrm{B}}+1)np(n) + (n_{\mathrm{B}}+1)(n+1)p(n+1)].
\end{aligned} \tag{9}$$

The first two terms in the rhs. of this equation describe the up-going transitions in the Fock states ladder corresponding to the creation of thermal photons in the cavity. The second line terms correspond to the down-going transitions due to thermally induced and "spontaneous" processes. The steady state for the photon number distribution can be obtained from this equation by a detailed balance argument [36]. This condition yields $p(n) = p(0)\exp(-n\hbar\omega/k_BT)$, where $p(0)$ is a normalization coefficient. This is obviously the Boltzmann distribution for an oscillator in thermal equilibrium at temperature T.

Let us mention finally that, in the $T = 0\mathrm{K}$ case, which applies to most optical experiments, and to low temperature millimeter-wave ones (below

1K), the relaxation of a Glauber coherent state takes a very simple form. A state with an initial amplitude $\alpha(0)$ remains a coherent state, whose amplitude circles around the origin in the complex plane, while being exponentially damped at a rate $\omega/2Q$:

$$\alpha(t) = \alpha(0)e^{-i\omega t}e^{-\omega t/2Q}. \tag{10}$$

1.2. A TWO LEVEL ATOM

A single atomic transition is resonant with the high-Q cavity mode. Assuming that the atomic levels are non degenerate, the atom is a two level system (upper and lower levels $|e\rangle$ and $|g\rangle$ respectively). The transition from e to g has the angular frequency ω_0, close to the cavity mode one: the atom-cavity detuning $\delta = \omega - \omega_0$ is assumed to be small compared to ω_0.

One may describe the atom as a spin $1/2$. The relevant operators are

$$D^3 = \frac{1}{2}[|e\rangle\langle e| - |g\rangle\langle g|]; \quad D^+ = |e\rangle\langle g|; \quad D^- = |g\rangle\langle e|. \tag{11}$$

D^+ and D^- are the atomic raising and lowering operators ($D^+|g\rangle = |e\rangle$ and $D^-|e\rangle = |g\rangle$), which obey, with D^3, angular momentum commutation rules. The atomic hamiltonian reads: $H_A = \hbar\omega_0 D^3$. The electric dipole on the $e \rightarrow g$ transition is: $\boldsymbol{d} = (D^+ + D^-)d\boldsymbol{\epsilon}$, where $\boldsymbol{\epsilon}$ is the unit vector describing the transition polarization, and d the dipole matrix element, which we assume in the following to be real and non zero.

In Rydberg atoms experiments, the atomic system relaxation is negligible, but it should be included in the description of most optical domain experiments. The two level model we are using does not accomodate easily a relaxation to other levels than e and g. In order to keep the algebra simple, we will restrict to internal relaxation between states e and g. This relaxation may be due to non radiative processes, or to spontaneous emission into radiation modes other than the cavity one [25]. This relaxation is described by a rate equation for the atom density matrix ρ_A:

$$\frac{d\rho_A}{dt} = \frac{1}{i\hbar}[H_A, \rho_A] + \Lambda_A\rho_A, \tag{12}$$

where the Liouvillian operator Λ_A is given by:

$$\Lambda_A\rho_A = -\frac{\gamma}{2}[D^+D^-, \rho_A]_+ + \gamma D^-\rho_A D^+. \tag{13}$$

γ is the damping rate of the upper state population. Note the similitude between this equation and Eq. (8) with $n_B = 0$ describing the field relaxation when thermal radiation is neglected. Thermal effects could also be included here, but they are usually negligible in the optical domain.

1.3. ATOM–FIELD COUPLING

In the dipole approximation the atom-field interaction hamiltonian can be written as:

$$H_{AF} = -\boldsymbol{d} \cdot \boldsymbol{E}. \tag{14}$$

In order to make the algebra simple, we assume that the dipole transition polarization and the mode polarization at the atom location are well matched:

$$\boldsymbol{\epsilon} \cdot \boldsymbol{f}(\boldsymbol{r}) = f(\boldsymbol{r}), \tag{15}$$

where f is real. We treat the atom motion in the cavity classically. $\boldsymbol{r}$ is then a number (and not an operator). Finally, we perform the rotating wave approximation. It amounts to neglecting in H_{AF} non resonant terms corresponding for instance to the emission of a photon and a simultaneous transition of the atom from g to e.

With these conditions, the interaction hamiltonian writes:

$$H_{AF} = -\hbar\Omega f(\boldsymbol{r})[aD^+ + a^\dagger D^-]. \tag{16}$$

The atom-field coupling constant is:

$$\Omega = \frac{d\mathcal{E}_0}{\hbar}. \tag{17}$$

We will see in the next sections that it characterizes the rate of the energy exchange between the atom and the field: 2Ω is the one-photon Rabi angular frequency.

1.4. ORDERS OF MAGNITUDE

The most interesting cavity QED effects are observed in the strong coupling regime, with the atom-field coupling Ω much larger than the reciprocals of all other characteristic times: atom-cavity interaction time t_{int}, atomic levels lifetime t_{rad} and cavity field energy damping time t_{cav}. Rydberg atoms transitions correspond to these conditions. Let us consider for instance a transition between two neighbouring rubidium circular Rydberg states [41], with principal quantum numbers $n = 50$ and $n = 51$. The transition frequency is $\omega_0/2\pi = 51.099$ GHz, and the electric dipole matrix element is $d = 1250$ atomic units. These levels decay only through millimeter-wave transitions toward the closest less excited circular state. Their radiative lifetime t_{rad} is thus extremely long, about 30 ms. This lifetime corresponds to the flight of 10 m/s atoms across a 30 cm long apparatus. These orders of magnitude correspond to experiments being presently made or planned [14]. Let us mention also that Rydberg atoms can be detected in a sensitive and selective way. A moderate electric field is enough to ionize these atoms. The resulting electrons can be easily detected. The fast variation of the ionizing electric field with the atom's binding energy makes it possible to detect selectively adjacent states.

A low order mode cavity resonant with this transition has a typical 1 cm size, and an effective volume of the order of 100 mm^3. The field per photon $\mathcal{E}_0$ reaches then 10^{-3} V/m. The one photon Rabi precession angular frequency Ω is typically 10^5 rd/s, much greater than the reciprocal of the atom-cavity interaction time t_{int}, ranging from a few tens of μs to 1 ms. With niobium superconducting cavities used at very low temperatures, the mode quality factor is in the $10^9 - 10^{10}$ range, corresponding to cavity energy damping times t_{cav} of the order of 10 to 100 ms. All the conditions for a very strong coupling regime are therefore fullfilled.

Strong coupling regime can also be achieved in the optical domain. Millimeter-size high finesse Fabry-Perot resonators [42] or whispering gallery modes at the surface of small dielectric spheres [27] offer very high atom-field couplings (Ω reaches values in the 10^8/s range in the latter case), and quite long cavity damping times (in the 1-10 μs range). The latter resonators will be described in more details in the last section of this paper.

1.5. THE DRESSED LEVELS

Instead of describing the atom-field system evolution in the basis of the eigenstates of $H_A + H_F$, it is often simpler to use the so-called "dressed atom basis" [35], corresponding to the eigenstates of the complete system hamiltonian H.

Besides the ground state $|g, 0\rangle$ corresponding to a deexcited atom in an empty cavity, all the eigenstates of $H_A + H_F$ are grouped in equidistant manifolds of nearly degenerate levels: $|e, n\rangle$ and $|g, n + 1\rangle$ ($n \geq 0$). The energy difference between these two levels is $\hbar\delta$ (δ: atom-cavity detuning), while the energy difference between manifolds is $\hbar\omega \gg \hbar\delta$. The atom-field interaction H_{AF} couples $|e, n\rangle$ to $|g, n + 1\rangle$, and does not connect different manifolds. Diagonalizing H_{AF} amounts therefore to separate diagonalizations of 2×2 matrices.

Using Eq.(16), one gets easily the expressions of the eigenstates of H. They are simple linear combinations of $|e, n\rangle$ and $|g, n + 1\rangle$:

$$
\begin{aligned}
|+, n\rangle &= \cos\theta_n |e, n\rangle \quad - \sin\theta_n |g, n + 1\rangle \\
|-, n\rangle &= \sin\theta_n |e, n\rangle \quad + \cos\theta_n |g, n + 1\rangle,
\end{aligned}
\tag{18}
$$

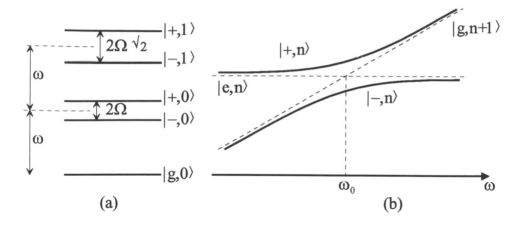

Figure 1. (a) Lowest energy dressed levels at exact atom-cavity resonance and at cavity center. (b) Dressed energy levels as a function of the cavity frequency.

with the mixing angle θ_n defined as:

$$\tan 2\theta_n = -\frac{2\Omega f(r)\sqrt{n+1}}{\delta}, \qquad \text{with } 0 \le \theta_n \le \pi/2. \tag{19}$$

The energies of these states are given by:

$$E_{\pm,n} = (n+1)\hbar\omega \pm \frac{\hbar}{2}\sqrt{4\Omega^2 f^2(r)(n+1) + \delta^2}. \tag{20}$$

Note that $|g,0\rangle$ is also an eigenstate of H with energy $\hbar\delta/2$. The positions of the first dressed states in the resonant case ($\delta = 0$) are sketched on Figure 1(a). The degencracy between $|e,n\rangle$ and $|g,n+1\rangle$ at resonance is removed by the interaction, and the separation between the two dressed states is $2\hbar\Omega f(r)\sqrt{n+1}$. The $\sqrt{n+1}$ factor expresses that the coupling is proportional to the field amplitude, which scales as the square root of the photon number. Note that Eq.(18) simplifies considerably in this case, corresponding to $\theta_n = \pi/4$:

$$|+,n\rangle = \frac{1}{\sqrt{2}}[|e,n\rangle - |g,n+1\rangle]$$

$$|-,n\rangle = \frac{1}{\sqrt{2}}[|e,n\rangle + |g,n+1\rangle] \tag{21}$$

Figure 1(b) represents the variation with the cavity frequency of two dressed levels belonging to a manifold. For large negative detunings ($\delta <$

$0; |\delta| \gg \Omega$), the dressed states nearly coincide with the uncoupled atom-field states: $|+, n\rangle$ merges into $|e, n\rangle$ and $|-, n\rangle$ into $|g, n + 1\rangle$. While the uncoupled states cross at resonance ($\omega = \omega_0$), the dressed states anticross, the width of the anti-crossing being $2\hbar\Omega\sqrt{n + 1}$. The dressed levels merge again into uncoupled states for very large positive detunings: $|+, n\rangle$ tends towards $|g, n + 1\rangle$ and $|-, n\rangle$ towards $|e, n\rangle$.

The dressed levels energies are functions of both the atom-field detuning and the atomic position r inside the cavity. This dependance is associated to a force acting on the atom. This force has very interesting effects, but is usually very small, and we neglect it in the next section. We assume that the atomic motion does not depend upon the cavity field and that the atoms cross the cavity along a straight line at constant velocity.

1.6. THE DRESSED ATOM AT WORK: VACUUM RABI SPLITTING

Many Quantum Electrodynamics effects can be interpreted very easily in the dressed atom basis. A complete review can be found in [35] and is out of the scope of this chapter. We will only discuss the "vacuum" or "normal mode" splitting effect. This effect has now been observed in a wide range of situations, from optical atomic transitions [25] to Rydberg states transitions [13] or excitonic transitions in integrated microcavities [32].

A single, initially deexcited atom, is placed in an empty cavity. The excitation spectrum of the atom-cavity system is recorded by coupling it to a weak probe field, coupled either to the cavity mode (for instance by transmission through the mirrors) or directly to the atom. This weak probe connects the ground state of the system ($|g, 0\rangle$) to the first two excited states ($|+, 0\rangle$ or $|-, 0\rangle$) (see figure 1(a)). The excitation of the system can be detected either on the cavity subsystem, by monitoring, for instance, the probe transmission through the mirrors, or on the atom, by monitoring its state at the exit of the cavity. The first technique is well adapted to optical transitions [25], the second to microwave transitions in Rydberg atoms [13].

The spectrum consists then of two lines, at $\omega_0 \pm \Omega$. The width of these lines is the average of the atom and cavity damping rates ($\omega/Q + \gamma)/2$ [37]. They are therefore well resolved only in a strong coupling regime. Let us stress also that this effect can be observed only with very weak probes. As soon as the population of the first dressed levels becomes important, transitions to higher manifolds occur, which wash out the splitting. One should note also the analogy of this "vacuum Rabi splitting" effect with the Mollow triplet observed in the fluorescence spectrum of an atom driven by an intense laser field [43], resulting from a radiative cascade in high lying multiplicities of the dressed atom.

Though the interpretation of the Rabi splitting effect is simple in the dressed atom picture, it should be noted that this effect is not basically different from the classical "normal mode splitting" observed for two degenerate weakly coupled mechanical harmonic oscillators. Instead of a single resonant frequency, one observes two lines, associated to an antisymmetric and a symmetric modes of oscillation, whose frequencies differ by an amount corresponding to the frequency of the energy exchange between the two oscillators. The only difference for the atomic splitting is the requirement of a very weak probe. While the two oscillators system is linear, with eigenfrequencies independant of the level of excitation, the atomic system is easily saturated, and the level structure should be probed weakly to yield a simple mode splitting effect. This interpretation relates also strongly the vacuum Rabi splitting with the polariton splitting observed in semi-conductors (see the lecture notes by A. Quattropani in this book).

Another classical picture can be given for the vacuum Rabi splitting. The atom constitutes a dispersive medium, albeit a microscopic one, placed in a resonator. Taking into account the atomic index in the cavity transmission, one obtains also a split line. This point of view has been developped in [44]. Note that this interpretation is not fundamentally different from the coupled harmonic oscillators one: the dispersion relations express the presence of an atomic resonance coupled to the cavity one. Note that this point of view also requires a very weak probing of the system, since the linear dispersion holds only as long as the atomic transition is not strongly saturated. This point of view is well adapted to the solid state physics experiments, with a large number of resonant excitons in the cavity (see for instance the lecture notes by C. Delalande).

2. Spontaneous Emission in a Cavity

2.1. SINGLE ATOM SPONTANEOUS EMISSION: THE TWO REGIMES

Let us focus on the simplest, widely encountered, cavity QED problem: the spontaneous emission of a single atom in an empty cavity. We show in this section how the spontaneous emission regime changes when the cavity relaxation is increased, from an oscillatory behavior, closely linked to the vacuum Rabi splitting studied in the preceding section, to an irreversible decay, much closer to the ordinary picture of spontaneous emission.

We neglect, for sake of simplicity, the thermal field and the atomic relaxation (a more detailed treatment can be found in [36]). The atom is initially (time $t = 0$) prepared in its upper state $|e\rangle$ in an empty cavity. The initial atom-field state is thus $|e, 0\rangle = (|+, 0\rangle + |-, 0\rangle)/\sqrt{2}$. Let us consider first an infinite Q cavity. The probability of finding the atom in the initial state $P_e(t)$ at time t is then:

$$P_e(t) = (1 + \cos 2\Omega t)/2. \tag{22}$$

The cavity modifies so strongly the spontaneous emission that it becomes a reversible process oscillating at 2Ω. The cavity stores the photon emitted by the atom long enough for it to be reabsorbed and then reemitted. These oscillations in $P_e(t)$ are the time-domain counterpart of the vacuum Rabi splitting observed in the frequency domain.

If we include, in a more realistic description, cavity relaxation, the atom-field system is described by its density matrix ρ. Only three uncoupled levels can be populated in the process: $|e, 0\rangle$, initial state, $|g, 1\rangle$, reached by atomic emission, and the ground state $|g, 0\rangle$, reached from $|g, 1\rangle$ by field relaxation. In the uncoupled state basis, the system of equations describing the 3×3 density matrix evolution can be reduced to a three equations differential system for the two excited states populations and their coherence [36, 38].

The evolution of $P_e(t)$ is ruled by the eigenfrequencies of this system:

$$-\frac{\omega}{2Q} \quad \text{and} \quad -\frac{\omega}{2Q} \pm \frac{\omega}{2Q}\sqrt{1 - \frac{16\Omega^2 Q^2}{\omega^2}}. \tag{23}$$

The system behavior depends thus drastically upon the sign of the quantity under the root in Eq.(23). One should distinguish two regimes:

2.1.1. The Oscillatory Regime

When $4\Omega > \omega/Q$, two of the three eigenfrequencies in Eq.(23) have an imaginary part. For very high Q values ($\Omega \gg \omega/Q$), these imaginary parts are $\pm 2\Omega$. The evolution of the system is therefore a damped oscillation at frequency 2Ω, with a damping rate $\omega/2Q$. Note that the relaxation rate is $\omega/2Q$ since the energy is only half of the time stored in the field subsystem. In the spectral domain, these damped oscillations are associated to a broadening of the vacuum Rabi splitting lines, their width being in this case also $\omega/2Q$.

2.1.2. The Overdamped, Irreversible Regime

When the cavity damping is large enough ($4\Omega < \omega/Q$), the three eigenvalues in Eq.(23) are real, and the evolution towards the ground state is irreversible. Let us focus here on the strongly damped regime $\Omega \ll \omega/Q$. The evolution of $P_e(t)$ is then almsot exponential with a rate:

$$\Gamma = \frac{4\Omega^2 Q}{\omega}. \tag{24}$$

It is interesting to compare Γ with the free space emission rate Γ_0, given by:

$$\Gamma_0 = \frac{d^2 \omega_0^3}{3\pi\varepsilon_0 \hbar c^3}. \tag{25}$$

One gets:

$$\Gamma = \eta\Gamma_0 \text{ with: } \eta = \frac{3}{4\pi^2} \frac{\lambda^3}{\mathcal{V}} Q. \tag{26}$$

The spontaneous emission rate in the cavity is thus generally greater than the free space one by a factor η, proportional to the cavity quality factor (as long as the overdamped regime conditions are fullfilled), and inversely proportional to the mode effective volume $\mathcal{V}$.

In this approach, we have increased the relaxation rate of a single radiation mode until the oscillatory regime characteristic of a relaxation free system turned into an irreversible process. It is also possible to consider the cavity resonance, much wider than the atomic line in this strong relaxation regime, as a continuum and to get the spontaneous emission rate through a Fermi Golden rule argument. The Fermi golden rule can be written:

$$\Gamma = \frac{2\pi}{\hbar^2} |\langle V \rangle|^2 \varrho(\omega_0), \tag{27}$$

where $\langle V \rangle$ is the atom-field coupling $\hbar\Omega$, and $\varrho(\omega_0)$ the field mode density per unit frequency at the atomic transition. Since we have a Lorentzian resonance spanning the frequency interval ω/Q, the mode density is $2Q/\pi\omega$. One thus retrieves the spontaneous emission rate given by Eq.(24).

Note that, when cavity relaxation becomes very large, nearby modes of the cavity overlap and the single mode approximation for the atom-field coupling fails. Fermi Golden rule arguments can be used to derive the spontaneous emission rate in this case. As this matter is treated in great details in other chapters of this book and in preceding reviews, we will not enter in this topic any further. Let us mention only that the main problem is to identify the cavity modes in order to calculate the mode density and the coupling of the atom to these modes (which have different spatial structures). In many situations, a mere classical calculation of the radiation reaction damping in presence of the boundary conditions, which can be made by the electric images picture, is enough to calculate the enhancement factor η [37, 45, 46].

Observed for the first time on a microwave transition between Rydberg atoms [4], the spontaneous emission enhancement has since then been reported in a wide range of situations [21, 22, 23].

2.2. THE N ATOMS CASE: COLLECTIVE SPONTANEOUS EMISSION AND COLLECTIVE VACUUM RABI SPLITTING

In this section, we give a brief description of a many-atoms system interacting with a single high-Q mode. The collective emission of this atomic sample has been extensively studied in the Rydberg atom context [7, 8, 36, 38]. Though it is no longer relevant to the present Rydberg atom experiments, which deal mainly with single atom systems, it is of importance to describe the semiconductor experiments, which deal always with large ensembles of radiators.

2.2.1. *The Atomic System*

We consider now a set of N identical two-level atoms (levels $|e\rangle$ and $|g\rangle$), located at positions $r_1, \ldots, r_N$ in the cavity mode. Each of these atoms is described by the spin-1/2 operators D_i^3, D_i^+, D_i^-. We assume that the atoms interact only through their coupling to the same radiation field. We neglect all the instantaneous dipole-dipole interactions between the atoms which play a role only for quite high densities [47]. The atomic hamiltonian, as well as the atom-field interaction one, are thus mere sums of individual atomic energies and couplings to the field.

Stated in these terms, the problem is extremely complex: there are 2^N highly degenerate energy levels. The problem becomes much simpler if all atoms are coupled in the same way to the field (for instance if they are all close to a same antinode of the cavity field). This condition is easily realized with Rydberg atoms in cavities, since the wavelength is large. It is relevant also to describe quantum wells emission in resonant structures, since the size of these emitters is much less than the optical wavelength.

The atomic hamiltonian and the atom-field interaction may then be expressed in terms of collective operators defined by;

$$D^3 = \sum_i D_i^3; \quad D^+ = \sum_i D_i^+; \quad D^- = \sum_i D_i^-. \tag{28}$$

The atom-field interaction hamiltonian is:

$$H_{AF} = -\hbar\Omega[aD^+ + a^\dagger D^-], \tag{29}$$

where Ω is defined by Eq.(17). The atomic hamiltonian is simply:

$$H_A = \hbar\omega_0 D^3 \tag{30}$$

The collective operators D^3, D^+ and D^- obey angular momentum commutation rules. This symmetric superposition of N spin 1/2-like systems is a spin $N/2$. The atomic hamiltonian H_A has therefore $N+1$ non degenerate energy levels (Dicke states [48]) $|J, M\rangle$, with $J = N/2$ and $-J \leq M \leq J$.

They are obtained by repeated action of the raising operator D^+ on the ground state $|G\rangle = |J, -J\rangle = |g, g, \ldots, g\rangle$, until the fully inverted state $|J, J\rangle = |e, e, \ldots, e\rangle$ is reached. In all these states, the excitation is symmetrically shared by all atoms. The first excited state $|J, -J + 1\rangle = |E\rangle$, for instance, is obtained by action of D^+ on $|G\rangle$:

$$D^+|G\rangle = \sqrt{N}|E\rangle. \tag{31}$$

Note that this expression is, besides the extra factor $\sqrt{N}$, identical to the one giving the action of the raising operator on a two level atom.

2.2.2. Collective Vacuum Rabi Splitting

Assume that an initially deexcited atomic sample is placed in the empty cavity. A weak probe, coupled either to the atom or to the field, is used to probe the energy levels of the complete system. We consider here, for sake of simplicity, exact atom-field resonance ($\omega = \omega_0$). The problem, as in the single atom case, amounts to determining the first eigenstates of the complete atom–field hamiltonian.

The first uncoupled atom-field state is the ground state $|G, 0\rangle$, describing a fully deexcited atomic sample in an empty cavity. This state is also the ground level in the collective dressed states ladder. Two degenerate states bear a single excitation: $|G, 1\rangle$, describing all the atoms in the ground state in the presence of one photon, and $|E, 0\rangle$, describing a collectively excited atomic system in an empty cavity. These two states are coupled by atom-field interaction. The first two dressed states are therefore the antisymmetric and symmetric superpositions of these uncoupled states, exactly as in the one-atom case:

$$\begin{aligned}
|+, 0\rangle &= \frac{1}{\sqrt{2}} [|E, 0\rangle - |G, 1\rangle] \\
|-, 0\rangle &= \frac{1}{\sqrt{2}} [|E, 0\rangle + |G, 1\rangle].
\end{aligned} \tag{32}$$

The weak excitation spectrum of the system consists in two lines, symmetrical with respect to the atomic frequency. Instead of 2Ω, the splitting is here $2\Omega\sqrt{N}$, due to the $\sqrt{N}$ factor in Eq. (31). There is therefore no fundamental difference between the vacuum Rabi splitting in the one atom case and in the N atoms case. The collective Rabi splitting is however much more easily observed, since it is increased by a factor $\sqrt{N}$.

As in the single atom case, this splitting must be studied with extremely weak probes. The structure of the higher dressed manifolds is quite different from the simple two level structure of the first manifold. As the atomic system can hold more than one elementary excitation, there are three degenerate uncoupled atom-field states in the second manifold :$|G, 2\rangle$, $|E, 1\rangle$

and $|J, -J + 2, 0\rangle$ which describes a doubly excited atomic system with no photons. There are therefore three equally spaced dressed levels in this manifold, 4 levels in the next one, a.s.o. The spectra corresponding to multiple excitations of the system are quite complex.

The collective vacuum Rabi splitting and its linear dependance versus $\sqrt{N}$ have been observed in various experiments, in the microwave domain [13] as well as in the optical one [25], and in the solid state physics context [32].

We should quote here that the symmetrical coupling hypothesis is not essential for describing the vacuum Rabi splitting, as long as there are many atoms in the system. Even with different couplings, there is a single ground state for the atomic system (all atoms deexcited), and a single atomic state with only one excitation, $|E'\rangle$, given by:

$$|E'\rangle = \left[\sum_i (\boldsymbol{\epsilon} \cdot \boldsymbol{f}(\boldsymbol{r}_i)) D_i^+ \right] |G\rangle. \tag{33}$$

There are therefore only two dressed levels in the first manifold, with a spacing

$$2\Omega \sqrt{\sum_i |\boldsymbol{\epsilon} \cdot \boldsymbol{f}(\boldsymbol{r}_i)|^2}. \tag{34}$$

When much more than one atom is involved, this splitting can be written as

$$2\overline{\Omega}\sqrt{N}, \tag{35}$$

where $\overline{\Omega}$ is the rms averaged coupling of the atoms to the field. Besides a numerical scaling factor, all the conclusions are the same as in the symmetric coupling case.

2.2.3. Collective Spontaneous Emission

Let us now turn to the spontaneous emission by this system. All the atoms are initially excited and the cavity empty, with a symmetrical coupling. Even for an undamped cavity, the problem is complex since $N + 1$ energy eigenstates are accessible.

Much simpler results can be obtained in the frame of a semi-classical approximation. For large N values, the spin $N/2$ describing the atomic system is close to a classical vector called the Bloch vector [38]. The z component of this vector describes the population inversion, while the x and y components describe the atomic medium polarization. The length of the vector is constant $(N/2)$, and only the polar angles ϑ (the atomic energy is $N/2 \cos \vartheta$) and φ of the vector evolve.

The equations of motion for ϑ and φ can be deduced from the equations of motion for the field and atomic operators [38]. For exact atom-cavity resonance, φ is constant, and ϑ obeys:

$$\frac{d^2\vartheta}{dt^2} + \frac{\omega}{2Q}\frac{d\vartheta}{dt} - \Omega^2 N \sin\vartheta = 0, \tag{36}$$

the field amplitude being proportional to the time derivative of ϑ.

This equation is identical to the one describing a damped pendulum in a gravitational field. The small amplitude oscillation frequency of this pendulum is $\Omega\sqrt{N}$. We recognize here the collective vacuum Rabi splitting frequency. Depending on the relative magnitudes of the pendulum damping rate and oscillation frequency, we get an irreversible or an oscillatory evolution for the pendulum around its equilibrium position, which corresponds obviously to deexcited atoms ($\vartheta = \pi$) in an empty cavity.

The initial condition (all atoms excited) corresponds to the unstable equilibrium point ($\vartheta = \dot{\vartheta} = 0$). In this semi-classical picture, the evolution would never start. In fact, the quantum field fluctuations (or the blackbody induced effects) trigger the evolution. The Bloch vector, when driven by these fluctuations, experiences a kind of Brownian motion which takes it away from its equilibrium position. This Brownian motion can be treated in a quantum approach, since it takes place in an almost fully inverted system, for which the Heisenberg equations of motion can be linearized [38]. The bottom line of this treatment is that everything happens "as if" the Bloch vector had been initially slightly tilted with respect to the vertical position, by some random "tipping angle". This random variable obeys to a Gaussian statistics, which can be explicitly determined.

The further evolution depends upon the system parameters. For a high Q cavity ($\Omega\sqrt{N} \gg \omega/Q$), damped oscillations of large amplitude are obtained. This is nothing but a collective Rabi precession of the atoms in the field they have themselves emitted in the cavity.

For a strongly damped cavity, on the other hand ($\Omega\sqrt{N} \ll \omega/Q$), the pendulum evolves in a very viscous medium, and reaches its equilibrium position after a monotonous evolution. This regime is quite similar to the irreversible spontaneous emission process of a single atom. Note however that the evolution of the atomic energy is far from being exponential. The velocity of the pendulum (and hence the radiated field) is maximum when it is horizontal. The emitted radiation is thus pulse shaped. This pulse, emitted after some delay, is very reminiscent of the superradiance phenomenon [49].

The transition between these two regimes has been observed [8] with a sample of Rydberg atoms radiating in a cavity with a moderate Q factor. Depending upon the value of N, the system can be either in the oscillatory

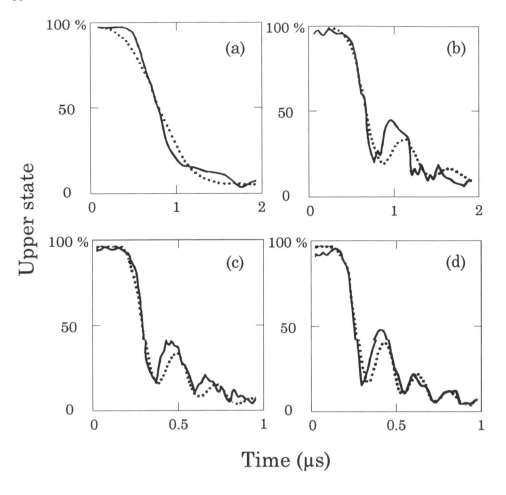

Figure 2. Average atomic energy versus time for a sample of N Rydberg atoms radiating in a cavity ($Q/\omega = 200$ ns, $\Omega = 1.3\ 10^5$ s^{-1}). Solid and dotted lines are experimental and theoretical respectively. The numbers of atoms are $N = 2000, 19000, 27000$ and 40000 in boxes (a), (b), (c) and (d). Curve (a) corresponds to irreversible superradiant emission. The oscillatory regime is clearly apparent in boxes (b), (c) and (d) (adapted from [8]).

or in the overdamped regime. By recording the atomic energy versus time, these two regimes have been demonstrated. Figure 2 shows a recording of the average atomic energy versus time for various populations. The onset of the oscillatory regime, as well as the $\sqrt{N}$ dependence of the small amplitude oscillations are clearly apparent.

This collective emitting system is a kind of transient maser. The microscopic nature of the atomic sample and the high cavity Q makes it very different from an ordinary maser. The spontaneous emission enhancement

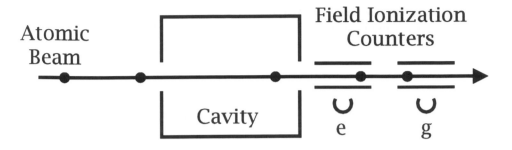

Figure 3. Scheme of a micromaser experiment. A weak beam of Rydberg atoms initially prepared in the upper level $|e\rangle$ crosses a high Q superconducting cavity, and is detected downstream in field ionization counters. The atoms release part of their energy in the cavity, and a "macroscopic" field builds up while there is, on the average, less than one atom in the cavity.

experiments are the one atom limit of this maser emission. Using the extraordinary features of Rydberg atoms in cavities, it is possible to realize a continuous-wave version of these transient masers, the micromaser.

2.3. THE MICROMASER

The principle of a Rydberg atom micromaser [9, 10, 11], is sketched in Fig.3. The weak beam of Rydberg atoms, prepared in the upper level $|e\rangle$ of the transition, crosses the high Q resonant superconducting cavity. The atoms are detected downstream by field ionization counters. Due to oscillatory spontaneous emission in the cavity, each atom has a large probability to emit a photon. If the atomic flux is high enough, the gain due to atomic emission overcomes the losses due to cavity dissipation, and a "macroscopic" field eventually builds up.

Though the basic principle of this maser oscillator is not very different from the one of the first ammonia maser, the orders of magnitude involved are radically different. Three important time scales have to be considered. The cavity relaxation time t_{cav} is the photon lifetime. The atom flux is measured by the average time between atoms, t_{at}, reciprocal of r_{at}, pumping rate of the micromaser (average number of atoms crossing the cavity per unit time). The last important time scale is the mean atom-cavity interaction time $\overline{t_{\text{int}}}$. If the atomic velocity is not selected, the interaction time t_{int} is a random variable whose statistics is determined by the Maxwellian velocity distribution in the beam. If a velocity selection is used [9, 10], all atoms have the same t_{int}.

Micromaser operating conditions correspond to the simultaneous full-fillment of two inequalities:

$$t_{\text{cav}} \gg t_{\text{at}} \gg \overline{t_{\text{int}}}. \tag{37}$$

The left one states that the mean time between atoms is much shorter than the cavity damping time. If each atom has a non zero probability to emit a photon, this condition states that the field does not decay before the next atom enters the cavity. This is a threshold condition, expressing that a macroscopic field can build up. The ratio of these two times,

$$N_{\text{ex}} = \frac{t_{\text{cav}}}{t_{\text{at}}} = r_{\text{at}}Q/\omega, \tag{38}$$

gives an indication of the photon number.

The right inequality in Eq.(37) means that the average time between atoms is much longer than the interaction time of each atom with the field. There is therefore, at any time, much less than one active atom inside the cavity. This is the main difference with an ordinary laser or maser. The dynamics of such a microscopic, nearly relaxation free, quantum oscillator exhibits many new features where the quantum fluctuations of the field play a central role.

Let us consider some orders of magnitude for the micromaser realized in Garching [9, 10]. The cavity, operating around 20 GHz, has a Q in the 10^9-10^{10} range, corresponding to t_{cav} in the 10-100 ms range. As t_{int} is of the order of 100 μs (crossing time of a centimeter-sized cavity at thermal velocity), there is a wide range of t_{at}'s corresponding to micromaser operation. The "weak intensity" limit corresponds to one atom crossing the cavity during each t_{cav} period. The mean photon number is then about unity, and the flux of excited atoms is between 10 and 100/s. This maser is above threshold with 10^{-3} atoms, on the average, in the cavity. The "high intensity" limit corresponds to $t_{\text{at}} \simeq t_{\text{int}}$. The excited atoms flux is of the order of 10^4/s, and the average photon number about 10^3. Even at highest pumping rate, the micromaser field remains rather weak.

The superconducting cavities damping rates are so low that it is even possible to operate a micromaser on a much weaker transition. A two-photon micromaser, on a transition between atomic levels with the same parity, has been operated in our laboratory [11, 50]. It was the first realization of a two-photon continuous wave oscillator, suggested long ago [51]. The orders of magnitude involved are quite comparable to the ones of the one-photon micromaser, though the spontaneous emission probability in an empty cavity is weaker, rising the threshold up to 10^5 atoms per second, corresponding to about one atom at a time in the cavity.

Developping the complete theory of micromaser operation would be far beyond the scope of this paper. The interested reader should refer to other papers for a complete description [37, 52, 53, 54]. We outline here only the main features.

The most striking differences with ordinary lasers occur when the atoms are velocity selected. Due to the Rabi precession, the atomic "gain" is then an oscillatory function of the photon number in the cavity, whereas it increases monotonically towards the saturated value for ordinary lasers. A simple gain/loss balance argument yields then that the loss curve, linear versus intensity, may have many intersections with the gain curve. The micromaser is therefore, far above threshold, a multistable device [53]. Quantum fluctuations generally make all the operating points but one unstable. Under specific conditions, two points may be equally stable. The maser then switches randomly between these two operating conditions. This switching behavior, recently observed experimentally [56], is quite reminiscent of the situations studied, in a different context, by Leggett [57], where a macroscopic system jumps between two equally stable conditions under the influence of quantum or thermal noise. Similar switching behavior has also been observed for the two photon maser start-up [58]. In this device, the low-signal gain is quadratic versus intensity, therefore always smaller than losses near the origin. Only quantum fluctuations can drive the maser out of this metastable zero intensity point.

The oscillatory gain behavior has also important consequences for the field statistics. The gain is, in some conditions, a decreasing function of the intensity around the operating point. This results in a stabilization of the photon number against fluctuations of the atomic beam intensity, which are the only cause of field amplitude fluctuations in this very simple oscillator. The field in the cavity is then highly sub-poissonian. This effect reflects on the statistics of the out-going atoms, as demonstrated experimentally in [10].

For some precisely defined atom-field interaction times, the gain may even vanish for some photon number. The micromaser field is then, in steady state, very close to a Fock state with this photon number. These so-called "trapping states" [53, 54] of the micromaser are very sensitive to dissipation and fluctuations, and have not been put in evidence yet.

3. Atom-Field Entanglement

An important point in the cavity QED experiments is that the atom and the field get correlated after their interaction. The global atom-field quantum state is an entangled one which cannot be cast, under any representation, as a tensor product of an atom and a field state. The two components of this correlated system may have a lifetime much longer than the interaction time t_{int} (tens of ms for circular Rydberg atoms and superconducting cavities, as compared to tens of μs). The atom-field state is thus a non-local quantum state, very similar to the correlated pairs of the famous EPR paradox [15].

This atom–field quantum correlation can be used to gain some information on the field state. This is at the heart of a Quantum Non Demolition (QND) measurement method for the field intensity [17, 18]. The measurement of a well chosen atomic observable allows one also to project the field onto an highly nonclassical state. "Schrödinger cats" of the field can be prepared in this way [18]. They are quantum superpositions of fields differing through their macroscopic attributes. The study of the relaxation of the coherence between their components due to the coupling with the outside world can help us bridging the gap between the quantum world, where superposition is the rule, and the macroscopic world, where it is never observed (there is not such thing as a living *and* dead cat, following Schrödinger's provocative wording [16]). This problem of the decoherence of macroscopic quantum superpositions is indeed at the heart of the quantum measurement theory. Hamiltonian interaction of a microscopic measured system with a macroscopic measuring apparatus should produce a measuring device in a coherent superposition of the possible outcomes. It is precisely because of the fast relaxation processes of such states that these coherent superpositions are never observed [59].

Tests of quantum non-locality based on Bell inequalities [60] could also be undertaken with this system. It is possible, as will be shown later, to prepare entangled states with more than two particles. Proposed for the first time by Greenberger et al. [61], these states provide tests of quantum non-locality much more stringent than the "ordinary" pairs. Finally, the quantum non-local correlated states could be used, in a quantum computer architecture [62], to perform some calculations exponentially faster than any classical computer. These different points will be briefly analyzed in the next paragraphs.

3.1. DISPERSIVE ATOM-FIELD ENTANGLEMENT

The more interesting situations for atom-field entanglement are encountered when the atom and the cavity are not exactly resonant. The interaction is then purely dispersive. No photon can be absorbed or emitted by the atom, provided the atom–cavity detuning δ is large enough. The effect of the coupling is then twofold.

First, the atomic transition is shifted by the cavity field. The value of these shifts can be easily derived from the dressed atom levels introduced in section 1. The energy shift of the upper level $|e\rangle$ at the center of a cavity containing n photons is $-(n + 1)\hbar\Omega^2/\delta$ [18]. The n term, linear versus the field intensity, accounts for the light shifts [35]. The "1" term is present even if the cavity is empty. It can be described as the Lamb shift of the upper level due to the interaction with the vacuum fluctuations. Level $|g\rangle$ experiences, in the same conditions, a shift $n\hbar\Omega^2/\delta$ (note that, within the approximations we use, $|g\rangle$ is not shifted in an empty cavity). The atomic transition frequency is therefore shifted by $\Delta\omega = \Omega^2(2n + 1)/\delta$ (note that the Lamb shift effect is equal to the light shift produced by half a photon). In parallel with the atomic levels modifications, the interaction shifts the cavity mode frequency. This level dependant shift (Ω^2/δ for an atom in level $|e\rangle$ and $-\Omega^2/\delta$ for an atom in level $|g\rangle$) can be attributed to the atomic index of refraction.

These light or Lamb shifts are accessible experimentally with Rydberg atoms. With the parameters of the experiment under progress in our laboratory [14] ($\Omega/2\pi = 25$ kHz and $\delta/2\pi = 100$ kHz —much larger than the cavity and atomic linewidths), the shifts are a few kHz per photon. The atom produces a cavity frequency shift also in the kHz range, which means that a single Rydberg atom changes the index of refraction in the cavity by about 1 part in 10^7, quite a large value.

The time-integrated energy level shifts for an atom crossing the cavity result in a wavefunction phase shift. For an atom in state e, the phase shift is $\varepsilon(n + 1)$, with $\varepsilon = -t_{\text{int}}\overline{\Omega}^2/\delta$. $\overline{\Omega}$ is the r.m.s. average of the atom-field coupling along the atom's trajectory. An atom in state $|g\rangle$ would experience a dephasing $-n\varepsilon$. Note that, for $\delta \gg t_{\text{int}}^{-1}$, the atom-cavity system follows then adiabatically the dressed state connected to the initial uncoupled level. The photon number is left unchanged and the atom returns to the initial state after the interaction

If one considers the external degree of freedom of the atom, this atom's wavefunction dephasing corresponds to a modification of the atom's position due to the slowing down or acceleration in the cavity, under the action of the force deriving from the position–dependant shifts [63, 64, 65]. This force is extremely weak (the associated potential well depth, $\hbar\Omega^2/\delta$, is of the order of 1μK only). It corresponds to a large modification of the atomic motion only for very slow atoms, such as the one produced by the laser cooling techniques [66]. For thermal velocity atoms, it results only in a modification of the atom's position in the picometer range. Since this is comparable to the De Broglie wavelength of the atom, this dephasing may be detected by an interferometric technique, as described in the next paragraph.

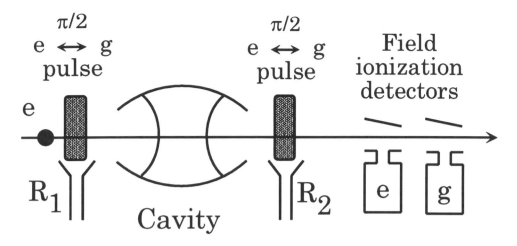

Figure 4.   Principle of the Ramsey interferometer

3.2. FROM LAMB SHIFTS TO LIGHT SHIFTS

To detect the Rydberg atom wavefunction dephasing due to the light shifts
(or Lamb shifts in an empty cavity), we make use of the Ramsey atomic
interferometry method [67]. As sketched on Figure 4, the superconducting
cavity containing the quantized field is sandwiched between two zones, R_1
and R_2, where the atom interacts with a classical microwave field. The
zone R_1 is set to prepare a superposition with equal weigths of $|e\rangle$ and
$|g\rangle$. This superposition crosses the cavity, which dephases in different ways
the wavepackets associated to $|e\rangle$ and $|g\rangle$. The atomic coherence phase
modification is translated into a population information in zone R_2, which
mixes again the two levels. Depending upon the relative phase of the atomic
coherence and of the R_2 field, the effect of the second zone doubles or
cancels the one of R_1. The atomic energy is then a sinusoidal function
of the frequency $\nu_r = \omega_r/2\pi$ applied in R_1 and R_2, with a period equal
to the reciprocal of the transit time across the apparatus, modulating the
line profile associated to a single zone. The phase of these fringes reflects
directly the shifts experienced in the cavity. This signal reveals a quantum
interference between two undistinguishable paths: a transition from $|e\rangle$ to
$|g\rangle$, for instance, can occur either in R_1 or in R_2. Nothing in the final
outcome can be used to tell which path the atom has followed.

   The whole experimental set–up is contained in an Helium 4 croystat,
operating at 1.4 K. The low temperature allows one to use superconducting
cavities, and to get rid of the thermal radiation. The rubidium atoms are
first promoted, by a laser–diode excitation, in an "ordinary" low angular
momentum Rydberg state ($51F, m = 2$). A "circularization" process then

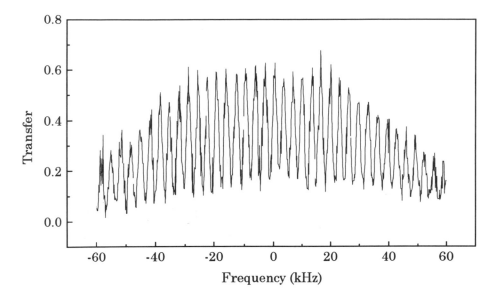

Figure 5. Ramsey fringes signal: $|e\rangle$ to $|g\rangle$ transfer rate versus the frequency difference between the fields in R_1 and R_2 and the atomic line. 45 fringes, 3.2 kHz apart, modulate the single zone interaction profile.

transfers most of the population to the circular state. This very efficient technique has already been described in details elsewhere [68] and will not be recalled here. Each circularization sequence produces about 400 circular atoms. The sequence is repeated at a 0.8 kHz rate, well adapted to the 600 μs transit time across the apparatus, the preparation time being known within 10 μs.

Circular atoms are stable only in a small electric field [69]. In free space, they are degenerate with many other high angular momentum levels. Stray fields can induce transitions between these levels. A very homogeneous, carefully controlled, 0.3 V/cm field is thus applied along the atom's trajectory with the help of 40 adjustable electrodes. At the exit of the interferometer, the atoms are detected by the field ionization technique. The association of the time resolved preparation and detection allows us to perform a passive velocity selection (relative velocity dispersion 1.5 %). As optical interferometric signals, the Ramsey fringes are better observed with a "monochromatic" atom source.

Two low-finesse cavities are used for the Ramsey zones. They are 9 cm apart. This corresponds, for 300 m/s atomic velocity, to a 3.2 kHz fringe spacing. In order to observe a large fringe number, we broaden the single zone line by applying the microwave in R_1 and R_2 for a short time, while the atoms are close to the center of the zones. The observed transfer rate as a function of the difference between ν_r and the atomic line frequency $\omega_0/2\pi$ is displayed in Figure 5.

The superconducting cavity C is an open Fabry–Pérot. This is the only design compatible with the static electric field needed for circular states preservation. It is made of two spherical niobium mirrors. They sustain two orthogonally polarized gaussian TEM_{900} modes, with 9 antinodes. The degeneracy of these modes is slightly lifted by geometric imperfections: the two frequencies are 146 kHz apart. The modes can be tuned simultaneously by translating one mirror with the help of a mechanical drive, and the other by a PZT stack.

Both modes have a quality factor $Q = 8.\ 10^5$. It is determined by monitoring the cavity transmission through two coupling holes drilled in the mirrors centers. The field damping time $t_{cav} = 2\ \mu s$ is thus much shorter that the atom–cavity interaction time $t_{int} = 25\ \mu s$. The field is renewed many times by the source during the interaction. The light shifts are then proportional to the average intensity, which is not quantized.

The average atom–field coupling at cavity center, $\Omega/2\pi$, is a little lower than the 25 kHz maximum value. This is due to a slight misalignment of the beam with respect to the cavity antinodes, and to the atomic beam size. The value deduced from a careful measurement of the beam position in the cold apparatus is 17(3) kHz.

Figure 6 presents the central fringes for an empty cavity and a cavity containing one photon (the field calibration will be discussed later). The noise on the signal is mainly statistical: only a few atoms are detected for each frequency channel. The thick line is a fit on a sine function. The precision on the fringes position, obtained through this fit, is about 25 Hz, corresponding to a $5.\ 10^{-10}$ resolution. The daily stability of the fringe pattern is also 25 Hz. A very careful screening of the magnetic field fluctuations, down to the 10 μG level, has been necessary to reach such a stability.

The fringes shift for a one photon field is 315 Hz, corresponding to $2\varepsilon = 0.6$ rd, for $\delta/2\pi = 150$ kHz. This corresponds to a few picometers displacement of the atomic wave packets in the cavity. We checked the linearity versus field intensity and the dispersive nature of the shift (proportional to $1/\delta$). The precision of the fringes position determination allows us to detect 0.1 photon fields, corresponding to a 125 fm retardation of the atom's wave packet!

With such a sensitivity, the Lamb shift is easily measured. Figure 7 shows, as open circles, the position of the fringes versus δ for an empty cavity. In order to interpret these data, two effects should be taken into account. First, the two modes of the cavity contribute to the shift (for the light shift experiment, the narrow bandwith source is coupled to a single

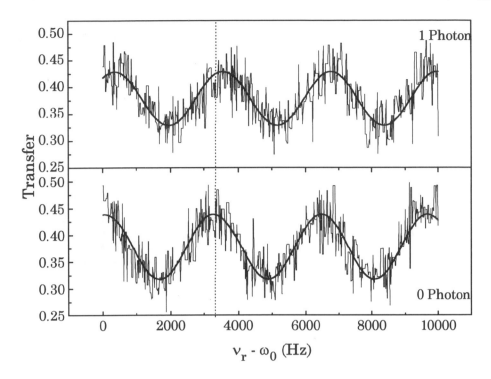

Figure 6. Ramsey fringes for a cavity containing zero or one photon ($\delta/2\pi = 150$ kHz). The fringes dephasing due to the light shift in the cavity is clearly apparent.

mode). One should also take into account the residual thermal field. The number of blackbody photons per mode has been found in an auxiliary experiment to be 0.32, corresponding to a 1.7 K radiation temperature. The contribution of the light shifts induced by this field have been removed from the raw data to obtain the pure Lamb shift effect, depicted by solid circles on Fig. 7. The solid line presents a fit on the theoretical values. The only adjustable parameter is the atom–field coupling. The obtained value, $\Omega = 16$ (0.5) kHz is in good agreement with, and more precise than, the one deduced from the beam position. It gives an excellent and independant calibration of the field intensity in the light shift experiment.

Lamb shifts are of course well known, and have been, in some cases, measured with excellent precision. Ordinarily, however, they are due to the coupling of the atomic levels with a whole continuum of field modes. This experiment is, to our knowledge, the first one which singles out the effect of only one field mode and studies the continuous transition between Lamb shifts and light shifts.

This preliminary experiment clearly demonstrates that our circular Rydberg atoms interferometer is sensitive to single photon fields. By minor

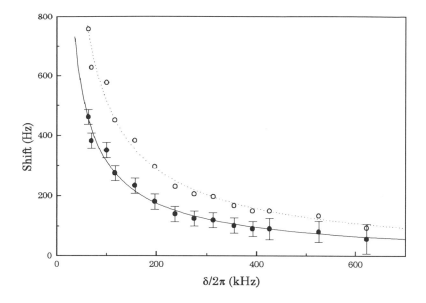

Figure 7. Shifts of the fringe system in an empty cavity versus atom–cavity detuning. Open circles: raw data. Full circles: data corrected for the residual thermal radiation. Solid line: theoretical fit.

improvements of the experiment (optimizing the atomic beam position in the cavity mode pattern and selecting slower atoms), it would even be possible to achieve a π phase shift of the Ramsey fringes for a single photon. The atom-field entanglement could then be realized and used, provided the photon lifetime in the cavity is much longer than the experiment duration. We have recently made considerable progresses in this direction. By improving the surface quality of the mirrors (which is the main Q limiting factor in this open geometry), we have been able to reach Q values in the 10^9 range, corresponding to photon lifetimes of the oder of 5 ms, much longer than the atomic transit time trough the apparatus. This realization opens the way to many experiments discussed in the next paragraphs.

3.3. QND MEASUREMENT OF THE CAVITY-FIELD INTENSITY

The simplest experiment, in progress now, uses the atom–field correlations to measure, without absorption, the intensity of a field containing 0 or 1 photon [18]. The interferometer is tuned so that $\varepsilon = \pi/2$ (π phase shift of the Ramsey fringes for a single photon). By a proper tuning of ω_r, the atom (incoming in $|g\rangle$) is certainly found in $|g\rangle$ if the cavity is empty (at least if we assume a perfect contrast for the fringes). It is also found in $|g\rangle$ if the cavity contains an even photon number. At variance, the atom is found in $|e\rangle$ if the cavity contains odd photon numbers. Assuming that the initial field has a low probability of containing 2 or more photons, the detection of a single atom is enough to pin down the field intensity and to count the photon number (a small bunch of atoms would be needed if the fringes visibility is not one).

A field containing either 0 or 1 photon can be easily prepared by using single atom spontaneous emission. A first atom, prepared in R_1 in an arbitrary superposition $ic_e|e\rangle + c_g|g\rangle$ by a classical field with proper phases and amplitude, is sent through the cavity. Stark effect tunes the atom in exact resonance with the initially empty mode for such a time that the oscillatory spontaneous emission undergoes an exact π pulse. An atom in e leaves a photon in the cavity, while an atom in g does not evolve. The final state of the atom-cavity system is thus $|g\rangle(c_e|1\rangle + c_g|0\rangle)$. A second atom, interacting dispersively with the cavity, measures then the photon number.

At variance with most photodetections, this measurement does not change the photon number: the non-resonant atom cannot emit or absorb in the cavity. It is therefore a Quantum Non Demolition (QND) measurement [70, 71, 72, 73]. If one neglects relaxation, the photon left in the cavity after detection of an atom in $|e\rangle$ will remain forever, and all following atoms will exit the cavity in $|e\rangle$. This correlation between repeated measurements is one of the main features of such processes and can be used to check the QND character. Taking into account cavity relaxation, the photon has a small chance to be absorbed or diffused out of the cavity between two measurements. Assuming that the sources are switched off, one should observe a sequence of atoms in $|e\rangle$ (photon present) followed by an infinite sequence of atoms in $|g\rangle$ (cavity empty). This quantum jump, monitoring in real time the photon death, is very reminiscent of the quantum jumps observed on the fluorescence of a single trapped ion [74]. The ordinary quantum mechanics prediction (exponential decay of the energy) is recovered only as a statistical average over many individual realizations of the experiment. Such features illustrate the properties of quantum mechanics applied to single systems, and not to large ensembles.

Note that this QND method can be generalized to arbitrary photon numbers [18, 75, 76]. By using an optimized measurement scheme, it is possible to determine an abitrary photon number with only $\log_2 N_{max}$ atoms, where N_{max} is the maximum possible photon number [76]. This experiment, using the minimal number of atoms allowed by information theory, would realize an ideal analog/digital converter operating at the photon level. In an easier experiment [18], it is enough to detect atoms with different velocities in the Maxwell distribution. Different velocities correspond to different

Ramsey fringes patterns and to different informations on the field. Once a few tens of atoms have been recorded, a field intensity up to about ten photons is efficiently measured.

3.3.1. *GHZ Non-Local Correlations*

The states of atoms performing successive QND measurements of 0 or 1 photon field intensity, as discussed above, are strongly correlated. This feature can be used to prepare triplets of correlated atoms [61], exhibiting in a striking way the nonlocal properties of quantum mechanics.

Following Mermin [77, 78], let us consider a set of three spins 1/2 (eigenstates $|+\rangle$ and $|-\rangle$ along a quantization axis Oz). The system is prepared, via a common interaction, in the entangled state:

$$(|+,+,+\rangle - |-,-,-\rangle)/\sqrt{2}. \tag{39}$$

Once the spins have been spatially separated, a measurement of their components along an axis Ox, orthogonal to Oz, is performed. Quantum mechanics predicts that the product of the three outcomes, $m_x^1 m_x^2 m_x^3$, should be equal to -1. On the other hand, any local theory, based upon the concept of "elements of reality" (hidden variables theory), predicts an outcome of $+1$. There is a striking difference between the two predictions, and, at least with an ideal set–up, a single realization of the experiment would be enough to confirm or contradict quantum mechanics. This is a much more stringent test than the ones based on particle pairs and Bell inequalities [60], which involve statistics performed on a large number of experiments.

To prepare the state described by Eq.(39), we make use of the QND technique. A first atom, prepared in R_1 in state $(|e\rangle - |g\rangle)/\sqrt{2}$ crosses the initially empty cavity. It is tuned to resonance with the mode for such a time that it undergoes a π pulse in spontaneous emission. The final atom–cavity state is thus $|g\rangle(-i|1\rangle - |0\rangle)/\sqrt{2}$. The first atom's state is not correlated to the cavity one, and its detection may be used as a triggering event.

The second and third atoms then cross the Ramsey set-up in the non-resonant conditions needed for a QND measurement (see above). The interferometer is tuned so that the atoms exit cavity in state e if it contains 1 photon, in state g if it is empty. The final atom-cavity state is thus $(-i|1, e, e\rangle - |0, g, g\rangle)/\sqrt{2}$, an entangled state of the two atoms and of the cavity. To destroy the cavity entanglement, a fourth atom is sent trough the cavity. Initially in state g, it is tuned at exact resonance so that it undergoes a π rotation if the cavity contains one photon. The final state of the last three atoms is thus:

$$\frac{1}{\sqrt{2}}\left(|e, e, e\rangle - |g, g, g\rangle\right), \tag{40}$$

a three atoms correlated state identical to the spin state described by Eq.(39). In order to measure the analog of the x components of the spins, each atom is submitted to a $\pi/2$ Rabi pulse before measuring the atomic state. The parity of the excited atom number is then different for the predictions of quantum mechanics and the ones of any local theory. Let us stress that the method could obvioulsy be generalized to an arbitrary number of atoms (by adding additional atoms in the QND measurement sequence).

The main experimental difficulty is to make sure that four atoms only have crossed the cavity. Since the detection efficiency is not one, an additional atom may escape detection, and perturb the parity measurement. The only practical possibility to reduce this effect is to deal with four preparation events yielding, on the average, much less than one atom. Retaining only the seldom cases where exactly four atoms are detected, one makes sure that the probability of having a spurious atom is low. Very long acquisition times would be associated to this procedure.

3.3.2. *Quantum Computing*

This QND process can be viewed as an operation on two two-level quantum systems (the field containing either 0 or 1 photon and the atom, with states e or 1 and g or 0). It is very close to a logic operation between two bits. With a proper set-up of the interferometer, the first bit (the field) is not affected in the process. The second one (the atom) changes state if the field is in in state 1, while it is unaffected if the field is in state 0. The "logic function" implemented by this system is then very close to a XOR logic gate. At variance with an ordinary logic gate, this one relies on an hamiltonian interaction between two two-level quantum systems. It should thus be able to process coherent quantum superpositions of the input states.

Such a "quantum logic gate" is the building block for a "quantum computer" [62]. Instead of ordinary bits of information, these computers operate on "q-bits", two-level systems indeed. The whole computing process is a reversible one, involving hamiltonian evolution of the computer processor and of the q-bits. A quantum machine has been shown to solve some problems much faster than any classical machine [62, 79]. The key of this rapidity is the massive parralelism a quantum machine allows. Since it processes quantum superpositions of q-bits values, it can process, in a single step, all the possible values of the input data. Instead of performing a single calculation on an N-bits word, it can compute in the same step all the 2^N possible outcomes. Carefully designed algorithms can take benefit of this massive quantum parralelism. The most famous one [79] factors an integer exponentially faster than any classical program. As many cryptographic systems are based on the difficulty of factoring large integers, this result has raised a lot of interest.

The building block of a quantum computer is a "quantum logic gate" [62]. This gate is a small computing machine, performing on two or more q-bits a well defined unitary transformation. When wired together to form a network, these gates allow in principle to realize any quantum computation. When a single gate type can be used to form the network, it is said to be universal. While universal gates are known since long for classical computers, universal gates for quantum computations have only been exhibited in the last months [80]. The "QND gate", also called "measurement gate" or "controlled not" in the quantum computing litterature is universal associated with single bit gates, amounting to a trivial phase factor on the q-bit state. It has recently been shown that a minor amendment of the technique would allow to realize a true universal gate [81, 83] with our setup.

The success of the QND experiment entails thus the successfull operation of a single quantum gate. Wether such gates could be wired together to form an useful quantum computer is a more difficult question. The main problem concerns relaxation. One may take advantage of the massive quantum parralelism only if the quantum coherence is maintained during the whole computation among all the elementary q-bits. This puts very severe constraints on the implementations of networks with thousands of gates and q-bits, as needed to factor large integers [82]. At least, it is very interesting to study experimentally this complex decoherence process of a large ensemble of quantum systems. The next section disusses another case where the decoherence of large systems could be studied.

3.4. "SCHRÖDINGER CAT" STATES

Let us consider now the QND experiment when the cavity initially contains a classical field with amplitude α, represented by a Glauber's coherent state $|\alpha\rangle$ [40]. The atomic index of refraction shifts transiently the cavity frequency and changes the field phase. With $\varepsilon = \pi/2$, an atom in state $|e\rangle$ dephases the field by $\pi/2$, changing the initial field $|\alpha\rangle$ into $|\beta\rangle = |i\alpha\rangle$. An atom in state $|g\rangle$ has an opposite effect, changing $|\alpha\rangle$ into $|-\beta\rangle$. The atom behaves in fact as a dielectric plunger. It changes transiently the cavity frequency. This results in a phase modification for the field stored in the cavity. At variance with an ordinary macroscopic plunger, the atom can be prepared in a quantum superposition of its two states, that is a quantum superposition of opposite indices of refraction. The field might thus be left in a quantum superposition of components with different classical phases.

In a QND measurement, the number of phase components increases with the number of atoms crossing the cavity. Finally, all initial phase information is lost and the phase distribution is uniform. This is clearly requested by the intensity-phase uncertainty relation: when the QND determination of the photon number is complete, the phase should be completely scrambled. The phase shifts discussed here are at the heart of this phase scrambling mechanism [18].

Let us analyze in more details the first step of this phase scrambling process, for an atom crossing the cavity with $\varepsilon = \pi/2$. The Ramsey zones R_1 and R_2 are set to achieve the transformations: $|e\rangle \to (|e\rangle - |g\rangle)/\sqrt{2}$ and $|g\rangle \to (|g\rangle + |e\rangle)/\sqrt{2}$. The atom enters then C in state $(|e\rangle - |g\rangle)/\sqrt{2}$ and the atom–field system is, after the interaction:

$$\frac{1}{\sqrt{2}}\Big(|e, \beta\rangle - |g, -\beta\rangle\Big), \tag{41}$$

where the first symbol inside the kets refer to the atom, the second to the cavity. This is clearly an entangled atom–field state. Performing at this stage a measurement of the atomic energy amounts in projecting the field state on $|\beta\rangle$ or $|-\beta\rangle$. In such a situation, the state of the atom in the cavity is known, and the interaction only produces a classical phase shift.

A more interesting situation is obtained by scrambling the information on the atomic state. This is achieved when the atom crosses R_2, which mixes both levels. The final atom–field state is then:

$$\frac{1}{2}\Big(|e\rangle[|\beta\rangle - |-\beta\rangle] - |g\rangle[|\beta\rangle + |-\beta\rangle]\Big). \tag{42}$$

Measuring the atomic energy projects the field onto one of the states:

$$|\Psi_\pm\rangle = \frac{1}{\mathcal{N}}(|\beta\rangle \pm |-\beta\rangle), \tag{43}$$

where $\mathcal{N}$ is a normalization factor, close to $1/\sqrt{2}$ when $|\beta|$ is large enough (this will be assumed in the following).

The $|\Psi_\pm\rangle$ states exhibit remarkable quantum features [18, 84]. They are quantum superpositions of two classical fields with opposite phases. The name "Schrödinger cat" is often coined out for such superpositions of macroscopically distinguishable states, in reference to the famous "paradox" stated by Schrödinger [16]. $|\Psi_+\rangle$, for instance, corresponding to an atom detected in $|g\rangle$, has no classical counterpart: the sum of two classical fields with opposite phases is zero. Moreover, it contains only even numbers of photons: the odd photon numbers probability amplitudes associated to the two components cancel out. This can be understood easily in terms of the QND measurement. The atom exits the interferometer in $|g\rangle$ only if the cavity contains an even photon number. Detecting $|g\rangle$ makes us sure that no odd photon number can exist in the cavity. In a similar way, $|\Psi_-\rangle$ contains only odd photon numbers. $|\Psi_+\rangle$ can be called an "even cat", and $|\Psi_-\rangle$ an "odd cat".

This peculiar photon number distribution explains why the "cats" are short lived. After a time of the order of t_{cav}/N (N is the average photon number in the cat), relaxation fills the gaps in the photon number distribution, and the cat turns into a mere statistical superposition of two fields. The larger the energy, the larger the "distance" between the two cat components, the faster the decoherence. For macroscopic objects, the same decoherence occurs in times so short that quantum superpositions are never observed. This explains why the needle of a detector, measuring a quantum system in a coherent superposition of states, is never found in a quantum superposition of the corresponding positions [59]. In the present system, the classical phase of the field is the needle indicating the state of the atom in the cavity. The size of this "measuring apparatus" can vary continuously from microscopic to mesoscopic, from single photon fields to large macroscopic fields. A study of the "relaxing cat" is thus a study of the decorrelation mechanisms at the heart of the quantum measurement theory, and an exploration of the subtle border between the quantum and classical worlds.

The coherent nature of the "cat" can be revealed by the quantum interference effects it provides [85]. Let us send a second atom, in the same conditions, through the cavity, a time T after the first (which has been detected in $|g\rangle$). If $T \ll t_{cav}/N$, this atom interacts with an intact cat, containing only even photon numbers. It exits therefore the apparatus in $|g\rangle$. When $T \gg t_{cav}$, this atom crosses an empty cavity, and exits also in $|g\rangle$. At variance, for $t_{cav}/N \ll T \ll t_{cav}$, the atom encounters a statistical superposition of coherent fields, containing odd and even photon numbers. The atom is then found in $|e\rangle$ or $|g\rangle$ with equal probabilities. The conditional probability of detecting two successive atoms in the same state thus exhibits a fast variation around $T = 0$, which reveals the decoherence. Let us stress that the orders of magnitude for the experimental realization are quite comparable to the ones for the QND measurement, as long as the "cat" is not too big.

Other breeds of Schrödinger cats can be prepared and studied in this system. Using, for instance, the atomic index of refraction to tune the cavity in resonance with a classical source, it is possible to control the flow of energy in the cavity trough the state of the atom. One realizes then a kind of "quantum switch" [19] which can be in a quantum superposition of its "open "and "closed" states. One can thus prepare quantum superpositions of the vacuum state and a large coherent component ("amplitude cat", by opposition to the "phase cat" described above). Making use of the same atom crossing successively two cavities, it is even possible to realize a superposition of quantum states corresponding to a large coherent field localized either in the first or in the second cavity. The quantum coherence of such non-local states can also be displayed in experiments making use of a second atom crossing the apparatus.

4. Mie resonances of dielectric microspheres

Though Rydberg atoms in superconducting cavities offer an almost perfect tool for fundamental CQED studies, it is of great importance to extend these experiments to the optical domain since the physics of single radiators interacting with highly confined electromagnetic fields in this frequency range is likely to open the way to interesting practical applications.

The strong single atom-field coupling has already been realized with alkali atoms in very high finesse Fabry-Pérot cavities [25]. However, the volume of these cavities cannot be very small. The intermirror separation is at least about 1 mm, while the mode waist is of the order of 50 μm [42]. This implies a mode volume of many thousands of λ^3, and a one-photon Rabi frequency in the few MHz range for an highly allowed transition, which does not exceed by far the atomic or cavity relaxation rates.

A promising alternative to these conventional cavities is offered by the Mie (or Morphology Dependent) resonances of small silica spheres [26, 27, 28]. These modes are discussed, in the context of small liquid droplets, in the lecture notes by R. Chang in this book. Liquid droplets can be produced easily, with very reproducible shapes, and have been extensively used for non-linear optics studies. However, the liquid evaporation makes the resonant frequencies evolve rapidly in time and the liquid intrinsic non-linearities (Raman effect, harmonic generation) are high enough to limit the Q factor to about 10^8, i.e. a cavity damping time t_{cav} of the oder of 30 ns. Following the pioneering work of Braginsky's group [26], we have decided to use instead silica microspheres. The high transparency of synthetic silica and its extremely low non-linear coefficients makes it a very promising material for the realization of very high Q microspheres.

The modes offering the higher spatial confinement are localized along the sphere's equator. One can give a very simple qualitative picture of these modes in terms of a ray repeatedly reflected by total internal reflection at the sphere surface, and orbiting in the equatorial plane. A more detailed picture, as given in R. Chang's lecture notes, assigns three "quantum numbers" to the mode. The radial one (n) gives the number of nodes (or of extrema) along a radius. The two other are the analogs of the orbital (ℓ) and magnetic (m) quantum numbers in atomic physics. We use here atomic physics notations for these quantum numbers. The reader should be aware that the notations in R. Chang's lecture notes are slightly different (n and ℓ are exchanged in the standard litterature on Mie resonances). The angular dependance of the mode is derived from the corresponding spherical harmonic. The equatorial modes which we are interested in correspond to low values of the radial quantum number, that is to say to modes extending only along a few λ's inside the sphere. They correspond also to the highest possible values for the photon angular momentum (of the order of the sphere's size parameter $x = 2\pi Na/\lambda$ where a is the sphere radius, N its index), and to the highest allowed value of the magnetic quantum number $|m| = \ell$. Their angular dependance then bears a strong analogy with a circular Rydberg atom wavefunction. The field is localized in a thin torus along the equator, with transverse dimensions of the order of λ. A small part of the mode extends outside the sphere, forming an evanescent wave near the surface. A sphere with a size parameter of a few hundreds (corresponding to a radius in the 50 μm range), has therefore a mode volume $\mathcal{V}$ in the $100 - 1000$ λ^3 range. Such a small volume corresponds to extremely large single-photon fields, of the order of 10 kV/m! An atomic transition with a dipole of one atomic unit would experience in such a field a single photon Rabi precession at about 100 MHz. These orders of magnitude illustrate clearly the interest of the high field confinment achieved by these modes.

For a perfect non-absorbing dielectric, the quality factors of these low n modes should reach extremely high values. The losses due to imperfect total internal reflection are quite negligible for these modes corresponding to grazing incidence in the ray optics picture, as soon as the diameter of the sphere exceeds a few microns. A more realistic Q limit is given by the bulk absorption of the dielectric. At 780 nm wavelength (used in our experiments), the attenuation of the best opical fibers is 2 dB/km, corresponding to maximal Q values in the 10^{10} range (at 1.55 μm, the quality factor could reach values up to 10^{11}). Such Q values correspond to cavity damping times of a few μs.

An electron microscope picture of one of the microspheres used in our experiments is presented in figure 8. The sphere is obtained by melting, with the help of two counterpropagating CO_2 laser beams, the end of a very pure silica fiber (with a diameter ranging from 10 μm to 50 μm). Provided the beams intensities are properly balanced to avoid radiative forces, the high surface tension of melted silica gives a very good sphericity around the equator (the measured eccentricities are in the 10^{-3} range). The sphere remains attached to the fiber stem, making its manipulation easy. As the modes of interest are highly localized around the equator, this stem does not perturb the quality factor. In order to avoid surface contamination, the sphere fabrication and handling should take place in a clean atmosphere or under vacuum.

Figure 9 (a) presents the experimental arrangement to observe the sphere resonances. We obtain a good coupling efficiency by placing the

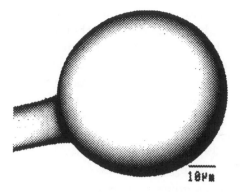

Figure 8. Electron microscope photograph of a silica microsphere.

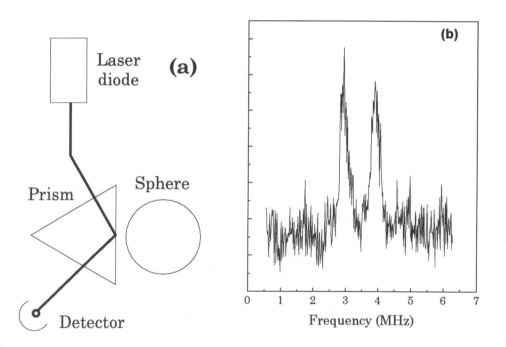

Figure 9. (a) Experimental arrangement to observe the spheres resonances. (b) A typical high Q resonance doublet.

sphere in an evanescent wave produced by total internal reflection of a highly focused laser beam on the hypothenuse of a high index prism [27]. When the laser incidence is close to the critical angle for the prism/sphere reflection, the k-vector of the prism evanescent wave is well matched with that of the whispering gallery mode outside the sphere. The coupling efficiency varies exponentially with the sphere-prism gap. The scale of this variation is $\lambda/2\pi \simeq 100$ nm. A very good control of the sphere-prism distance is thus needed, allowing to tune the coupling efficiency. While it may reach values up to 30%, the highest quality modes are observed only for a vanishingly small coupling (gap of the order of 1 μm).

A quite efficient coupling may be obtained also by placing the sphere near the core of a monomode silica fiber. By optically polishing the fiber's cladding, the evanescent wave around the core can be exposed to air. Since its k-vector is naturally matched with the evanescent part of the sphere mode, large coupling efficiencies (up to 10%) can be obtained in an easy-to-align arrangement [86]. Even when the sphere is in contact with the fiber, relatively high Q modes may be observed, opening the way to compact integrated systems.

We make use of high spectral quality laser diodes to measure the mode width. A diode laser with grating feedback is stabilized on an external Fabry-Pérot cavity. The resulting linewidth is of the order of 100 kHz, well suited to the measurement of Q's up to a few 10^9. For higher Q values, time decay measurements can be used [87]. Since the observation of high Q's implies a low coupling efficiency, we make use of a frequency modulation technique to improve the signal to noise ratio [27].

A typical high Q resonance is depicted on figure 9(b). It appears as a doublet, with a 1 MHz separation, and a width of the components of 270 kHz, corresponding to $Q = 2.\,10^9$. This high Q value is quite remarquable, when one keeps in mind the small volume of the mode. It corresponds to a finesse of the order of 10^7, much higher than the best ones achieved with multidielectric mirrors. The cavity damping time t_{cav} is 0.8 μs, and should be well adapted to the realization of CQED experiments.

The doublet structure, observed systematically on high enough quality modes, is attributed to a lifting of the Kramers degeneracy between the two modes with same n and ℓ, but opposite magnetic quantum numbers, corresponding to two counter-propagating modes [87] (the degeneracy of modes with different m values is completely lifted by the small eccentricity of our sphere). Light backscattering on small imperfections can couple these two modes. As for any situation of coupled degenerate oscillators (see for instance the vacuum Rabi splitting in section 1), the coupling lifts the degeneracy and the splitting is equal to the rate of energy exchange between the two modes. A simple model explains correctly this effect and shows why

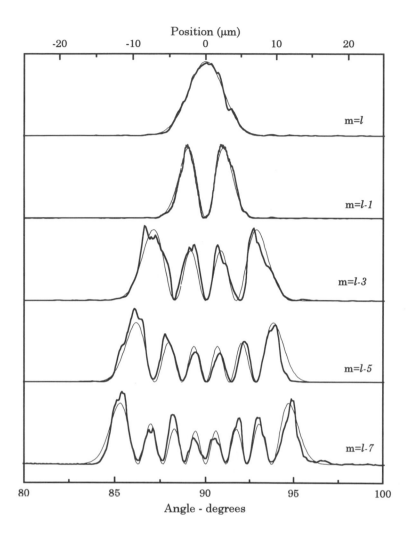

Figure 10. Spatial structure of the sphere's modes recorded by near field optical microscopy. The thick curves display the recorded intensity as a function of the tip position along a meridian. From top to bottom, the curves correspond to increasing $\ell - |m|$ values. The thin lines correspond to a theoretical fit.

the quality factor degradation due to scattering by these defects is much smaller than the backscattering efficiency.

To determine the modes quantum numbers, a necessary step to calculate the mode effective volume and the atom-field coupling, we have explored the spatial dependance of the mode by near-field optical microscopy [88]. A very thin tip, eroded out of an optical fiber, is scanned along a meridian of a

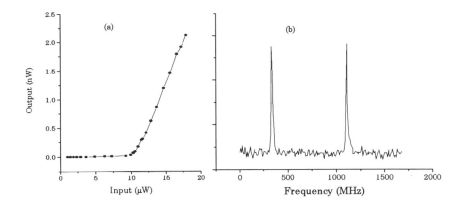

Figure 11. (a) Output power versus pump power for a microsphere Nd laser (cold cavity $Q = 3.8 \ 10^6$). The threshold (10 μW) is clearly apparent. (b) Fary Perot scan for the same oscillator. The operation is monomode, the mode width (17 MHz), much smaller than the cold cavity one, being limited by the Fabry-Perot finesse.

sphere. The sphere is coupled, on the other side, to a monomode fiber, and the exciting diode laser is electronically locked on a mode resonance. The signal emerging from the tip is displayed on figure 10 as a function of the tip position. The fit with the theoretical prediction, also displayed on figure 10 is quite good, and allows us to get a value for the orbital quantum number in good agreement with the expected one. This implies that we effectively excite modes with low n's. Moving the tip along the equator will allow us in near future to measure directly m, hence ℓ, and to get the precise value of n from the mode frequency, completing the mode attribution. Note also that the measurement of the mode position as a function of m for a given n and ℓ gives a precise value for the sphere's eccentricity, in good agreement with the visual observation.

The next step in these experiments is to couple atoms to the sphere's resonance. We went in this direction by observing a laser effect in neodymium doped spheres [89]. To make these spheres, we made use of a fiber with a large core (20 μm diameter), doped with 0.1 % of neodymium. The fiber cladding is first removed by immersion in HF. The exposed core is then melted to form the sphere. The pumping light at 810 nm is coupled to a resonant mode of the sphere. Due to the neodymium ions absorption, the quality factor of the pump mode is rather low. When laser action occurs on a mode near 1.06 μm (peak of the doped fiber fluorescence), the light is coupled out of the sphere to the prism. The output beam is filtered in a monochromator to remove the pump light reflected off the prism. Figure 11 (a) presents the output power as a function of the pump power coupled into the sphere. A clear threshold effect is observable. One should note the relatively low value of this threshold, in the μW range. The cold-cavity quality factor for the lasing mode was only 3.8 10^6, limited by the coupling. We have checked that the threshold varies almost linearly with Q^{-1}. With higher Q's, the threshold could be reduced to sub-μW values. In order to check that the emission is monomode, we have sent the output light to a Fabry-Pérot analyzer. The transmitted signal is depicted on Figure 11 (b). A single peak appears in the FSR. The width of this peak is limited by the Fabry-Pérot finesse to about 17 MHz. The laser linewidth is thus certainly much smaller than the cold-cavity bandwith (of the order of 100 MHz).

A more detailed analyzis of this laser operation is under way. It is already clear that the threshold and the required neodymium concentrations could be considerably reduced by getting higher Q's for the cold cavity, and by reducing the homogeneous linewidth of the transition. Using microspheres in a low temperature environment, it should be possible to realize micro-lasers with only a few active ions resonant with the cavity. This would realize a very interesting CQED situation, since the ions are permanently coupled to the mode.

5. Conclusion

Rydberg atoms in superconducting cavities offer an unprecedented test system for matter-field interaction. One of the most striking features is the simplicity of both the atomic system (a mere collection of two level atoms) and the field state (a geometrically well confined single mode of the radiation field). At the same time, the extremely long lifetimes of these systems makes relaxation negligible for most purposes. The outer world complexity (for instance the thermal bath to which the field is coupled) plays only a secondary role in the system evolution. This makes possible a detailed theoretical analysis, in terms of the most basic ingredients of quantum mechanics. Of course, this conceptual simplicity of the system is paid at the price of huge difficulties to actually perform the experiments.

The experiments realized on this system are direct illustrations (and even direct tests) of our understanding of quantum mechanics. They are very close to the Gedankenexperiments which were used by the founders of quantum mechanics to test the Copenhagen interpretation. One of the most challenging fields for the experimenters is certainly the one of coherent macroscopic superpositions. Close-to-ideal quantum measurements on a nearly relaxation free single quantum system could yield more insight into the most conceptually difficult part of quantum mechanics.

The extension of these very fundamental experiments to other systems, or even to other domains of physics, is certainly a fascinating challenge. The realization of very high quality optical resonators made with mere dielectric spheres is certainly encouraging. The strong, permanent, coupling of a single optically active ion to a single mode of the sphere could allow the realization of interesting CQED situations. Moreover, such systems could be easily integrated and offer various applications.

References

1. Purcell E.M., Spontaneous emission probabilities at radio frequencies, *Phys. Rev.* 69:681 (1946).
2. Casimir H.B.G., and Polder D., The influence of retardation on the London-van der Waals force, *Phys. Rev.* 73:360 (1948).
3. Drexhage K.H., Interaction of light with monomolecular dye layers, in: "Progress in Optics XII", Wolf E. ed., p. 163, North Holland, Amsterdam (1974).
4. Goy P., Raimond J.M., Gross M., and Haroche S., Observation of cavity-enhanced single atom spontaneous emission, *Phys. Rev. Lett.* 50:1903 (1983).
5. Hulet R.G., Hilfer E.S., and Kleppner D., Inhibited spontaneous emission by a Rydberg atom, *Phys. Rev. Lett.* 55:2137 (1985).
6. Sandoghdar V., Sukenik C., Hinds E., and Haroche S., Direct measurement of the van der Waals interaction between an atom and its images in a micron-sized cavity, *Phys. Rev. Lett.* 68:3432 (1992).
7. Raimond J.M., Goy P., Gross M., Fabre C., and Haroche S., Statistics of millimeter-wave photons emitted by a Rydberg atom maser: an experimental study of fluctuations in single mode superradiance, *Phys. Rev. Lett.* 49:1924 (1982).
8. Kaluzny Y., Goy P., Gross M., Raimond J.M., and Haroche S., Observation of self-induced Rabi oscillations in two-level atoms excited inside a resonant cavity, *Phys. Rev. Lett.* 51:1175 (1983).
9. Meschede D., Walther H., and Klein N., One-atom maser, *Phys. Rev. Lett.* 54:551 (1985).
10. Rempe G., Schmidt-Kaler F., and Walther H. , Observation of sub-Poissonian photon statistics in a micromaser, *Phys. Rev. Lett.* 64:2783 (1990).
11. Brune M., Raimond J.M., Goy P., Davidovich L., and Haroche S., Realization of a two-photon maser oscillator, *Phys. Rev. Lett.* 59:1899 (1987).
12. Rempe G., Walther H., and Klein N., Observation of quantum collapse and revival in a one-atom maser, *Phys. Rev. Lett.* 58:353 (1987).
13. Bernardot F., Nussenzveig P., Brune M., Raimond J.M., and Haroche S., Vacuum Rabi splitting observed on a microscopic atomic sample in a microwave cavity, *Euro. Phys. Lett.* 17:33 (1991).
14. Brune M., Nussenzveig P., Schmidt-Kaler F., Bernardot F., Maali A., Raimond J.M. and Haroche S., From Lamb shift to light shifts: vacuum and subphoton cavity fields measured by atomic phase sensitive detection, *Phys. Rev. Lett* 72:3339 (1994).
15. Einstein A., Podolski B., and Rosen N., Can quantum mechanical description of physical reality be considered complete?, *Phys. Rev.* 47:777 (1935).
16. Schrödinger E., Die gegenwärtige situation in der quantenmechanik, *Naturwissenschaften* 23:807, 823 (1935).
17. Brune M., Haroche S., Lefèvre V., Raimond J.M., and Zagury N., Quantum non-demolition measurements of small photon numbers by Rydberg atom phase sensitive detection, *Phys. Rev. Lett.* 65:976 (1990).
18. Brune M., Haroche S., Raimond J.M., Davidovich L., and Zagury N., Manipulation

of photons in a cavity by dispersive atom-field coupling: quantum non demolition measurements and Schrödinger cat states, *Phys. Rev.* A45:5193 (1992) .

19. Davidovich L., Maali A., Brune M., Raimond J.M., and Haroche S., Quantum switches and non-local microwave fields, *Phys. Rev. Lett.* 71:2360 (1993).
20. Jhe W., Anderson A., Hinds E.A., Meschede D., Moi L., and Haroche S., Suppression of spontaneous decay at optical frequencies: test of vacuum field anisotropy in confined space, *Phys. Rev. Lett.* 58:666 (1987).
21. de Martini F., Innocenti G., Jacobowitz G.R., and Mataloni P., Anomalous spontaneous emission time in a microscopic optical cavity, *Phys. Rev. Lett.* 59:2955 (1987).
22. Yokoyama H., Suzuki M., and Nambu Y., Spontaneous emission and laser oscillation properties of microcavities containing a dye solution, *Appl. Phys. Lett.* 58:2598 (1991).
23. Heinzen D.J., Childs J.J., Thomas J.E., and Feld M.S., Enhanced and inhibited visible spontaneous emission by atoms in a confocal resonator, *Phys.Rev. Lett.* 58:1320 (1987).
24. Heinzen D.J., and Feld M.S., Vacuum radiative level shift and spontaneous emission linewidth of an atom in an optical resonator, *Phys. Rev. Lett.* 59:2623 (1987).
25. Thompson R.J., Rempe G., and Kimble H.J., Observation of normal mode splitting for an atom in an optical cavity, *Phys. Rev. Lett.* 68:1132 (1992).
26. Braginsky V.B., Gorodetsky M.L. and Ilchenko V.S., Quality factors and non-linear properties of optical whispering gallery modes, *Phys. Lett.* A 137:393 (1989).
27. Collot L., Lefèvre V., Brune M., Raimond J.M., and Haroche S., Very high Q whispering gallery mode resonances observed on fused silica microspheres, *Euro. Phys. Lett.* 23:327 (1993).
28. Mabuchi H. and Kimble H.J., Atom galleries for whispering atoms: binding atoms in stable orbits around an optical resonator, *Opt. Lett.*, 19:749 (1994).
29. Yablonovitch E., Gmitter T.J., and Bhat R., Inhibited and enhanced spontaneous emission from optically thin AlGaAs/GaAs double heterostructure, *Phys. Rev. Lett.* 61:2546 (1988).
30. Yokoyama H., Nishi K., Anan T., Yamada H., Brorson S.D., and Ippen E.P., Enhanced spontaneous emission from GaAs quantum wells in monolithic microcavities, *Appl. Phys. Lett.* 57:2814 (1990).
31. Björk G., Machida S., Yamamoto Y., and Igeta K., Modification of spontaneous emission rate in planar dielectric microcavity structures, *Phys. Rev.* A44:669 (1991).
32. Weisbuch C., Nishioka M., Ishikawa A., and Arakawa Y., Observation of the coupled exciton-photon mode splitting in a semiconductor quantum microcavity, *Phys. Rev. Lett.* 69:3314 (1992).
33. Machida S., and Yamamoto Y., Observation of sub-poissonian photoelectron statistics in a negative feedback semiconductor laser, *Opt. Comm.* 57:290 (1986); Richardson W.H., Machida S. and Yamamoto Y., Squeezed photon number noise and sub-poissonian electrical partition noise in a semi-conductor laser, *Phys. Rev. Lett* 66:2867 (1991).
34. Loudon R., The quantum theory of light, Oxford University Press (1983).
35. Cohen-Tannoudji C., Dupont-Roc J., and Grymberg G. "An Introduction to Quantum Electrodynamics" and "Photons and atoms", Wiley, New York (1992).
36. Haroche S., Rydberg atoms and radiation in a resonant cavity, *in:* "New Trends in Atomic Physics, Les Houches Summer School Session XXXVIII", Grymberg G., and Stora R. eds., North Holland, Amsterdam (1984).
37. Haroche S., Cavity Quantum Electrodynamics, *in:* "Fundamental Systems in Quantum Optics, Les Houches Summer School, Session LIII", Dalibard J., Raimond J.M., and Zinn-Justin J., eds.,North Holland, Amsterdam (1992).
38. Haroche S. and Raimond J.M., Radiative properties of Rydberg states in resonant cavities, *in:* "Advances in Atomic and Molecular Physics Vol XX", Bates D., and Bederson B. eds., Academic Press, New York (1985).

39. Haroche S. and Raimond J.M., Manipulation of non classical field states in a cavity by atom interferometry, *in* "Cavity Quantum Electrodynamics, Special Issue of Advances in Atomic and Molecular Physics", Berman P. ed., Academic Press, New York, p123 (1994).

40. Glauber R.J., Optical coherence and photon statistics, *in:* "Quantum Optics and Electronics, Les Houches Summer School", de Witt C., Blandin A., and Cohen-Tannoudji C. eds., Gordon and Breach, London (1965).

41. Hulet R.G., and Kleppner D., Rydberg atoms in "circular" states, *Phys. Rev. Lett* 51:1430 (1983).

42. Rempe G., Thompson R.J., Kimble H.J., and Lalezari R., Measurement of ultra low losses in an optical interferometer, *Opt. Lett.* 17:363 (1992).

43. Mollow B.R., Power spectrum of light scattered by two-level systems, *Phys. Rev.* 188:1969 (1969).

44. Zhu Y., Gauthier D.J., Morin S.E., Wu Q., Charmichael H.J., and Mossberg T.W., Vacuum Rabi splitting as a feature of linear dispersion theory: analyzis and experimental observation, *Phys. Rev. Lett.* 64:2499 (1990).

45. Cohen-Tannoudji C., Introduction to quantum electrodynamics, *in:* "New Trends in Atomic Physics, Les Houches Summer School Session XXXVIII", Grymberg G. and Stora R. eds., North Holland, Amsterdam (1984).

46. Dalibard J., Dupont-Roc J., and Cohen-Tannoudji C., Vacuum fluctuations and radiation reaction: identification of their respective contributions, *J. Phys. (Paris)* 43:1617 (1982).

47. Raimond J.M., Vitrant G., and Haroche S., Spectral line broadening due to the interaction between very excited atoms: the dense Rydberg gas, *J. Phys. B. Lett.* 14:L655 (1981).

48. Dicke R.H., Coherence in spontaneous radiation processes, *Phys. Rev.* 93:99 (1954).

49. Gross M., and Haroche S., Superradiance, an essay on the theory of collective spontaneous emission, *Phys. Rep.* 93:302 (1982).

50. Davidovich L., Raimond J.M., Brune M., and Haroche S., Quantum theory of a two-photon micromaser, *Phys. Rev.* A36:3771 (1987).

51. Sorokin P.P., and Braslau N., Some theoretical apsects of a proposed double quantum stimulated emission device, *IBM J. Res. and Dev.* 8:177 (1964); Prokhorov A.M., Quantum electrodynamics, *Science* 149:828 (1965).

52. Filipowicz P., Javanainen J., and Meystre P., Theory of a microscopic maser, *Phys. Rev.* A34:3077 (1986).

53. Filipowicz P., Javanainen J., and Meystre P., Quantum and semi classical steady states of a kicked cavity mode, *J. Opt. Soc. Am.* B3:906 (1986).

54. Rempe, G. and Walther H., Sub-poissonian atomic statistics in a micromaser, *Phys. Rev.* A42:1650 (1990).

55. Meystre P. and Wright E.M., Measurements-induced dynamics of a micromaser, *Phys. Rev.* A37:2524 (1988).

56. Benson O., Raithel G. and Walther H., Quantum jumps of the micromaser field: dynamic behavior close to phase transition points, *Phys. Rev. Lett.* 72:3506 (1994).

57. Leggett A.J., Chakravarty S., Dorsey A.T., Fisher M.P.A., Garg A. and Zwerger W., Dynamics of the dissipative two-state system, *Rev. Mod. Phys.* 59:1 (1987); Caldeira A.D., and Leggett A.J., Quantum tunneling in a dissipative system, *Ann. Phys. (N.Y.)* 149:374 (1983).

58. Raimond J.M., Brune M., Goy P., and Haroche S., Micromaser à deux photons, *Ann de Physique, Coll, Paris* 15:17 (1990).

59. Zurek W. Decoherence and the transition from quantum to classical, *Physics Today*, Oct 1991, p. 36 .

60. Bell, J.S., Speakable and unspeakable in quantum mechanics, Cambridge University press, Cambridge, (1987).

61. Greenberger, D.M., Horne, M.A., Shimony, A., and Zeilinger, A., Bell's theorem without inequalities, *Am. J. of Phys.* **58**, 1131 (1990).

62. Deutsch D., Quantum theory, the Church-Turing principle and the universal quantum computer, *Proc. Roy. Soc. London* A400:97 (1985); Deutsch D., Quantum computational networks, *Proc. Roy. Soc. London* A425,73 (1989); Deutsch D. and Josza R., Rapid solution of problems by quantum computation, *Proc. Roy. Soc. London* A439,553 (1992).

63. Haroche S., Brune M., and Raimond J.M., Trapping atoms by the vacuum field in a cavity, *Euro. Phys. Lett.* 14:19 (1991).

64. Englert B.G., Schwinger J., Barut A.O. and Scully M.O., Reflecting slow atoms from a micromaser field, *Europhys. Lett.* 14:25 (1991).

65. Ivanov D., and Kennedy T.A.B., Photon number measurements with cold atoms, *Phys. Rev.* A47:566 (1993).

66. Cohen-Tannoudji C., Atomic motion in laser light, *in:* "Fundamental Systems in Quantum Optics, Les Houches Summer School, Session LIII", Dalibard J., Raimond J.M., and Zinn-Justin J. eds., North Holland, Amsterdam (1992).

67. Ramsey N.F., " Molecular Beams", Oxford University Press, New York (1985).

68. Nussenzveig P., Bernardot F., Brune M., Hare J., Raimond J.M., Haroche S.and Gawlik W., Preparation of high principal quantum numbers circular states of rubidium *Phys. Rev.* A48:3991 (1993).

69. Gross, M. and Liang, Is a circular Rydberg atom stable in a vanishing electric field?, J., *Phys. Rev. Lett.* **57**, 3160 (1986).

70. Braginsky V.B., and Khalili F.Y., *Zh. Eksp. Theor. Fiz.* 78:1712 (1977) [Quantum singularities of a ponderomotive meter of electromagnetic energy, *Sov. Phys. JETP* 46:705 (1977)].

71. LaPorta A., Slusher R.E., and Yurke B., Back-action evading measurements of an optical field using parametric down conversion, *Phys. Rev. Lett.* 62:28 (1989).

72. Levenson M.D., Shelby R.M., Reid M., and Walls D.F., Quantum non demolition detection of optical quadrature amplitudes, *Phys. Rev. Lett.* 57:2473 (1986).

73. Grangier P., Roch J.F. and Roger G., Observation of backaction-evading measurement of an optical intensity in a three level atomic non-linear system, *Phys. Rev. Lett.* 66:1418 (1991).

74. Nagourney W., Sandberg J. , and Dehmelt H., Shelved optical electron amplifier: observation of quantum jumps, *Phys. Rev.Lett.* 56:2797 (1986).

75. Haroche S., Brune M., and Raimond J.M., Manipulation of optical fields by atomic interferometry: quantum variations on a theme by Young, *Appl. Phys.* B54:355 (1992).

76. Haroche S., Brune M., and Raimond J.M., Measuring photon numbers in a cavity by atomic interferometry: optimizing the convergence procedure, *Journal de Physique II, Paris,* 2:659 (1992).

77. Mermin, N.D., What is wrong with these elements of reality? *Phys. Today,* **43**, 9 (1990).

78. Mermin N.D., Extreme entanglement in a superposition of macroscopically distinct states, *Phys. Rev. Lett.* **65**, 1838 (1990).

79. Shor P.W., Algorithms for quantum computation: discrete log and factoring, in *Proceedings of the 35th annual symposium on the foundations of computer science* IEEE Computer Society Press, Los Alamitos, CA p.124 (1994).

80. DiVicenzo D.P., Two-bit gates are universal for quantum computation, *Phys. Rev. A* 51:1015 (1995).

81. Sleator T. and Weinfurter H., Realizable quantum logic gates, *Phys. Rev. Lett.* 74:4087 (1995).

82. Unruh W.G., Maintaining coherence in quantum computers, *Phys. Rev. A,* 51:992 (1995).

83. Domokos P., Raimond J.M., Brune M. and Haroche S., Simple cavity-QED two-bit universal quantum logic gate: principle and expected performances, *Phys. Rev. A* to be published.

84. Yurke, B. and Stoler, D., Generating quantum mechanical superpositions of macro-

scopically distinguishable states via amplitude dispersion, *Phys. Rev. Lett.* **57**, 13 (1986); Yurke, B., Schleich, W., and Walls, D.F., Quantum superpositions generated by quantum non-demolition measurements, *Phys. Rev. A* **42**, 1703 (1990); Milburn, G., Quantum and classical Liouville dynamics of the anharmonic oscillator, *Phys. Rev. A* **33**, 674 (1986).

85. Haroche, S., Brune, M., Raimond, J.M., and Davidovich, L., Mesoscopic quantum coherences in cavity QED, in *Fundamentals of Quantum Optics III*, ed. Ehlotzky, F., Springer-Verlag, Berlin, (1993).

86. Dubreuil N., Knight J.C., Leventhal D.K., Sandoghdar V., Hare J. and Lefèvre-Séguin V., Eroded monomode optical fiber for whispering gallery mode excitation in fused silica microspheres, *Opt. Lett.* 20:813 (1995).

87. Weiss D.S., Sandoghdar V., Hare J., Lefèvre-Séguin V., Raimond J.M. and Haroche S., Splitting of high Q Mie modes induced by light backscattering in silica microspheres, *Optics Lett.* to be published.

88. Knight J.C., Dubreuil N., Sandoghdar V., Hare J., Lefèvre-Séguin V., Raimond J.M. and Haroche S., Mapping whispering gallery modes in microspheres using a near-field probe, *Opt. Lett.* to be published.

89. Lu B., Wang Y., Li Y., Lu Y., High order resonance structures in a Nd-doped glass microsphere, *Opt. Comm* 108:13 (1994).

PHOTON STATISTICS, NON-CLASSICAL LIGHT AND QUANTUM INTERFERENCE

J.G.RARITY AND P.R.TAPSTER
Defence Research Agency
St Andrews Rd, Malvern,
Worcestershire WR14 3PS
UK

1. Introduction

In this chapter we hope to give an overview of studies in non-classical photon statistics and interference phenomena. We start by studying the generic optical measurement as illustrated in figure 1. The light is created in an emission process typically involving the transition of an atom from the excited state to the ground state with consequent emission of a 'photon' which is a manifestly quantum event. Photo-detection is also manifestly quantum in that a single electron is excited into the conduction band by the absorbtion of a 'photon' from the electro-magnetic field. The photo-electron can subsequently be amplified and detected in a pulse discriminator as a single 'photon' detection. With such a description the reader could be led to believe that the light propagating between the source and detector can be described simply as a stream of these particle like photons. In the first part of this article (section 2) I will describe measurements where interference effects can be largely ignored. The photodetection statistics are essentially those of a Poisson point process arising from the lack of correlation between individual emission events (section 2.1). When the source shows classical fluctuations these statistics are broadened. However our experimental goal in recent years has been to reduce the random fluctuations to below the Poisson value thus producing non-classical light with applications in low noise measurement. In section 2.2 we describe how parametric pair photon sources can be used to demonstrate non-classical effects such as antibunching, one-photon states and sub-poissonian light generation. We then extend the single photon counting picture to photocurrent detection with the in-

47

M. Ducloy and D. Bloch (eds.), Quantum Optics of Confined Systems, 47-73.
© *1996 British Crown. Printed in the Netherlands.*

troduction of the concepts of shot noise limited and sub-shot noise light (section 2.3). This is illustrated by experiments using parametrically generated twin beams and current noise limited solid state emitters (section 2.4).

The notion of particle like photons is dismissed in section 3 where we discuss experiments in which the particle and wave properties of photons are explored. Here a full description involves a probability amplitude for emission at a particular time where the amplitude is related to the field in a classical description of the experiment. The detection process involves the coherent summation of all possible probability amplitudes arising from the many paths from emitter to detector. The square modulus of this sum gives us the detection probability per unit time equivalent to the classical intensity. This is illustrated using experiments showing one photon interference in arrangements analogous to Young's slits (section 3.1). Here the simplicity is that the infinite number of paths between source and detector are reduced to two. This concept can be extended to two photon interference experiments where again interference occurs between two indistinguishable paths leading to the detection of two photons, sometimes at remote locations (section 3.3) . This leads us on to illustrate the non-local nature of these quantum mechanical objects we like to call photons. Application of such one- and two-photon interference effects in secure key sharing schemes (popularly known as Quantum Cryptography) is also described (section 3.2). We finish (section 3.4) by describing recent work where the two indistinguishable paths involve a spontaneous emission of a photon pair and discuss the resolution of philosophical questions in such experiments.

2. Photon Statistics

In this section I will try briefly to review photon statistics from the experimentalists view point. The first point to note is that photon statistics is a misnomer and we never measure more than the statistics of the individual photo-electrons (section 2.1) or the integrated photo-current (section 2.3). For space reasons, for much of the detailed derivations I will refer the student elsewhere [1, 2].

2.1. PHOTO-ELECTRON STATISTICS

In the generic experiment (figure 1) a classical constant intensity (theorists should read 'coherent state' here, see [1]) source results in a Poisson random train of photo-electron pulses at the detector. The Poisson process reflects the uncorrelated nature of the emission/detection events. Photodetection at a time t does not change the probability of detection at any other time which remains proportional to the mean detection rate multiplied by the sampling

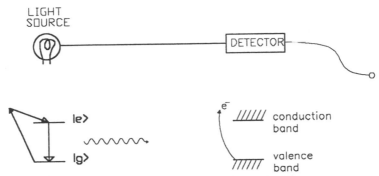

LIGHT
SOURCE

DETECTOR

|e>

|g>

conduction
band

valence
band

Figure 1. The generic optical experiment. Light is typically created by an atom passing from excited to ground state and detected when an electron is excited from a valence band to the conduction band. These are clearly quantum processes but the light can usually be treated classically in propagation.. In this chapter we explain experiments designed to show the limitations of these semi-classical models

interval. This Poisson process is well known as the source of (baseline) noise in all classical light measurement experiments and also in nuclear particle counting experiments. It is characterised by a variance equal to the mean $< n^2 > - < n >^2 = < n >$ and counting distribution

$$P(n, T) = \frac{< n >^n}{n!} \exp - < n >. \tag{1}$$

where $P(n, T)$ is the probability of counting n photo-events in the time interval T.

In general photo-event statistics can be characterised by the moments of this distribution and for light we often study the normalised factorial moments

$$n^{(r)} = \frac{\langle n(n-1)..(n-r+1) \rangle}{\langle n \rangle^r} \tag{2}$$

where the angle brackets define the ensemble average,

$$\langle f(n) \rangle = \Sigma_{n=0}^{n=\infty} f(n) P(n, T) \tag{3}$$

The factorial moments are used in the study of photo-count statistics because all $n^{(r)} \equiv 1$ for Poisson light (easily checked for all moments). Hence factorial moments below unity are characteristic of light with fluctuations below the classical limit. For classical fluctuations the factorial moments are equivalent to the normalised moments of the classical intensity fluctuations $< I^r > / < I >^r$ thus are very useful measures.

To characterise the timescale of fluctuations the photo-count autocorrelation function is normally used

$$g^{(2)}(\tau) = \frac{< n(t)n(t+\tau) >}{< n >^2} \tag{4}$$

Although in theory $g^{(2)}$ is a continuous function in most experimental situations delay time is incremented in integer units of the sampling interval $\tau = jT$ and $n(t)$ represents the number of counts measured in the interval t to $t + T$. Due to lack of time correlation between individual photo-events of Poisson light the correlation function measured from our ideal constant intensity source will be unity at all non-zero times while at zero time $g^{(2)}(0)$ is obtained from the Poisson variance,

$$g^{(2)}(\tau) = 1 \qquad\qquad \tau \neq 0 \qquad\qquad (5)$$

$$g^{(2)}(0) = < n^2 > / < n >^2 \equiv 1 + 1/ < n >$$

For light with classical intensity fluctuations the correlation function is greater than unity and equivalent to the classical intensity correlation function $g^{(2)}(\tau) \equiv < I(0)I(\tau) > / < I >^2$. As a result $g^{(2)}$ is often used in low light level scattering experiments to provide velocity information and diffusion coefficients of particles.

2.2. PHOTODETECTION

To allow measurement of photon statistics we require an efficient photon counting detector. Until the early 80's the photomultiplier was the detector of choice for single photon counting. The photomultiplier consists of a series of electrodes (dynodes) in a vacuum tube fronted by a photosensitive and conducting window, the photocathode. A high voltage is applied between the photocathode and first dynode. Photoelectrons ejected from the photo-cathode by absorption of a light quantum are accelerated into this first dynode where they create a number ($\simeq$ 20) of secondary electrons. This amplification process is repeated as the secondary electrons are accelerated down the chain from dynode to dynode by smaller voltage steps. The gain of the whole process can lead to 10^7 electrons reaching the anode for each detected photoelectron. These macroscopic electron pulses can be detected in a discriminator circuit to give standardised pulses suitable for photon counting and timing apparatus. The noise in this gain process is quite low as the initial gain step between photo-cathode and first anode is high. The output pulse height distribution is narrow and suitably engineered phototubes can have very low background and dark noise. The design of an efficient and low noise photocathode is critical to the performance of the photomultiplier. The cathode has to be conducting yet near transparent so that photons are absorbed close to the surface and electrons are ejected into the vacuum and accelerating field. This limits the maximum efficiency to around 50% for conventional designs. Typical red extended photocathodes (S20) will have quantum efficiencies peaking around 40% in the region of 500nm and falling to less than 5% beyond 633nm.

In recent years the photomultiplier has been challenged by the avalanche diode biased beyond breakdown both passively[3] and actively quenched[4]. Although biased beyond breakdown the diode will not break down until an electron-hole pair is created in the avalanche region either by absorbtion of a quantum or by noise sources. The electron and hole are accelerated in the high field of the avalanche region and successive collisions create further electron hole pairs. Below breakdown the device has a finite but extremely noisy gain arising from such a random binary amplification process and the resulting pulse height distribution is extremely broad. As yet no photon counting devices operating below breakdown have proven practical. Above breakdown a macroscopic current flows until quenched by an external circuit and a narrow pulse height distribution results (although due to the noise in gain not all photodetection events lead to a breakdown). In the simplest arrangement this breakdown is quenched by a series resistor between the device and the voltage supply. When the current flows the voltage across the resistor rises and the voltage across the device drops to below the breakdown voltage. Improved device performance has been achieved in recent years in silicon where the hole to electron ionisation ratio K can be engineered to be very low[5]. Commercially available devices show dark counts below 100 counts per second (cps) when cooled to -20°C and quantum efficiencies greater than 70% in the near infra-red and reasonable efficiency from 400nm to 1μm wavelength. The compact size, lower operating voltages and order of magnitude improvement in quantum efficiency in the near infra-red when compared to photomultipliers has made the silicon avalanche diode the detector of choice in most recent experiments.

Beyond 1μm InGaAs or germanium avalanche diodes must be used. Material quality in these devices is much poorer and thus noise and dark counts dominate their performance near room temperature. Performance of Ge avalanche diodes at 77K [6] is summarized in figure 2 where we see efficiencies of order 10% with dark counts of tens of Kilohertz. Further cooling does not improve this performance. Commercially available InGaAs avalanche detectors operate at slightly higher temperatures (175K) but appear to have lower quantum efficiencies and higher dark counts than the Ge devices[7].

2.3. ANTIBUNCHING EXPERIMENTS

Our interest has been primarily in the creation and use of non-classical lights where factorial moments and correlation functions dip below unity. The first experiments showing light with such properties used the resonance fluorescence of single atoms [8, 9]. In principle an atom cannot emit more than one quantum of light at a time. The maximum rate at which emission

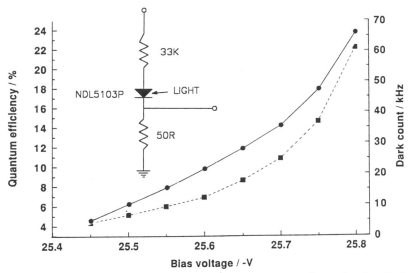

Figure 2. Performance of passively quenched photon counting Ge avalanche diodes at 77K. Inset show the passive quenching circuit. Bias voltage is applied to the 33 kOhm resistor and pulses are measured across the 50 Ohm. Filled circles with solid line show the quantum efficiency and filled squares joined by the dashed line shows dark count both plotted as a function of bias voltage. Breakdown voltage for this device is 25.32V. Further details in [6].

can occur is set by the natural lifetime of the atom coupled with the intensity of the pumping light. In spectral terms a strongly pumped atom shows a split emission spectrum with the splitting being the Rabi frequency. The probability of seeing a photon a short time after an emission event is reduced below the classical value and this leads to a measured autocorrelation function suppressed below unity at near zero delay time.

A second means to control the quantum noise of light has been through feedback methods. Initially closed loop experiments were carried out using the detected output fluctuations negatively fed back to a modulation means in the beam [10, 11]. Although this can produce light with non-classical statistics at the detector the closed loop feedback can always be described semi-classically [12, 13] and no measurement gain can be achieved in this way. One way around this problem is to create two beams with identical quantum statistics and use measurement of one beam to correct fluctuations in the other. To do this we use parametric downconversion in ($\chi^{(2)}$) non-linear birefringent crystals pumped by short wavelength polarized (laser) light (see [14]). The birefringence of the crystal allows phase and energy conservation to be satisfied for creation of a cone of downconverted light of different polarization. Pair beams with identical quantum statistics can be selected from this cone with frequencies summing to that of the pump beam

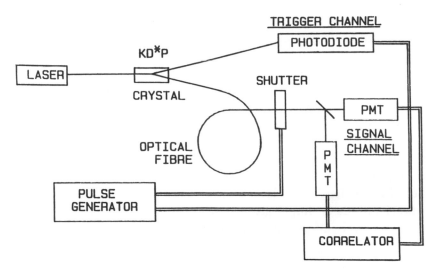

Figure 3. Apparatus used to optically gate out single photon states. See text and [18] for details

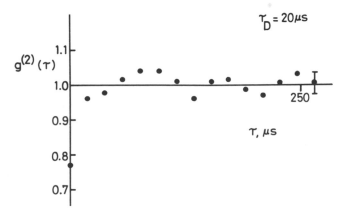

Figure 4. Correlation function $g^{(2)}(\tau)$ as a function of delay time τ. Measured across the signal beamsplitter as shown in figure 3

(energy conservation) and at angles satisfying $\vec{k}_0 = \vec{k}_1 + \vec{k}_2$ (momentum conservation) where $\vec{k}_i$ are the wave vectors of pump (0) and downconverted beams (1,2).

Twin beams of this type were first identified some twenty years ago by photon counting coincidence measurements[15] . However it was not until the early eighties that it was realised that the detection of photons in one beam could be used to modify the statistics of the other. This led to the demonstration of one photon (Fock) states[16], antibunching[17] and sub-Poissonian light[18]. These and other anti-bunching experiments are reviewed in [2].

In this chapter we give a brief description of one of them [18] where we used optical gating to produce anti-bunched and sub-Poissonian light. The apparatus is illustrated schematically in figure 3. A short wavelength laser (HeCd 325nm) illuminated a crystal of deuterated potassium dihydrogen phosphate (KD*P). The crystal was cut so that pair beams (at 650nm) could be selected from the downconversion cone at angles around ±3 degrees to the pumping beam. A high efficiency avalanche diode detected a train of photo-electron pulse in a "trigger" beam while the other "signal" beam was measure using two photomultipliers looking through a 50/50 beamsplitter. The rate of coincident detections between one of the photomultipliers and the avalanche diode is simply given by the rate of emission of photon pairs r multiplied by the lumped channel efficiencies η_s, η_t

$$< C >= r\eta_s\eta_t \tag{6}$$

In this experiment the aim was to create a light source with non-classical statistics. To do this the trigger detections were used to open an optical shutter (accousto-optic switch) for a short time t thus allowing only one photon of signal light through. To allow time for the electronic delays the signal beam passed through 170m of multimode fibre before the accousto-optic switch. The trigger pulses form a Poisson random time sequence thus the gated single photons show similar statistics. To regularise this Poisson train a dead time τ_D is introduced after every trigger pulse. This produces an anti-bunched and slightly sub-poissonian train of signal photons. To measure the intensity correlation function of the output light without distortion due to detector afterpulsing we analyse the cross correlation between two detectors measuring across a 50/50 beamsplitter. With classical light in a beamsplitter the intensity at the outputs are equal and the probability of seeing a photo-electron pulse in each output is dependent simply on that intensity. We find that the normalised cross-correlation function is unity and equivalent to the auto-correlation function for all non-zero delays independent of the intensity. In our case we have gated out single photons. We cannot then split these quanta and detect half a photon in each detector. The result then is an anti-correlation between the detectors at zero time extending out to delays equal to the dead time. This translates to a cross-correlation function suppressed below unity at near zero delay. In a rigorous analysis we can show that the cross correlation function is again equivalent to the auto-correlation function at non-zero times and equivalent to the second factorial moment at zero delay. The result (figure 4) shows a correlation with early points suppressed which confirms anti-bunched and sub-poissonian behaviour. By ratioing the coincident counts (eq 6) to the mean counting rate in each detector ($\eta_t r, \eta_s r$) we can measure the effective detector efficiencies [19]. The efficiency of the trigger channel was about 10% while that of the signal channel was 0.06%. This low efficiency was due to optical losses in the delay and shutter (10% transmission) and the use of photomultipliers as detectors (2% efficiency at this wavelength) as well as the 50% loss at the beamsplitter. If we had a true source of single-photon states the suppression of the first channel in figure 4 would be down to zero. Here we are primarily corrupted by dark counts in the photomultipliers (10 counts per second) and the low efficiencies. When we correct for the random coincidences between dark counts and signal counts in the two photomultipliers the intercept is reduced to $n^{(2)} = 0.42$ from a measured value 0.77. Ideally the intercept should drop to zero but the remaining counts arise simply from the fact that we have a finite shutter open time $t \simeq 0.2\mu s$ and thus a low but significant probability of two signal detections (one in each photomultiplier) within this time window.

Due to these limited efficiencies the variance of this source which is strongly antibunched is only slightly sub-poissonian. This is made obvious when we write the identity

$$\frac{<n^2> - <n>^2}{<n>} = 1 - <n>(1 - n^{(2)}) \tag{7}$$

and consider that $<n> \simeq 10^{-4}$ in the sample time of $20\mu s$ used here.

2.4. PHOTOCURRENT DETECTION AND SHOT NOISE

To get around the efficiency problems of photon counting experiments we can use high efficiency of near IR silicon and InGaAs PIN photodiodes without internal amplification. Here, each photodetections creates one electron pulse. At quite low powers (typically nA detected current) the pulse decay time is longer than the interevent time. The resulting pulse pile up leads to a continuous analogue photocurrent. The noise in this photocurrent arises from the inherent Poisson randomness of the underlying individual photoelectrons and is known as shot-noise. In continuous analogue measurements it is easier to talk of noise powers in the various frequency bands making up a signal. We can relate the power spectrum of the photocurrent noise $S_i(f)$ to its autocorrelation function $C_i(\tau) =< i(t)i(t+\tau) >$ using the Weiner-Khintchine theorem which states

$$S_i(f) = \int_{-\infty}^{\infty} C_i(\tau)e^{i2\pi f\tau}d\tau \tag{8}$$

Given that $i(t) = en(t)/T$ and writing the mean current $I =< i >= e < n > /T$ we can write

$$C_i(\tau) = I^2 g^{(2)}(\tau) \tag{9}$$

This then allows us to calculate the spectral density using equation 8 in the limit $T \to 0$.

$$S_i(f) = 2Ie \tag{10}$$

This compares with the thermal fluctuations in the number of electrons passing through a resistor value R

$$S_i(f) = \frac{4kT}{R} \tag{11}$$

where kT is the Boltzmann energy. In an idealised detector the photocurrent is measured as a voltage across a resistor of value R. The ratio of thermal (Johnson) noise to shot noise is thus

$$\frac{S_{Electron}}{S_{Shot}} = \frac{2kT}{eV} \tag{12}$$

where $V = IR$ is the voltage produced across the resistor. As kT/e is 25mV at room temperature we need to use a resistor large enough to drop a voltage greater than 50mV to reduce electronic noise below the shot noise level.

2.5. SUB-SHOT-NOISE SOURCES AND MEASUREMENTS

From the above we see that to get above the electronic amplifier noise we need higher downconversion powers or measurements across a high resistance. The highest practical resistance one can use is around a GigaOhm limiting the frequency response with a few pF stray capacitance to a few kilohertz. Strong noise correlations between twin beams created in parametric downconversion have been measured at nanowatt powers (1 kilohertz) by us[20], at milliwatt powers[21, 22] in optical parametric oscillators (10 MHz) and in parametric amplifier experiments using high peak power pulsed beams [23]. Photocurrent correlations of order 0.9 of the ideal (unity) are measured at milliwatt power levels[22, 23]. In all cases it is possible to use the fluctuations measured on one beam to correct the fluctuations on the other leading to sub-Poissonian or sub-shot-noise light sources[24, 25].

One projected use for twin beam light is in improving the accuracy of transmission measurements (in absorbtion spectroscopy for example[26, 20]). In certain biological or light sensitive samples a maximum absorbed dose limit will exist, limiting the accuracy with which a measurement of the absorption can be performed. In figure 5 we show the principle of noise reduction by subtraction of the identical shot noise of quantum identical beams. Here we describe a simple demonstration experiment where the signal to noise ratio could be improved by up to 4dB in a measurement of turbidity in a weakly scattering sample [20].

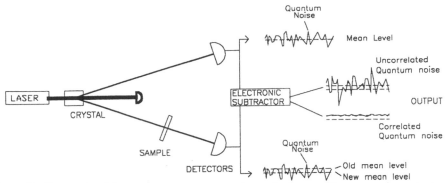

Figure 5. Improving Measurement accuracy using twin beams. From a parametric down-conversion source we select out identical photon trains correlated in emission time and thus with identical shot-noise. When we detect these beams and subtract photocurrents the identical noise components cancel (lower output). This allows measurement of the small change in the mean intensity due to insertion of the sample. The upper output trace shows the classical result when shot noise in the two beams are uncorrelated and the small change in intensity cannot be detected.

In the experiment an amplitude stabilised Krypton ion laser producing 400mW of 413.4nm wavelenth light was weakly focussed into a Lithium Iodate non-linear crystal cut so that phase matched 826.8nm light was emitted in a cone of half angle 10°. Correlated beams were selected from this cone by apertures subtending an angle of $\simeq 1°$ at the crystal. The resulting beams were broad band (aproximately 60nm bandwidth). Both beams are focussed using anti-reflection (AR) coated lenses through AR coated colour glass filters (cut-on > 750nm) onto PIN photodiodes connected to low noise transimpedance amplifiers with gains around 1GOhm and. The current noise in the amplifiers was close to the Johnson noise for $R = 1GOhm$ and thus was small compared with the typical shot noise levels when photocurrents of order 4nA were measured. The DC level and noise outputs of the two detector amplifiers were separately digitised in a computer programmed to perform repeated fast fourier transform spectrum analysis to allow long time averaging of the noise.

This apparatus was used to demonstrate the potential signal to noise gains in simple transmission measurements. A sample consisting of a liquid crystal cell with mean transmission coefficient $\alpha = 0.85$ was placed in front of one detector (see figure) and its transmission was weakly modulated at 480Hz and 960Hz by applying a low voltage sine wave across the cell.

The computer calculated noise voltage spectra with frequency bin width δf

$$< \Delta V_I^2 >= R^2 S_i(f)\delta f \qquad (13)$$

58

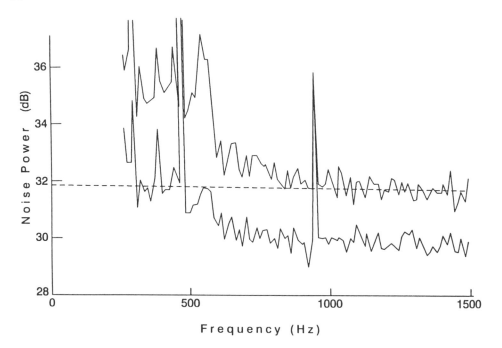

Figure 6. Results of a measurement of modulated absorption as schematically shown in figure 5. The peak at 980Hz is the signal arising from weak modulation in the transmission of a liquid crystal cell. The surrounding background is a measure of noise power as a function of frequency. When we detect the light in a single detector we obtain the upper curve S_{I1}, which is close to the shot noise level as shown by the dashed line. The lower curve shows S_{I1-I2} (equation 14) where the background is clearly reduced to below the shot noise level. The vertical scale is referenced to 0dB$\equiv 1\mu V^2$/Hz across a 1GOhm resistor. For more details see [20]

and the voltage difference spectra

$$< (\Delta V_1 - k\Delta V_2)^2 > = R^2 S_{I1-I2}(f)\delta f \tag{14}$$

between the two detector amplifier outputs (labelled 1 & 2) with feedback (more correctly feedforward) parameter k. For optimum quantum noise reduction[20] $k \simeq \alpha\eta_1$ where η_1 is the effective efficiency (including optical losses) of channel 1. In this case the noise referenced to the single beam shot noise limit is given by

$$\frac{S_{I1-I2}}{2eI_1} = (1 - \alpha\eta_1\eta_2) \tag{15}$$

The result of a measurement with detected photocurrents of $I \simeq 3.4$nA and using $k = 0.66$ is shown in figure 6 along with the uncorrected voltage noise spectrum which reached the shot noise limit above 800Hz. Clearly the background noise level of the difference spectrum drops some 2dB below the shot noise limit for measurements at 960Hz. When there is also significant classical noise (such as at the 480Hz signal) it is more realistic to use full feedback $k \simeq 1$ and to compare with the sum of the noise variance in both beams (S_{I1+I2}). In this hypothetical situation 4dB of measurement improvement beyond the dual beam shot noise limit could be acieved.

Improved detectors (to 99.5% efficiency[27]), a higher sample transmission and lower noise amplifiers could improve the measurement to give near 10dB noise reduction. It must be noted that the maximum noise reduction even with ideal detection will be ultimately limited to $1 - \alpha$ which is the fraction of light lost in the sample.

In this section I have omitted the subject of phase squeezing [28, 29, 30] partly because this subject involves phase and interference effects which I have avoided in this section. I give a brief word picture in the following and assume a more detailed coverage elsewhere in these proceedings. In degenerate parametric downconversion the pair beams cannot be separated by polarisation or momentum. Downconversion can be viewed as a parametric amplification process. As in any parametric amplifier there can be gain g when the signal is in one quadrature (a cosine wave for instance) and de-amplification $1/g$ in the other (the sine wave component in the simple example). This property means that light created in strongly pumped degenerate parametric downconversion will be phase squeezed. This non-classical property is measured when the light is mixed at a beamsplitter with a coherent local oscillator beam in phase with either of these quadratures. When in-phase with the deamplified quadrature the output beams from the beamsplitter are quantum correlated which can be measured as a reduced difference noise in detected photocurrents below the normal shot noise limit. The difference noise is greater than shot-noise when the other quadrature is selected (by phase shifting the local oscillator beam by $\pi/2$). Such experiments can be configured to show improved measurement accuracy in both phase and transmission measurements [31, 32].

This section has also ignored the direct generation of sub-shot noise light in high efficiency constant current driven semiconductor LED's [33] and lasers [34] which should be addressed in more detail in the lectures of Yamamoto (in this book).

3.   Quantum Interference

3.1.   ONE-PHOTON INTERFERENCE

The first evidence of one photon interference was the experiments of G I Taylor[35] in 1909. Taylor used extremely faint light sources in a standard

Young's slit interference apparatus (see figure 7). In his experiments the source emitted much less energy than a single quantum ($\hbar\omega$) in the time it took for light to pass from source to screen. Thus the probability of more than one 'photon' being between source and screen at any one time was extremely low, yet after integrating long enough interference fringes still appeared at the detector (a sensitive photographic plate). This was the first experimental evidence supporting Dirac's famous statement 'Each photon then interferes only with itself'[36]. Of course this experiment could in principle be explained by semi-classical means without recourse to a fully quantised field. A definitive experiment came later[37] using a pair photon source based on correlated emissions from three level atoms. We show a schematic of such an experiment based on parametric downconversion in figure 7. A one-photon state can be created (see section 2.3) by electronic gating on the detection of its partner. This gated 'one-photon' state passes through a Mach-Zehnder interferometer and collection of many one-photon events at various interferometer settings will show the expected cosinusoidal interference fringes. At the same time the anticorrelation across the output ports of the second beamsplitter of the interferometer continues to confirm the presence of a non-classical field.

In a classical description of a simple Mach-Zehnder interferometer the incoming field I_{in} is split equally between the two arms and recombines at the second beamsplitter. The field arriving at the detector is the sum of two fields $E_{in}/2$ with phase difference determined by the path length difference

$$E_{tot} = \frac{1}{2}E_{in}(1 + e^{i\phi})e^{i\omega t} \tag{16}$$

and the resulting intensity distribution is given by the square modulus of the resultant field

$$I_{tot} = \mid E_{tot}\mid^2 = \frac{1}{2}\mid E_{in}\mid^2 (1 + \cos\phi) \tag{17}$$

In the quantum description of the experiment (figure 7) we associate a probability amplitude 1/2 for passage of the quanta via a particular arm (a or b) of the interferometer and add amplitudes at the detectors to produce an effective wavefunction

$$\Psi c/d = \frac{1}{2}\left[\mid 1>_a + e^{i\phi}\mid 1>_b\right] \tag{18}$$

We denote the modes of the interferometer arms as $\mid >_a, \mid >_b$ and in the experiments described above these modes are occupied by a single photon at a time. When these modes are made indistinguishable as viewed from the detector we can remove the subscripts and see interference in the probability

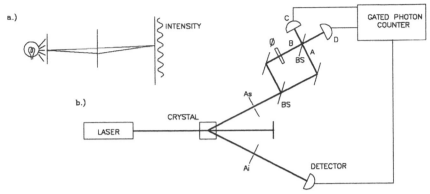

Figure 7. a.) Young's slit experiment carried out by Taylor using extremely weak light sources.. b.) True 1 photon interference apparatus where we gate out single photons in the signal channel (As) using a time correlated partner events in the idler channel (Ai).

of detection of a single photon at the interferometer output. This probability for detector D is given by

$$P_D(1) = < \Psi | \hat{a}_d^\dagger \hat{a}_d | \Psi > \tag{19}$$

where $\hat{a}_d^\dagger, \hat{a}_d$ are the creation and annihilation operators respectively for the detected mode $| >_d$. This leads to a probability of detection given by

$$P_D(1) = \frac{1}{2}(1 + \cos \phi) \tag{20}$$

for each single photon state input into the interferometer. A corresponding equation with the plus sign replaced by a minus holds for the other output mode $| >_c$ so that the total probability of detection remains unity (in this idealised example).

3.2. ONE-PHOTON INTERFERENCE AND CRYPTOGRAPHY

In equation 20 above we see that the probability of seeing a photon in a particular interferometer output can be adjusted from 0 to unity simply by adjusting the phase ϕ. Of course the anticorrelation is retained and when the probability is zero at one output the probability at the other is unity. This suggest a means of interferometric coding where we we identify the outputs of the interferometer with 0's and 1's. Furthermore we can use this system to create a shared secure key at remote locations[38, 39, 40].

In a variant of these one photon interference experiments the feasibility of a simple key sharing scheme with a range of 10km and more has been demonstrated [41, 42, 43]. The apparatus is shown schematically in figure 8.

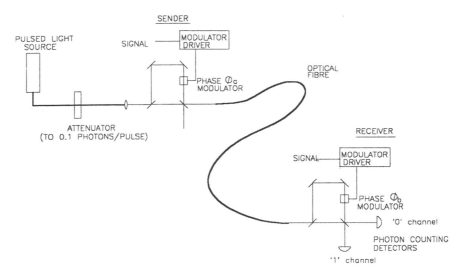

Figure 8. 10km long faint pulse interferometer using two time delay interferometers. Interference is seen between the two equal path-length routes through the system (either through the long then short path OR the short then long path). The phases can be adjusted using phase-plates ϕ_a, ϕ_b in the long arms of the interferometers.

Two identical time division interferometers are connected in series. A light source emitting short pulses of light is attenuated until the average photon number per pulse is about 0.1. The system works as an extended single Mach-Zehnder interferometer. The pulses are split in time in the first interferometer which has a path length difference much longer than the pulse width and the two pulses so formed propagate to the receiving interferometer. When the leading pulse takes the longer path and the trailing pulse takes the shorter path the pulses arrive simultaneously at the final beam-splitter and interference is seen. Events where the pulse takes the long (or short) path in both interferometers will not show interference but can be gated out as they will arrive later (or earlier) when compared to a timing reference originating in the source. Having gated out the non-interfering events the probability of seeing a photon in the '1' or '0' outputs for a single pulse is then given by

$$P_{1/0} = \overline{m}\alpha\eta/4\left[1 \pm cos\left(\phi_a - \phi_b\right)\right] \tag{21}$$

where ϕ_a is the transmitter phase, ϕ_b the receiver phase, α the lumped transmission, η the detection efficiency and $\overline{m} \ll 1$ the mean number of photons per pulse.

Clearly $\phi_a - \phi_b = 0$ can be used to direct gated pulses to the '0' channel and $\phi_a - \phi_b = \pi$ for the '1' channel. A simple scheme to establish identical random numbers (keys) at sender and receiver would use a fixed receiver phase, $\phi_b = 0$ for instance, and randomly switch the transmitter phase between $\phi_a = 0$ ('0') and π ('1'). The receiver will obviously receive only a small fraction of the sent bits when $\alpha\eta$ is small but the valid bits can be selected by sending information on detection times to the transmitter. A simple eavesdropping scheme here would involve an eavesdropper measuring all pulses with an interferometer identical to that at the receiver. As the mean number of photons per pulse is about 0.1 they cannot be detected by both eavesdropper and receiver. The eavesdropper thus measures all the pulses and creates copies to send to the receiver. To guarantee absolute security the receiver phase can be switched between $\phi_b = 0$ and $\phi_b = \pi/2$ with transmitter phases randomly selected from $\phi_a = 0, \pi$ and $\phi_a = \pi/2, 3\pi/2$ coding '0's and '1's. 100% correlation only occurs when the phase difference is 0 or π thus sender and receiver must communicate and discard all received pulses where the phase difference was $\pi/2$ or $3\pi/2$. After all uncorrelated bits are discarded the transmitter and receiver are left with near identical random bit strings to be used as a key. An eavesdropper must now guess which receiver phase was used $\phi_b = 0$ or $\pi/2$ and will resend pulses with errors 25% of the time (50% of the time they will use the wrong phase which leads to random detection in the '0' and '1' channels.). These errors can be detected by openly comparing a small fraction of the key bits (which are then discarded).

In the experimental test of this time-division interferometry interference visibilities greater than 0.95 corresponding to error rates below 3% with detection rates of a few kilo-counts per second (Kcps) at ranges of 10km were measured[41]. These experiments were carried out using standard telecommunications fibres and interferometers operating at 1.3μm with Germanium APD's operated in Geiger mode (see section 2.2) as detectors. Later work at British Telecom has demonstrated that key sharing at a few hundred bits per second can be performed of a similar system[43]. Other groups have made similar demonstrations using polarisation as the coding variable rather than interferometric phase[44, 45].

Obviously these experiments with faint pulses recreate the experiment of G I Taylor with a weak classical source which are in principle described semi-classically. To show truly non-classical behaviour we need to use one-photon states as created in pair photon gating experiments (see section 2.3) or by using engineered high efficiency LED's[33, 46].

4. Two-Photon Interference

Non-classical two photon interference effects were first predicted for a two photon field falling on a screen[47] and then in the interference of two 'one-photon' states input into opposite ports of a beamsplitter[48, 49]. In the simple particle picture used in section 2.3 a single photon input into a beamsplitter will appear randomly in one or other of the beamsplitter outputs. Extending this to one photon in each port we naively expect be no correlation in the output taken by each photon of the pair and thus a 50% probability that one photon will appear in each output port. A quantum mechanical picture includes the one-photon probability amplitudes $|1\rangle_1, |1\rangle_2$ and the phase shift between transmission amplitude t and reflection ir. If the input photons are indistinguishable from measurements at the beamsplitter outputs the beamsplitter output state can be represented by

$$|\Psi\rangle = 2^{-1/2}\left[|t^2|1\rangle_3|1\rangle_4 - r^2|1\rangle_3|1\rangle_4 + irt|2\rangle_3 + irt|2\rangle_4\right] \qquad (22)$$

We extend equation 19 to by including two sets of creation and annihilation operators for the two modes (3 and 4) and obtain the probability for seeing photons simultaneously in both beamsplitter outputs. We see a destructive interference effect can occur

$$P_{34} = t^4 + r^4 - 2r^2t^2 \qquad (23)$$

In fact when $r^2 = t^2 = 1/2$ (a 50/50 beamsplitter) $P_{34} = 0$ and the photons always appear as **pairs** in random outputs of the beamsplitter.

In section 2.3 we discussed the generation of photon pairs in parametric down-conversion. These pair beams can be made identical in wavelength simply by selecting them at equal (and opposite) angles to the pump beam and filtering through identical interference filters. Directing the beams into opposite input ports of a beamsplitter allows the above effect to be experimentally demonstrated[50, 51] as a reduction of the pair detection rate at the beamsplitter outputs. By scanning the path length difference from the crystal to the beamsplitter inputs one can show that the pairs are coincident to within their coherence length which can be much less than 100fs. This time is several orders of magnitude less than the fastest resolving time of typical photon counting detectors.

In an extension of this experiment the fact that the pairs need not be localised objects has been explored[52]. For communications applications of non-local correlations however, schemes suited to fibre optic implementation would be favoured. In the following we describe a fibre optic pair-photon interference experiment based on superpositions of emission times using out of balance interferometers[55, 56, 57].

4.1. LONG RANGE PAIR PHOTON INTERFEROMETRY

Time superposition interference experiments were first suggested by Franson[59] using atomic cascade pair photon sources. A schematic of a parametric downconversion based version using fibre-optic interferometers is

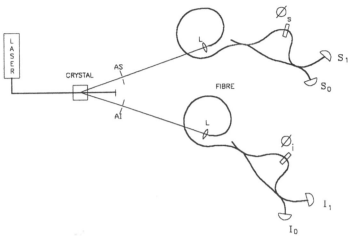

Figure 9. Schematic of the emission time superposition apparatus. Filters A_s and A_i select energy (and phase) matched photon pairs. They are launched into optical fibres and propagate through identical out of balance fibre-optic interferometers to signal $S_{0,1}$ and idler $I_{0,1}$ photon counting detectors. Coincidences between signal and idler detectors are measured as a function of phase shift ϕ_i and ϕ_s.

shown in figure 9. Pair beams created in parametric downconversion are launched into single mode fibres and propagate through identical fibre interferometers.

There are thus two routes from the crystal to the detector for each photon of the pair. A simplistic wavefunction for the two photons viewed from the detectors (but normalised at the crystal) assuming 50/50 beamsplitters is thus

$$\Psi_{si} = \frac{1}{4}[|1_{long}>_i |1_{long}>_s e^{i(\phi_i+\phi_s)}e^{i\Delta T(\omega_i+\omega_s)} + |1_{short}>_i |1_{short}>_s \quad (24)$$

$$+|1_{short}>_i |1_{long}>_s e^{i\phi_s}e^{i\Delta T\omega_s} + |1_{long}>_i |1_{short}>_s e^{i\phi_i}e^{i\Delta T\omega_i}]$$

where $|1>_{s,i}$ denotes a one-photon state in the s (for signal) and i (for idler) arm of the apparatus, the subscripts *long* and *short* denote propagation via the long and short paths through the interferometer and the small phase shifts $\phi_{i,s}$ incorporate any phase changes occurring on reflection while ΔT represents the mean time delay incurred on taking the long path in either interferometer. Containing a superposition of a *long, long* two-photon with a *short, short* two-photon this wavefunction directly expresses the uncertainty in time of emission of the photon pair. Coherence between the two possible emission times is ensured when the pump laser has a constant frequency ω_0 (ie a long coherence length as compared to ΔT) and when energy conservation is strictly applied in the relationship $\omega_i + \omega_s \equiv$

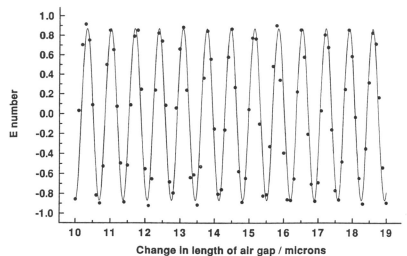

Figure 10. Coincidence correlation (E number) as a function of phase shift ϕ changed by extending the length of the air gap in the 830nm fibre interferometer. The sinusoidal fitting function (solid line) has a visibility of 87%. Reproduced from [61]

ω_0. Near coincident detections between *short, long* and *long, short* do not interfere because they are distinguishable. If detector time resolution is better than ΔT we can exclude these non-interfering events and evaluate the probability $P_{si}(j,k)$ of coincident detection in detectors indexed $j, k = 0, 1$ with efficiencies $\eta_{s,i}$

$$P_{si}(j,k) = \frac{1}{8}\eta_s\eta_i[1 + (-1)^{j+k}V \cos(\phi_i + \phi_s + \omega_0\Delta T)]. \qquad (25)$$

The interference has a maximum visibility of $V = 1$ and is clearly non-locally controlled by the phases ϕ_s, ϕ_i. In a realistic experiment the visibility is reduced from unity by dispersion effects in the fibres, changes of polarisation state in the interferometers and path length differences between the interferometers.

The similarity between equations 25 and 21 suggests a subtle key sharing scheme based on these non-local correlations[60]. In an ideal system we see 100% correlation when the phase sum is zero but by the very nature of the experiment the photodetections appear randomly in the '0' and '1' detectors and identical random numbers can be collected from coincidence counts between the remotely placed interferometers.. In a similar way to the faint pulse case security can be guaranteed by switching interferometer phases between $0, \pi/2$ and $0, -\pi/2$ and deleting all coincident bits when the phase sum is non-zero.

To confirm the feasibility of such key sharing schemes we have demonstrated that these correlations extend to kilometre separations [61]. We have adopt a hybrid approach using non-degenerate wavelengths of 820nm and $1.3\mu m$ for the separate beams. The $1.3\mu m$ beam then passes through a 4Km coil of low loss communication fibre and through an all fibre interferometer to a photon counting Germanium APD (see section 2.2). The 820nm light is detected locally using high efficiency silicon detectors after passing through an interferometer with an air gap path-length difference to reduce dispersion effects. The coincidence correlation coefficient ($V \cos \phi$) as a function of phase shift obtained by selecting only the interfering events is shown in figure 10. The visibility V is above 87% confirming the fully quantum mechanical nature of these correlations out to 4km spatial separation[61]. Application of this to key sharing [60], however, is limited as the present visibility represents too great an error rate.

4.2. INTERFERENCE IN SPONTANEOUS EMISSION

Modification of the spontaneous emission of an atom by its surrounding environment was first suggested some years ago [62] but it is only recently that the phenomenon has been investigated in various elegant experiments involving atoms close to mirrors[63, 64]. In a simple description, light propagating from an atom placed close to a mirror can reach a detector either directly or via reflection in the mirror. When the two paths cannot be distinguished, constructive and destructive interference effects occur between reflected and direct spontaneous emission. This interference effect decays away with increasing atom-mirror distance due to the limiting bandwidth (hence coherence length) associated with the emission process or predetection filter. Parametric downconversion in the weak pump limit is itself a spontaneous process. In this case there are three modes involved, the pump, signal and idler modes and there is a finite probability of spontaneous conversion of a pump photon to a signal-idler photon pair satisfying energy and momentum conservation rules. We can enhance and suppress this spontaneous emission process by suitably arranging three mirrors such that there are two indistinguishable routes for the creation of the downconverted photon pairs[65] Single atom experiments can be described by semi-classical models where the light can be thought of as a classical field interacting with a quantum mechanical oscillator, the atomic transition. In contrast, suppression of two-photon spontaneous emission is a truly nonclassical interference effect because the correlation of detection events in pairs allows us to gate detection of one-photon states conditionally on detection of partner photons while still observing the effect.

Consider the process of parametric down-conversion as shown in figure 11. A UV pump beam is reflected back on itself to pass through a non-linear crystal two times. This provides two possibilities for spontaneous emission of a pair of red photons into modes propagating in opposite directions. We

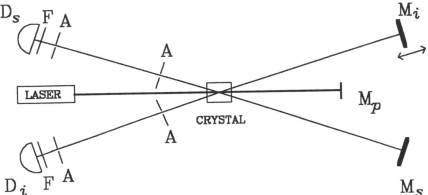

Figure 11. Schematic of the two-photon spontaneous emission experiment. Two possible ways of creating a photon pair in a non-linear crystal arise because the pump laser beam is retroreflected at mirror M_p. When these two possibilities are made to overlap with a further pair of mirrors (M_s, M_i) we see interference between them. The diaphragms (A) serve to define single spatial modes and filters (F) select out energy matched pairs. Pairs are detected in single photon counting detectors (D_s, D_i)

then reflect one set of signal-idler modes back through the crystal such that they overlap with the other signal-idler modes and detect pairs in photon counting detectors D_s, D_i. Viewed from the detectors there are two indistinguishable routes for any event to occur which leads to an interference effect.

In the weak pump limit the wavefunction can be written

$$|\Psi\rangle = \alpha \left(e^{i\phi_p} + e^{i(\phi_s + \phi_i)} \right) |1\rangle_s |1\rangle_i \qquad (26)$$

where $|\rangle_s |\rangle_i$ represent the signal and idler modes, ϕ_s and ϕ_i are the phases accumulated by the first signal and idler, respectively, to their mirror and back, ϕ_p is the phase accumulated by the reflected pump and $\alpha \ll 1$ is the probability amplitude for creation of a signal-idler pair for each passage of the pump beam (although the wavefunction above appears un-normalised we have omitted the near unit probability amplitude for zero photon pairs $|0>_i |0>_s$). Equation 1 makes the remarkable prediction that the probability for seeing signal P_s, idler P_i and (as a result) coincidence P_{si} counts all vary as

$$P_{si} \propto P_i = P_s = \eta \alpha^2 \left\{ 1 + \cos(\phi_i + \phi_s - \phi_p) \right\} \qquad (27)$$

Clearly signal idler and coincidence counting rates vary in an identical way. Thus varying, say the position of the idler mirror, varies all rates with period given by the idler wavelength. This can be interpreted as enhancement and suppression of two-photon spontaneous emission.

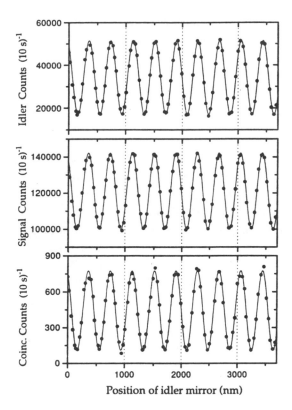

Figure 12. Simultaneous measurement of signal count rate, idler count rate and co-incidence count rate as a function of the displacement of the idler mirror. The phase agreement of these oscillations shows clearly that the boundary condition controlling the emission of both photons in the entangled state can be controlled by changing the position of just one of the mirrors. Reproduced from [65]

A more complete theoretical approach has to take into account the finite coherence lengths of the pump beam (meters) and of the down-converted radiation ($\approx 260\mu$m). Energy conservation confines the sum of signal and idler photon energies to that of the pump light. When considered in unison signal-idler photon pairs have a coherence length equal to that of the pump beam thus the signal and idler mirrors can be placed a large distance from the crystal without destroying the effect. We must only *equalise* the crystal-mirror path lengths of signal and idler beams to within their 260μm coherence lengths. Similarly the crystal-mirror distance of the pump, and signal/idler mirrors should not differ by more than the pump coherence length.

The results of an experiment optimised to overlap modes from both forward and backward pump direction are shown in figure 12. The signal count rate, idler count rate and coincidence count rate measured as a function of the idler mirror position. show fringes in agreement with equation 27. The effect may be fully understood as interference between two possible ways for emission to occur rather than interference of the photons themselves. The fact that the mirrors can be placed a large distance from the crystal allows the conception of some interesting 'Welcher Weg' experiments[66]. The question: "Are there any photons between the crystal and the mirrors when the detectors see no photons due to suppression of spontaneous emission", can be addressed by placing a fast optical switch just before one of the mirrors to switch the beam into a detector. The question arises because if photons (and the tendency is to imply particles) do exist between the crystal and mirror then they must be reabsorbed at the crystal (in some non-linear upconversion process) otherwise they would pass on to register in the detector. If there is nothing between the mirror and crystal until the mirror is removed then a we expect a delay before measurement of a first photon equivalent to the time for light to propagate to the crystal and back. In both these scenarios we have taken an oversimplistic view of the meaning of the word "photon". This is an interference effect between two equivalent paths leading to the creation (or suppression of creation) of a photon pair propagating to the detectors. In common with all interference phenomena a field is always present between the crystal and mirror. Hence there is always a probability amplitude for instantaneous detection of a photon.

5. Conclusion and Perspective

We began these lectures with a simplistic picture of photons as fuzzy balls that could be selected by opening shutters at predetermined times (section 2.2). We have extended that picture to the situation where noise in a beam of these 'photons' can be reduced by feedback from measurements performed of a quantum identical beam (section 2.3). The image that we were tempted to build from these initial studies is quickly destroyed when we describe the experiment where a one photon state created in the apparatus of section 2.2 is passed through a Mach-Zehnder interferometer to interfere with itself (section 3.1) echoing Dirac's famous words. To describe higher order interference effects and quantum interference in general we must go beyond the narrow interpretation of Dirac's statement. Quantum interference arises when we combine the various indistinguishable paths (modes) through the apparatus to a specific detection event (be it a single or multiply coincident photodetection). The probability amplitudes for passage via a particular path are proportional to the field in that path and instantaneously placing a detector in any one path will obviously produce an instantaneous probability for detection.

Here we have shown that the basic nature of quantum interference can itself be used to allow sharing of random numbers at remote locations. This is a simple example of a more general quantum processor capable of outperforming a classical Turing machine for specific applications including factoring[67]. Such a machine would code all possible inputs into an N-fold superposition state and manipulate that state using coherent/interferometric logic devices. However before we can build such a machine using photons we must note some of the obvious technical hurdles to be overcome. The first simple problem is the limited efficiency of existing photon counting detectors and the efficiency of single/multiple photon sources as the probability of seeing an N-fold coincidence will fall as η^N where η is the lumped efficiency of creation and detection. A second is the requirement for coherent logic gates that will switch phases by amounts of order π while retaining high visibility interference dependent on the presence or absence of a single photon. Over the past few years we have seen some progress towards ideal detectors and in generation of various one-and two-photon states but experiments to demonstrate N-photon entanglement are still at an early stage. To then take these N-states and demonstrate some logic function involving single particle Non-linearity is still some way off.

Acknowledgements

Pippa Owens for her contribution to work in section 2.2 on Germanium APD's. The work in section 3.2 was carried out in collaboration with Paul Townsend of BT research laboratories, Martlesham Heath, UK. The work in section 3.4 was carried out in collaboration with Anton Zeilinger, Thomas Herzog and Harald Weinfurter of the Institut für Experimentalphysik Innsbruck.

References

1. Loudon R. (1983) *The Quantum Theory of Light*, Oxford University Press, ch. 6.
2. Teich M.C. and Saleh B.E.A. (1988) Photon Bunching and Antibunching, *Progress in Optics* **XXVI**, 1-104.
3. Brown R.G.W. Rarity J.G. and Ridley K.D. (1986), *Applied Optics* **25**, 4122.
4. Brown R.G.W. Jones R. Rarity J.G. and Ridley K.D. (1987), *Applied Optics* **26**, 2383.
5. MacIntyre R.J. (1985) Recent developments in silicon avalanche photodiodes, *Measurements* **3** 146-152.
6. Owens P.C.M. Rarity J.G. Tapster P.R. Knight D. and Townsend P.D. (1994), Photon counting with passively quenched germanium avalanche photodiodes, *Applied Optics* **33**, 6895-6901.
7. Owens P.C.M. Rarity J.G. and Tapster P.R., (1995) in preparation.
8. Short R. and Mandel L. (1983) *Phys. Rev. Lett.* **51** 384
9. Kimble H. J. Dagenais M. and Mandel L. (1977) Phys. Rev. Lett. **39** 691

10. Walker J. G. and Jakeman E. (1984) *SPIE Proc ECOOSA'84* **492**, 274.
11. Machida S. and Yamamoto Y. (1986), *Opt. Commun.* **57**, 290.
12. Shapiro J.H. Saplacoglu G. Ho S.T. Kumar P. Saleh B.E.A. and Teich M.C. (1987) *J. Opt. Soc. Am. B* **4**, 1604.
13. Khoroshko D.B. and Kilin S.Ya. (1994), *JETP* **79**, 691.
14. Yariv A. (1988) *Quantum Electronics*, John Wiley NY. Press (1983).
15. Burnham D. C. and Weinberg D. L. (1973) *Phys. Rev. Lett.* **25**, 84.
16. Hong C K and Mandel L, (1986) *Phys. Rev. Lett.* **56** 58.
17. Walker J G and Jakeman E (1985) *Opt. Acta* **32** 1303.
18. Rarity J G, Tapster P R and Jakeman E (1987) *Opt. Commun.* **62** 201.
19. Rarity J.G. Tapster P.R. and Ridley K. (1987) *Applied Optics* **26**, 4616.
20. Tapster P R, Seward S F and Rarity J G (1991) *Phys. Rev. A* **44** 3266.
21. Heidmann A, Horowicz R J, Reynaud S, Giacobino E and Fabre C (1987) *Phys. Rev. Lett.* **59**, 2555
22. Debuisschert T, Reynaud S, Heidmann A, Giacobino E and Fabre C (1989) *Quantum Opt.* **1**, 3
23. Aytur O. and Kumar P. (1987) . *Phys. Rev. Lett.* **65** 1551.
24. Tapster P R, Rarity J G and Satchell J S (1988) Phys. Rev. A **37**, 2963.
25. Mertz J, Heidmann A, Fabre C, Giacobino E and Reynaud S (1990) *Phys. Rev. Lett* **58**, 2897.
26. Jakeman E and Rarity J G (1986) *Opt. Commun.* **56** 219.
27. Owens P.C.M. Rarity J.G. and Tapster P.R., (1996), in preparation.
28. Slusher R E, Holberg L W, Yurke B, Mertz J C and Valley J F (1985) *Phys. Rev. Lett.* bf 55, 2409
29. Wu L-A, Kimble H J, Hall J L and Wu H (1986) *Phys. Rev. Lett.* **57**, 2520
30. Slusher R E, Grangier P, La Porta A, Yurke B and Potasek M J (1987) *Phys. Rev. Lett.* **59** 2566.
31. Xiao M., Wu L-A. and Kimble H.J. (1988) *Opt. Lett.* **13** 476.
32. Nabors C.D. and Shelby R.M. (1990) *Phys. Rev. A* **42**, 556.
33. Tapster P.R. Rarity J.G. and Satchell J. (1987), *Europhys. Lett.* **4**, 293.
34. Richardson W.H. Machida S. and Yamamoto Y. (1991), *Phys. Rev. Lett.* **66**, 2867.
35. Taylor G.I. (1909), *Proc. Cambridge Philos. Soc.* **15**, 114 (1909).
36. Dirac P.A.M. (1930), *The Principles of Quantum Mechanics*, Oxford University Press, Oxford, p9.
37. Grangier P. Roger G. and Aspect A. (1986), *Europhys. Lett.* **1** 173.
38. Bennett C.H. *Phys. Rev. Letts.* **68**, 3121 (1992).
39. Rarity J.G. Tapster P.R. and Owens P.C.M. (1994), *J. Modern Opt.* **41**, 2435.
40. For a recent review of quantum cryptography see: *J. Modern Opt.* **41** December 1994.
41. Townsend P.D. Rarity J.G. and Tapster P.R. (1993), *Electron. Lett.* **29**, 634.
42. Townsend P.D. Rarity J.G. and Tapster P.R. (1993), *Electron. Lett.* **29**, 1291-1292.
43. Townsend P.D. (1994), *Electron. Lett.* **30**, 809-811.
44. Muller A., Breguet J. and Gisin N. (1993), *Europhys. Lett.* **23**, 383-388.
45. Franson J.D. and Jacobs B.C. (1995), *Electron. Lett.* **30**, 809-811.
46. Schnitzer I. Yablonovitch E. Caneau C. Gmitter T.G. and Scherer A. (1993), *Appl. Phys. Letts.* **63**, 2174.
47. Ghosh R. and Mandel L. (1987), *Phys. Rev. Lett.* **59**, 1903.
48. Fearn H. and Loudon R. (1987), *Opt. Commun.* **64**, 485-490.
49. Ou Z.Y. Hong C.K. and Mandel L. *Opt. Commun.* **63**, 118-122 (1987).
50. Hong C.K. Ou Z.Y. and Mandel L. *Phys. Rev. Lett.*, **59**, 2044 (1987)
51. Rarity J.G. and Tapster P.R. (1988), in 'Photons and Quantum Flutuations', E R Pike and H Walther eds., Adam Hilger, p122.; (1989), *J Opt Soc Am B*, **6**, 1221.
52. Rarity J.G. and Tapster P.R. (1990), *Phys. Rev. Letts.* **64**, 2495.
53. Bell J.S. (1964), *Physics* **1**, 19.
54. Rarity J.G. Tapster P.R. Jakeman E. Larchuk T. Campos R.A. Teich M.C. and

Saleh B.E.A. (1990), *Phys. Rev. Letts.* **65**, 1348.

55. Rarity J.G. Tapster P.R. (1992), *Phys. Rev. A* **45**, 2054.
56. Brendel J. Mohler E. and Martiennsen W. (1992), *Europhys. Letts.* **20**, 575.
57. Kwiat P.G. Steinberg A.M. and Chiao R.Y. (1993), *Phys. Rev. A* **48**, R867.
58. Rarity J.G. Burnett J. Tapster P.R. and Paschotta R. (1993), *Europhys. Lett.* **22**, 95.
59. Franson J.D. *Phys. Rev. Lett.* **62**, 2205 (1989).
60. Ekert A.K. Rarity J.G. Tapster P.R. and Palma M. (1992) *Phys. Rev. Letts.* **69**, 1293.
61. Tapster P.R. Rarity J.G. and Owens P.C.M. (1994), *Phys. Rev. Letts.* **73**, 1923.
62. Purcell E. (1946), *Phys. Rev.* **69**, 681.
63. Drexhage K.H. (1974), *Progress in Optics* **12**, 163.
64. Deppe D.G. Campbell J.C. Kuchibhotla R. Rogers T.J. and Streetman B.G. (1990), *Electron. lett.* **26**, 1666.
65. Herzog T. Rarity J.G. Weinfurter H. and Zeilinger A. (1994), *Phys. Rev. Lett.* **72**, 629.
66. Fearn H. Cook R.J. and Milloni P.W. (1995), *Phys. Rev. Lett.* **74**, 1327.
67. Shor P.W. (1994), in *Proc. 35th Annual Symp. on FOCS*, ed Goldwasser S. (IEEE Comput. Soc. Press, Los Alamos), 124.

NONLINEAR OPTICS IN MICRO-METER SIZED DROPLETS

Richard K. Chang, Gang Chen, and Md. Mohiuddin Mazumder
Department of Applied Physics and Center for Laser Diagnostics
Yale University, New Haven, Connecticut 06520-8284, USA

INTRODUCTION

Many groups are actively studying various aspects of microcavity physics and optics; hence the topic of this NATO ASI. One of the subfields is dye lasing and nonlinear optical effects in microdroplets (since 1984) and, more recently, in liquid jets and liquid in thin capillary tubes. Although dye lasing and nonlinear optical effects in an optical cell are well studied and much understood, when dye lasing and nonlinear optical experiments are conducted in the spherical, nearly spherical, and cylindrical liquid microstructures, inevitably "surprises" occur because of the microstructure morphology (size, shape, and index of refraction).

There are several attractions for the Yale University group to continue conducting research in dye lasing and nonlinear optical effects of microdroplets. First and foremost, the topic serves as a good training platform because of its multi-disciplinary nature, requiring that we learn from quantum mechanics, quantum and nonlinear optics, electromagnetics, hydrodynamics, thermodynamics, as well as the use of lasers, spectroscopic instruments, and detectors. Second and more challenging, the topic enables us to bridge the basic-research and application domains. During the past decade, many of the basic-research studies were driven by the need to provide some optical diagnostic techniques that have the potential of serving as a nonintrusive, *in-situ*, and real-time probe to determine the chemical species, physical properties, and temperature of multicomponent liquid droplets. Droplets and sprays are central to modern internal combustion engines.

In this chapter, we will review some of the dye-lasing and nonlinear-optics in microdroplet. The interplay between basic and applied research will become apparent to the reader. After a brief summary of the microcavity resonances, namely to introduce the notation used by the electromagnetic scattering community, we will confine our review to four topics: dye lasing, stimulated Raman scattering (SRS), SRS in a dye lasing medium, and third-order sum frequency generation. These four topics were selected to emphasize the symbiotic relationship of how the need for a diagnostic probe has driven the need to understand microcavity effects more thoroughly.

75

M. Ducloy and D. Bloch (eds.), Quantum Optics of Confined Systems, 75-99.
© 1996 *Kluwer Academic Publishers. Printed in the Netherlands.*

CAVITY RESONANCES

Liquid droplets exhibit high-Q cavity resonances, with size parameter $X = 2\pi a/\lambda$ > 10 and with index of refraction ratio of the liquid and the surrounding air $m(\omega) > 1.1$. These resonances have been called by a variety of names: morphology-dependent resonances (MDR's); whispering gallery modes (WGM's); structure resonances; ripples modes; and cavity resonances. In this chapter, we will refer to them as MDR's.

From the ray picture, total internal reflection occurs when a ray of light, inside the droplet, is incident on the droplet-air interface at an angle greater than the critical angle. However, because of diffraction, some light leaks out. The retained rays can traverse along the rim. If the traversed path length is an integral multiple of wavelength, successive round trips around the rim result in constructive interference and MDR's occur.

From the electromagnetic wave model, each MDR needs to be labeled by three mode indices: ℓ, n, and m, which together designate, inside the droplet, the electric field distribution with the standard spherical coordinates. Details of the vector spherical harmonics and spherical Bessel and Hankel functions can be found in standard textbooks and review articles [1-3].

Suffice it to summarize here the properties of the ℓ, n, and m -MDR's. For a cubical cavity, each of the three mode indices designates one of the three orthogonal field distributions in the x, y, and z direction. However, for a spherical cavity, the ℓ, n, and m indices are interrelated and these indices together designate the orthogonal field distributions in the r, θ, and ϕ direction. Nevertheless, some insights of the ℓ, n, and m indices can be individually summarized below.

The mode order or the radial mode order, ℓ, specifies the number of peaks in the angle-averaged radial intensity as **r** is increased from 0 to **a**. For low-order ℓ's, all the intensity peaks are localized just within the rim. From the ray picture, rays emanating from this region strike the liquid-air interface at an angle much greater than the critical angle. Hence, the low-order ℓ-MDR's, have short penetration depth into the air and have very high-Q values. For high-order ℓ MDR's, the intensity peaks are distributed more toward $a/m(\omega)$. Usually, the largest intensity peak (among the ℓ number of peaks) is situated farthest from the rim. Rays emanating from location of the largest intensity peak strikes the liquid-air interface at an angle closer to the critical angle. Hence, the high-order ℓ-MDR's have longer penetration depth into the air and have lower Q values than the low-order ℓ-MDR's.

The mode number or the angular mode number, n, specifies the number of MDR intensity peaks in the equatorial plane as the azimuthal angle ϕ is varied from $0°$ to $180°$. The equatorial plane is relative to the z-axis, which is defined for a sphere to be along the plane-wave propagation direction or for a spheroid to be along the axisymmetric axis. For MDR's to have sufficiently high Q values, the range of n values must be within $2\pi a/\lambda \leq n \leq m(\omega)2\pi a/\lambda$. The angular mode n can be considered as the "angular momentum (L)" of the MDR, where the $L^2 = n(n+1)$ is conserved for a sphere but not conserved for a spheroid.

The angular mode number n affects the radial distribution of the MDR. From a mathematical point of view [4], when the radial wave equation is written without any explicit angular dependence, the effect of the conservation of the "angular momentum" is implicit in a centrifugal potential term, $n(n+1)/r^2$. This centrifugal potential together with the discontinuity of the index of refraction at the liquid-air interface [m(ω) for $0 \leq r \leq a$, and 1 for $r > a$] form a "potential well" which traps the MDR's in the radial direction. Out-going radial waves must first tunnel through the index of refraction barrier and the outer centrifugal barrier, before the radial waves become an oscillatory propagating wave, which is the "leaky" wave from the microparticle.

By using the potential well model, the following properties of the MDR's can be derived. The potential well is high, when $n \rightarrow m(\omega)2\pi a/\lambda$ and, hence, can have long lived states or high-Q MDR's. The potential well is low, when $n \rightarrow 2\pi a/\lambda$ and, hence, would have shorter lived states or lower Q MDR's. For a fixed n value, the low ℓ's correspond to trapped states deep in the potential well, and, hence, high Q values. The higher ℓ's correspond to trapped states closer to the top of the potential well and, hence, lower Q values.

The azimuthal mode number, m, describes the intensity as a function of ϕ. The electric fields are described mathematically as exp($\pm im\phi$). For a fixed angular mode number n, the m values can range from $\pm n, \pm(n-1), \ldots., 0$. The double degeneracy of $\pm m$ is because clock-wise and counterclock-wise rotations are equivalent. In the ray picture, the m-MDR can be viewed as light circulating in an orbit, in the form of a thin band, with its normal inclined at $\theta = \cos^{-1}(m/n)$ with the z-axis. For a perfect sphere, the path length of all the orbits are the same, regardless of their inclination angle. Therefore, all the m-MDR's for a perfect sphere have the same frequency or $(2n+1)$ degenerate in frequency; hence, a MDR in a perfect sphere is described usually by two indices, just ℓ and n. For plane wave illumination of a sphere, only the $m = \pm 1$ MDR's can be excited, because the z-axis can always be defined to be along the propagation direction of the plane wave [1,2]. However, for the case of a focused beam illuminating the edge of the sphere, many more m-MDR's can be involved [5-8]. A focused beam is formed mathematically by a summation of plane waves propagating along a range of angles relative to a fixed z-axis. The relative amount of radiation that can be coupled into each m-MDR depends on the spatial overlap between the input beam and the spatial distribution of each m-MDR.

The azimuthal mode m is analogous to the "magnetic quantum number" or the projection of the angular momentum L along the z-axis, L_z. When the micro-object is a spheroid, with the z-axis defined along the axisymmetric axis, this angular momentum analog is particularly helpful. Then, only the $L_z = m = n \cos\theta$ is conserved and the L_x and L_y are not conserved. The latter leads to precession of m-MDR's. In the ray picture, for a slightly deformed sphere, in the form of a spheroid, the m-MDR can be viewed again as light circulating in an orbit, in the form of a thin band, inclined with its normal at θ with the z-axis (see Fig. 1). The path length of all the orbits are no longer the same and are now dependent on their inclination angle. Therefore, the m-MDR's for a deformed sphere have the different frequency, i.e., $(n + 1)$ discrete frequencies that are split from the $(2n + 1)$ degeneracy for a perfect sphere [9].

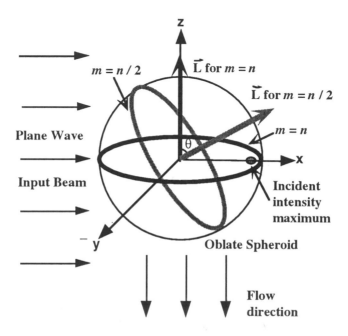

Figure 1. Schematic of a spheroidal droplet that is axisymmetric along the flow direction (z axis) and irradiated by the input beam along the x axis. Lasing is collected along the –y axis. The maximum intensity of the incident beam is shown as a shaded spot. The two MDR's (with azimuthal number $m = n$ and $m = n/2$) have their angular momentum L oriented at different θ's relative to the z axis.

DYE LASING

The dye lasing process in liquids is well understood. Since the first observation of dye lasing in microdroplets [10], pumped by a cw Argon-ion laser beam, there have been many interesting results from the several active research groups. In the original experiment, the energy fluence needed to reach the lasing threshold in droplets (with $\mathbf{a} \approx 30$ μm and 10^{-5} - 10^{-4} M Rhodamine 6G) was 35 W/cm^2, some three orders of magnitude lower than the threshold intensity needed for a commercial dye laser with dye-jet thickness greater than the droplet diameter and with external laser cavities. Dye lasing has now been achieved in cylindrical microstructures, i.e., free-flowing liquid cylindrical jets [11,12] and/or liquid inside capillary tubes [11]. Recently, even two-photon absorption pumped dye lasing (with a Q-switched laser pump pulse) has been observed [13] from microdroplets with similar radius and with Coumarin 460 and Rhodamine 6G.

Many studies have been conducted that use the lasing spectra and the lasing properties to further our understanding of the basic physics of microcavity, particularly the interesting QED effect in enhancing the Einstein A and B coefficients [14,15]. In order to quantify the QED enhancement, it is necessary to determine the Q values of the MDR's, which can be accomplished by adding a known amount of an absorber [16] or small Rayleigh-sized scatterers [17,18] to degrade the overall Q of a particular MDR with a known ℓ, n- values. Another important contribution to the microdroplet-research field has been the unique identification of the MDR's (in terms of the ℓ, n- values) in the lasing, fluorescence, and elastic scattering spectra [19]. At present, such unique identification can only be accomplished for smaller-sized droplets (e.g., with $\mathbf{a} < 10$ μm). Because the droplets are smaller, the MDR's are spectrally further apart and, hence, is less demanding on the spectral resolution of the spectrometer. However, dust particles in the liquid reservoir tends to clog the smaller droplet orifice more frequently. For unique identification of MDR's, it is important for an experimentalist to vary continuously the droplet $\mathbf{a}$. If the microdroplet is optically or electrodynamically levitated [20], one convenient way to continuously vary $\mathbf{a}$ is to allow evaporation to steadily decrease the droplet size. In the case for flowing microdroplets, generated by a Berglund-Liu vibrating orifice droplet generator (from Thermal Systems Incorporated), remarkably controlled variation of $\mathbf{a}$ can be achieved by sweeping continuously the oscillator frequency that is driving the PZT of the droplet generator [21]. The droplet radius varies as the cube root of the oscillator frequency.

In this chapter, we have chosen to review several applications of lasing spectra from microdroplets. For the past decade, we have attempted to use the lasing spectra as a potential optical diagnostic technique to determine the physical properties of droplets e.g., the droplet size change due to evaporation and the droplet shape change due to external perturbations. More recently, we have started exploring the possibility of deducing the droplet temperature change from the blue-shift of the lasing wavelength. In specific, we review below some of these attempts to apply the changes in the lasing MDR's as a consequence of droplet morphology perturbations or changes as well as droplet temperature variations.

1) Minute size changes, a radius decrease because of evaporation and a radius increase because of condensation, can be readily detected by measuring the wavelength shift of each lasing peak, even if the ℓ, n- values of the MDR are NOT known [22]. For a given radius change, Δa, there is no need to identify two MDR's with consecutive n values and then determine the wavelength change in their separation. The wavelength shift $\Delta\lambda$ of each lasing peak is related to Δa in the following manner: $\Delta a = (\Delta\lambda/\lambda)a$. The sign of $\Delta\lambda$ is informative, where a $+\Delta\lambda$ signifies a size increase and a $-\Delta\lambda$ signifies a size decrease. Assuming that $\Delta\lambda = 1$ cm^{-1} can be spectrally discerned of a MDR lasing peak at $\lambda = 20,000$ cm^{-1}, then $\Delta a/a$ can be determined to 5 parts in 10^5. By using this simple technique, evaporation rate of closely spaced (with center-to-center of 4a) and moving (at 10 m/s) droplets in a continuous stream as well as in a segmented stream have been determined [23]. Such evaporation rate data now serve to verify hydrodynamic computer codes and, recently, brought out the importance of evaporative cooling in modeling the evaporation rate of closely spaced moving droplets.

2) Determination of minute shape deformation (assuming a slightly distorted sphere into a spheroid) has been made possible after the seminal paper [9] that provided an analytical expression for the frequency shift of the m-MDR's. The frequency of the degeneracy-lifted m-MDR frequency of a spheroid is shifted from that of a sphere by an amount $\Delta\omega(m)$,

$$\frac{\Delta\omega(m)}{\omega_o} = -\frac{e}{6}\left[1 - \frac{3m^2}{n(n+1)}\right] \tag{1}$$

where ω_o is the frequency of a MDR of a perfect sphere [9]. The distortion amplitude is $e \equiv (r_e - r_p)/a \ll 1$, where r_e and r_p are the equatorial radius and the polar radius, respectively. The radius of the equivolume sphere is a. Note that the frequency shift [24] to the "red" or to the "blue" is dependent on the sign of e. Thus, the shape deformation (whether it is oblate or prolate) as well as the distortion amplitude can be spectroscopically determined. Note that $\Delta\omega(m)$ depends on m^2 and the $\pm$ m-MDR's (for different sense of rotation) have the same frequency. The frequency spacing between consecutive m-MDR's are much too close to be spectrally resolved by a spectrometer. The total frequency span of the $m = \pm n$ MDR and the $m = 0$ MDR is $\Delta\omega_{total} = \Delta\omega(n) - \Delta\omega(0) = e\omega_o/2$. If these individual $\pm m$-MDR's are NOT spectrally resolved, then each ℓ, n-MDR would be misconstrued as being broadened and it can lead to the wrong conclusion that slight shape deformation lowers the Q value to $\omega_o/\Delta\omega_{total}$. Whereas the perturbation theory [9] predicts that there is no lowering of the Q values for small distortion amplitudes, only second-order in |e|.

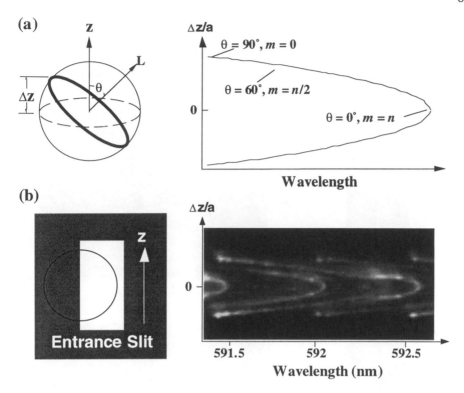

Figure 2. (a) Wavelength dependence of the light from different positions (with different θ and Δz) on the deformed droplet rim; various parts of the parabolic λ(Δz) curve can be related to different m-mode MDR's. (b) Actual CCD data recorded from a single droplet (with the right semicircle imaged onto the entrance slit) located at z = 10 mm downstream from the orifice. The ⊃-shaped λ(Δz) curve can be observed from four n-mode MDR's of the same radial mode order ℓ.

We were able to "beat" the spectral resolution limits of a spectrometer by using its other dimension, the dimension along the spectrometer slit [25]. By using a 2-dimensional CCD detector, we were able to disperse the frequency distribution (along the spectrometer horizontal axis) and displace the lasing radiation emerging from various physical locations along the droplet rim (projected along the vertical slit axis $\Delta z = a \sin\theta$). Recall that the different m-MDR's have a different inclination with respect to the droplet axisymmetric axis, z. The wavelength variation should form a parabola (see Fig. 2), starting from the droplet pole, and passing through the droplet

82

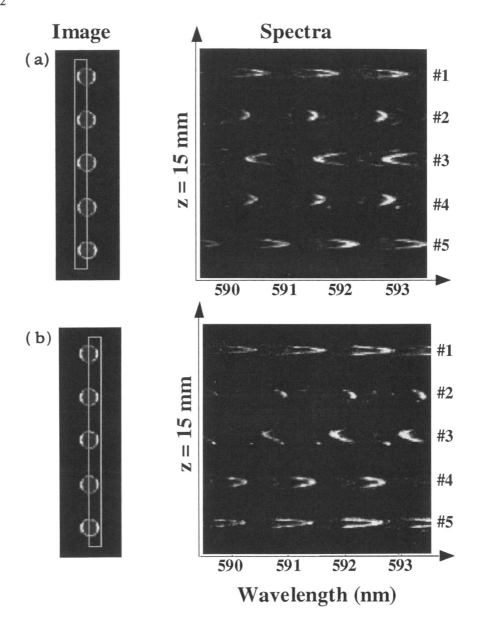

Figure 3. CCD images of the lasing droplets as they appear at the entrance slit of a spectrograph (left-hand column). For (a) the right-half rim and for (b) the left-half rim is imaged onto the slit. The spatially preserved wavelength-dispersed data captured by the CCD are shown on the right-hand side. Note the ⊃- and ⊂-shaped $\lambda(\Delta z)$ curves.

equatorial plane, and ending at the opposite pole. When the individual droplets are imaged onto the entrance slit of an image-preserving spectrograph (see left side of Fig. 3), the resultant lasing spectra should form a ⊃- and ⊂-shaped 2-dimensionally dispersed and displaced spectra. The resultant lasing spectra from five droplets are shown on the right side of Fig. 3. By observing the ⊃- and ⊂-shaped spectra, we can deduce that the droplet shape varies from prolate to nearly spherical to oblate, even though the actual photographs indicate that the droplets are spherical, at 15 mm down stream from the orifice of the droplet generator. Figure 2(b) shows the lasing spectrum from a single droplet, which is greatly magnified onto the entrance slit. The spatially preserved lasing spectrum forms a parabolic shape. We deduce from the ⊃-shaped spectrum that this droplet has a deformed shape of a oblate spheroid with a deformation amplitude of $e = 4 \times 10^{-3}$.

3)   Attempts to deduce the droplet temperature from the lasing spectra are underway [26]. The new idea is to add a temperature-dependent absorber, i.e., a thermochromic absorber into the dye lasing medium. To achieve lasing, the round-trip gain must overcome the total round-trip loss. The condition for lasing at any wavelength, λ, can be written as: $n_1 \sigma_e(\lambda) \geq \alpha_{leakage}(\lambda) + \alpha_{abs}(\lambda, T) + n_o \sigma_a(\lambda)$, where $\sigma_e(\lambda)$ and $\sigma_a(\lambda)$ are the stimulated emission and absorption cross-section (cm^2) of the fluorescent dye. The total number of fluorescent dye molecules per unit volume is $n_T = n_o + n_1$, where n_1 is the number density of molecules in the first-excited electronic single state and n_o is the number density of molecules in the ground-electronic single state. The radiation leakage loss out of the droplet cavity is $\alpha_{leakage}(\lambda)$ and the absorption loss of a temperature-dependent absorber, a thermochromic absorber [27], is $\alpha_{abs}(\lambda, T)$. Because $\alpha_{abs}(\lambda, T)$ increases with temperature, the lasing condition demands that $\sigma_e(\lambda)$ shifts toward its peak, which is more toward the blue. Experimentally, blue shift of the Rhodamine B lasing spectra has been observed when the microdroplet, containing the thermochromic absorber ($CoCl_2$) is heated. Much more work is needed before the absolute temperature of droplets can be deduced from the lasing spectra.

STIMULATED RAMAN SCATTERING (SRS)

The droplet morphology is again responsible for the low intensity threshold for first-order stimulated Raman scattering (SRS at $\omega_{1s} = \omega_L - \omega_{vib}$), where ω_{vib} is the molecular vibrational mode with the largest Raman gain, g_S. Therefore, the SRS spectra contain "finger print" information on the species within the microdroplet. In principle, the SRS intensity is related to the concentration of that species. The SRS process is the amplification of spontaneous noise. SRS threshold is reached when the round-trip gain at ω_{1s} exceeds the round-trip loss due to all possible mechanisms:

leakage, absorption, scattering, and depletion as a result of pumping other nonlinear processes. When the ω_L is off resonance with a MDR (off an input MDR), the internal intensity "hot spot" is just within the shadow face. Within this focal volume, the internal intensity creates QED enhanced spontaneous Raman scattering and provides gain for the "trapped" Raman radiation (on an output MDR). When the ω_L is on an input MDR, the entire rim can generate spontaneous Raman scattering and provide gain. In either case, the effective pumping length is dependent on the spatial overlap between the "hot spot" (or an input MDR) with the output MDR that is supporting the feedback for the SRS. When the input radiation is on an input MDR and because the SRS has to be on an output MDR, the condition is often referred to in the literature as "double resonance" [20,28].

Under double resonance conditions and using cw Argon-ion laser beam, cw-SRS from droplets has been reported [29]. This is a remarkable demonstration of the combined effect of the greatly enhanced internal field at ω_L, the QED enhanced Raman cross section [30, 31], and the high Q feedback provided by the MDR with good spatial overlap with the input-MDR. The use of cw laser beam to generate nonlinear optical waves in droplets opens up many new possibilities, among which is the ability to make more quantitative measurements.

Because the Raman gain profile is on the order of 10 - 100 cm^{-1} and the average frequency spacing among all high Q MDR's is ≈ 0.5 cm^{-1} (for droplet with $\mathbf{a} \approx 40$ µm), there will always be some MDR's to provide the needed feedback. Experimentally it is found [32,33] that the SRS spectra appear to select MDR's with one particular radial mode order ℓ. For example, in the SRS spectra from H_2O (Raman gain width of 400 cm^{-1}) and D_2O (Raman gain width of 300 cm^{-1}), a regular series of SRS peaks appeared, consisting of one fixed ℓ and consecutive series of n's. Upon increasing the pumping intensity, two series of MDR's with two ℓ's can appear [32].

Although it is difficult to predict a *priori* which particular ℓ the SRS process would select, we know that at least three factors need to be considered. First, the spatial overlap between the pumping field and the SRS-output MDR is important. When ω_L is off an input MDR, the spatial overlap with the "hot spot" (for the plane wave illumination) is better with higher ℓ's than that with edge illumination (with a focused beam aimed at the droplet rim). When ω_L is on an input MDR, the spatial overlap is best when the input and output MDR's have the same ℓ. Second, the Q factor of the SRS MDR must be large enough so that its photon lifetime τ_{MDR} be at least as long as the pumping time. The latter is equal to the incident pulse duration τ_L for the off-input MDR case and is equal to the $(\tau_{MDR})_{input}$ for the on an input MDR case. Because the Q factor increases with decreasing mode order number ℓ, there is no need to select ℓ for the output MDR's with Q much higher than needed. Once $(\tau_{MDR})_{output} > \tau_L$ or $(\tau_{MDR})_{output} > (\tau_{MDR})_{input}$, considerations regarding the spatial overlap between the input intensity and the output MDR are then more important. Lastly, the frequency of the output-MDR must be within the FWHM of the gain profile [34].

A large number of papers reporting on various aspects of SRS in microdroplets have appeared. Only some of these results, along with their implications, are mentioned below:

1) Cascade multiorder SRS has been seen [35], where the first-order SRS ($\omega_{1s} = \omega_L - \omega_{vib}$) pumps another first-order SRS to give the 2nd-order SRS ($\omega_{2s} = \omega_{1s} - \omega_{vib} = \omega_L - 2\omega_{vib}$). Figure 4 shows that up to 20th-order SRS ($\omega_{20s} = \omega_L - 20\omega_{vib}$) has been seen with CCl_4 droplets [36]. Even higher order SRS can be occurring at $\omega_{js} = \omega_L - j\omega_{vib}$, with $j > 20$. However, for $j > 20$ the Raman-shifted wavelength extended beyond the red sensitivity response of the CCD detector (beyond 1 μm). This result indicates that the internal SRS fields are high and the importance of spatial overlap between the pumping-SRS MDR and the resultant-SRS MDR. Figure 4 shows that even if ω_L is off an input-MDR, the pumping-SRS at ω_{js} are always on MDR's and, therefore, have excellent spatial overlap with the resultant-SRS at $\omega_{(j+1)s}$.

2) The temporally resolved data [37,38] demonstrate that the growth of the jth-order SRS is correlated with the decay of the (j – 1)th-order SRS and the decay of the jth-order SRS is correlated with the growth of the (j + 1)th-order SRS. This result further demonstrates that multiorder SRS is a cascade process.

3) The SRS may be directly pumped by the Q-switched laser (operating in a multi-mode with linewidth of 0.4 cm^{-1}) or pumped by the near backward ($\approx 180°$) circulating stimulated Brillouin scattering, SBS, (pumped by a single mode Q-switched laser with linewidth of 0.006 cm^{-1}). The SRS threshold for single-mode pumping is 3X lower than that for multi-mode pumping. The SBS threshold is not reached for multimode pumping, as the picosecond spikes are too short for the SBS process to reach its full steady-state gain. Direct evidence [39] that the SBS pumps the 1st-order SRS is from the temporally resolved data, which demonstrate that the decay of the SBS is correlated with the growth of the SRS.

4) In a binary mixture of fluids, the SRS of the majority species is more efficient in pumping the SRS of the minority species than the input laser [40]. The majority species having a lower threshold will build up first and then be partially depleted as a result of pumping the SRS of the minority species. This is further evidence that, besides the relative intensity of the two different pumps (with the input more intense than the SRS of the majority species), their spatial overlap with the field being pumped is an important factor in determining the overall pumping efficiency for the SRS.

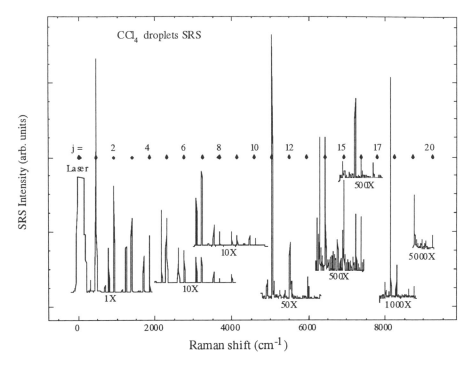

Figure 4. Detected SRS spectra from CCl4 droplets plotted as function of the Raman shift. The integer j (above the points) corresponds to the frequency $\omega_{js} = \omega_L - j\omega_{vib}$, where $\omega_{vib} = 459$ cm^{-1} is the frequency of the ν_1 vibrational mode of CCl4. Cascade multiorder Stokes SRS up to j = 20 is detected. Combinations and overtones of the ν_2 and ν_4 vibrational modes are also detected. The incident laser wavelength is $\lambda_L = 0.532$ μm, and the detector is a CCD.

5)   The measured SRS decay time τ_{SRS} decreases with increased input laser intensity [41,42]. This result demonstrates that besides MDR-related light leakage from the droplet there can be photon depletion, as the photons with ω_{1s} is partially converted to ω_{2s}. Thus the Q value deduced from $[Q = (\omega_{1s}\tau_{SRS})]$ can be less than $[Q = (\omega_{1s}\tau_{MDR})]$. Just above the SRS threshold, where there is no generation of the 2nd-order SRS, then τ_{SRS} should approach τ_{MDR}.

6)   The precession [43] of the SRS, restricted to be within a thin ribbon ring inclined with its normal at an angle θ, has been observed. The thin ribbon ring involves a small group of m-MDR's (with $m = n\cos\theta$),

determined by the spatial overlap of the focused beam, which is aimed at the droplet rim with $90° - \theta$ above or below the equator. For a slightly deformed spheroid, "angular momentum" in the z-direction ($m = n \cos\theta$) is conserved and the "angular momentum" in the x- and y-direction is not conserved and causes mixing among $m - 1$, m, and $m + 1$ MDR's. The mixing of m-MDR's of different frequencies leads to a precession about the axisymmetric z-axis with a precession frequency $\Omega_{precession} = \omega(m + 1) - \omega(m) = (d\omega/dm)$, which can be derived from the perturbation formula for a deformed sphere [9]. The precession of the group of m-MDR's is analogous to the group velocity of a linearly propagating wave $v_g = d\omega/dk$, where the linear momentum $\mathbf{k}$ is replaced by the MDR angular momentum m. Experimentally, the necessary condition to observe the precession is that the precession period $2\pi/\Omega_{precession}$ be shorter than τ_{SRS} and be longer than the response time of the detection system.

7)   Direct spectral observation of the frequency splitting among the m-MDR's has been made with a Fabry-Perot interferometer [44,45]. The predicted [9] frequency splitting is

$$\omega(m + 1) - \omega(m) = \frac{\omega_0|e|m}{n(n+1)} \approx \frac{\omega_0|e|(\cos\theta)}{n} \qquad (2)$$

where ω_0 is the MDR frequency, had the droplet been a perfect sphere, and $|e|$ is the magnitude of the distortion amplitude. Near the equator ($\theta \approx 0°$), the spacing between adjacent m-MDR's is nearly equal. Just above the threshold for SRS, only one m-MDR peak was detected by the Fabry-Perot interferometer. With increased input pumping, up to four m-MDR's were observed. From the frequency splitting, the shape distortion was deduced to be $|e| \approx 10^{-3}$. Experimentally, the necessary condition to observe the frequency splitting among the m-MDR's is that only one ℓ, n-MDR supports the SRS. Having more than one ℓ, n-MDR's in the SRS spectra will confuse the interferometer data, as the frequency difference between such MDR's are about 0.5 cm^{-1}, which exceeds the free-spectral-range of the Fabry-Perot interferometer. Consequently, the selectivity of the spatial overlap between MDR's and the relative narrow linewidth of the Raman gain profile (relative to the dye laser gain profile) were working in favor of this experiment. It is worth remarking that even though the wavelength splitting of the m-MDR's (a first-order dependence on $|e|$) were measured by the interferometer, the additional broadening (a second order dependence on $|e|$) was not detectable within the spectral resolution of the interferometer used.

8)   The fine structures in the angular distribution of MDR supporting SRS in a single droplet have been observed [46]. If the SRS is supported by a few

m-MDR's, then the far-field angular distribution of the SRS intensity, $I_{SRS}(\phi)$, emerging from the droplet equator should exhibit the expected angular distribution of $I_{SRS}(\phi) \propto \cos^2(m\phi + P_m)$. An arbitrary phase, P_m, can exist between the two counter-propagating waves circulating within the droplet equator. Near the equator, these m-MDR's have $m \approx n$. Experimentally, the necessary condition to observe such fine structure in the angular distribution is that only one ℓ, n-MDR support the SRS, otherwise there will be more than one "Fourier component" in the spatial distribution around the equator. Confirmation of such an angular dependence is reassuring that we are beginning to understand the MDR-mediated SRS process. The existence of $I_{SRS}(\phi)$ implies the following: (1) Even though the input beam is aimed at one edge, the SRS is a standing wave because its forward and backward gain are the same; (2) the m-value and hence the n-value of the MDR's (supporting the SRS from the equator) changed with edge illumination and center illumination. This change in n-values is consistent with the fact that higher Q with lower ℓ-MDR's have better spatial overlap with edge illumination. By using center illumination, no input-MDR's are excited and the spatial overlap with the internal "hot spot" favors a lower Q with higher ℓ values; and (3) the arbitrary phase factor, P_m, is stationary throughout the SRS growth and decay times. Had P_m been time varying, the angular distribution would be "washed out" during the time-integrated measurement with a CCD detector.

9)   The intense SRS leads to other nonlinear processes [47], such as third-order sum frequency generation, third-harmonic generation, stimulated anti-Stokes Raman scattering, and stimulated low frequency emission from anisotropic molecules.

10)   A heuristic model [48] for the growth and coupling of nonlinear four-wave mixing processes in droplets was attempted. At present, there does not exist an unified model for the SRS process in microdroplets. In the heuristic model, an attempt was made to incorporate some of the above mentioned experimental results. While qualitative features can be simulated by this heuristic model, particularly the coupling among the SBS and SRS as well as among the various multiorder SRS, this model lacks quantitative predictive capabilities.

SRS IN A FLUORESCENT AND LASING MEDIA

The addition of a lasing dye into a microdroplet provides some interesting interactions between SRS and lasing [49,50,51]. When the fluorescence profile overlaps with the Raman profile, SRS, fluorescence, and lasing can no longer be treated separately. In the absence of any such spectral overlap and of any external seeding by injecting radiation at the Stokes Raman-shifted wavelength, SRS builds up from spontaneous Raman scattering (usually referred to as noise). Such noise is generated at the internal "hot spot" (for off input-MDR case) and around the rim (for the on input MDR case). When there is spectral overlap between the fluorescence and Raman profiles, the fluorescence, being stronger than the spontaneous Raman noise, can be amplified. Furthermore, if there is some stimulated emission provided by the fluorescent dye, then it can provide "extra" Raman gain, even though the lasing gain alone may be insufficient to cause lasing. Evidence exists that with increased stimulated emission at ω_{1s}, the fluorescence at other wavelengths actually decreases with the increased SRS intensity.

Once the SRS intensity is sufficiently intense, the SRS can modify the lasing process. The stimulated emission rate will be selectively increased at the Raman-shifted wavelength, thereby, depleting the excited-state population. With increased input-laser intensity, the MDR peak within the Raman gain profile increases because both the Raman and lasing gain are simultaneously occurring. The other lasing peaks that are within the lasing gain profile, but outside the Raman gain profile, actually decrease because of the depletion of the excited-state population associated with the additional stimulated emission rate at ω_{1s}.

In an optical cell, the depletion of the pumping intensity (either the input laser or SBS) by the SRS buildup of the Raman mode with the largest gain prevents the significant growth of SRS of other Raman modes with weaker gain. This cell result does not forebode well for generating SRS from minority species in microdroplet. In a droplet, however, pump depletion is not so significant, because the spatial overlap of the pump and the particular MDR is only partial. The 1st-order Stokes SRS from several Raman modes with different Raman gain coefficients can grow with time.

The strongest-gain Raman mode of the minority species in a multicomponent liquid droplet can be selectively enhanced by several approaches. However, each of these techniques has some disadvantages. First, use internal seeding and some "extra gain" from a fluorescent dye (at low dye concentration, e.g., less than 10^{-6} M). The dye is chosen such that the fluorescence profile spectrally overlaps with the Raman shifted-wavelength of the minority species and have little or no spectral overlap with that of the majority species. Second, use internal seeding and the significant extra gain from a fluorescent dye (at a high dye concentration, e.g., greater than 10^{-4} M). The addition of a fluorescent dye in the first two approaches can introduce some contamination into the sample chamber, and complicates any attempts to quantify the SRS intensity, and to deduce the species concentration from the SRS intensity. Third, take advantage of the resonance Raman effect of the minority species in order to selectively increase the particular Raman gain. By combining the second and third techniques, the SRS of 10^{-3} M Rhodamine 6G has been observed in microdroplets. However, the resonance Raman effect of many minority species may occur near the same wavelength of that of the majority species. Even if it were possible to selectively reach the absorption band of the minority species, the resonance Raman effect is usually accompanied by absorption losses (at the Stokes Raman frequency) which affect the Q value of the MDR's. Lastly, there exists the possibility of external injection seeding at the Raman shifted wavelength of the minority species. However, the injection frequency must be tuned to the same MDR as the SRS of the minority species. The input coupling efficiency of the external radiation to that particular MDR is hard to control, when using the present approach of a focused external seed beam. Whereas in the internal seeding case with dye fluorescence, the spectral and spatial matching were essentially automatic.

THIRD-ORDER SUM FREQUENCY GENERATION (TSFG)

Third-order sum frequency generation (at $\omega_{TSFG} = \omega_a + \omega_b + \omega_c$) is one of the many four-wave mixing processes [47], which involve the product of three of the incident electric fields (at ω_a, ω_b, and ω_c) that induce a nonlinear polarization (at ω_{TSFG}), $P^{NLS} = \chi^{(3)}E(\omega_a)E(\omega_b)E(\omega_c)$. This nonlinear polarization acts as a source to generate a resultant $E(\omega_{TSFG})$. The third-order nonlinear susceptibility, $\chi^{(3)}$, is nonzero for a centrosymmetric medium, such as a liquid, while the second-order nonlinear susceptibility $\chi^{(2)}$ is identically zero within the droplet volume, although it is nonzero at the droplet interface. In the degenerate case, when $\omega_a = \omega_b = \omega_c$, third-harmonic generation (THG) results with $\omega_{THG} = 3\omega_a$.

In the plane-wave formalism for TSFG and THG, the spatial overlap (along the propagation direction) between the P^{NLS} and the resultant $E(\omega_{TSFG})$ give rise to the concept of phase matching and coherence length. For THG, the phase-matching condition requires that the phase velocity of the induced P^{NLS} (with wavevector $3k_a$) and the resultant $E(\omega_{THG})$ (with wavevector k_{THG}) be equal. That is, the phase velocity of the input and resultant E's must be equal Because liquids have normal dispersion [$m(3\omega_a) > m(\omega_a)$], the condition for perfect phase matching is not realizable in a bulk liquid which is isotropic and in its low absorption region. The typical coherence length in liquids for the THG is $\ell_{coh} = (\lambda_a/6)[m(3\omega_a) - m(\omega_a)]^{-1} \approx 10\ \mu m$.

The THG signal from a liquid cell is known to be weak. The intensity of TSFG and THG per unit volume of liquid in the form of a droplet can be greater than it is in an optical cell. Nevertheless, TSFG and THG is 4-6 orders of magnitude weaker than typical multiorder SRS signals from droplets. Consequently, it is not a commonly studied topic and requires the use of a laser with high intensity, low fluence, and high repetition rate, such as the output ($\lambda_L = 1.064\ \mu m$) from a cw-pumped mode-locked Q-switched Nd:YAG laser. The laser beam is focused to a beam radius of 1/3 **a** and is aimed at (or just outside) the edge of the droplet equator, in order to couple such incident laser pulses more efficiently into MDR's that are within the input laser linewidth (0.3 cm^{-1}, corresponding to the Fourier transform of 100 ps pulse duration).

THG of the focused incident beam is at $3\omega_L$. If no other nonlinear effects were simultaneously generated, then the detected spectra would consist of only the THG peak. Because the incident beam can also generate jth-order cascade SRS (at $\omega_{js} = \omega_L - j\omega_{vib}$), their electric fields can also produce a series of peaks corresponding to THG (at $3\omega_{js}$) and TSFG (at $\omega_{TSFG} = 3\omega_L - p\omega_{vib}$), where p is an integer [36].

Figure 5 shows the TSFG spectra from CCl_4 droplets at various droplet generator frequency, f_{osc}, which directly changes the droplet radius [36]. Accompanying the radius change, there are shifting of the MDR's within the laser linewidth, Raman gain profile, and the TSFG linewidth. New MDR's can be tuned into these three profiles. In Fig. 5, the p = 3, 6,....., and 15 TSFG peaks can not be interpreted as the THG of jth-order cascade SRS, i.e., not simply as THG of $3\omega_{1s}$, $3\omega_{2s}$,....., and $3\omega_{5s}$. For example, the p = 3 peak can have contributions from the following combinations: (1) $\omega_{1s} + \omega_{1s} + \omega_{1s}$; (2) $\omega_L + \omega_{1s} + \omega_{2s}$; and (3) $\omega_L + \omega_L + \omega_{3s}$. The number of different P^{NLS} contributing to a p peak in the TSFG spectra increases rapidly as p increases.

There are two noticeable features of Fig. 5. First, at a fixed f_{osc} or droplet radius, the intensity of the TSFG peaks do not follow any progression as a function of p. Second, at different f_{osc} or droplet radius, one or a few peaks "randomly" dominate the TSFG spectra. Both features would still be present even if $E(\omega_L)$ and all the $E(\omega_{js})$'s were equal and stationary. The observations that the TSFG peaks do not follow any progression and appear seemly random are manifestations of the phase-matching effect.

Phase matching considerations [52] in droplets are simplified if we restrict the discussion to THG. Also assume that all the E's of the generating and resultant fields are on MDR's. The generating fields $E(\omega_{js})$'s are always guaranteed to be on MDR's, because MDR's are needed for providing the feedback for SRS. There are bound to be some appropriate MDR's within the Raman gain profile with 1 to 10 cm^{-1} linewidth. Furthermore, it is safe to assume that $E(\omega_L)$ is on one or more MDR's, because of the 0.3 cm^{-1} linewidth of our 100 ps laser pulses and the slight amount of droplet shape distortion ($|e| \approx 10^{-3}$ to 10^{-2} with $a \approx 40$ µm). Thus, the three input fields can be considered to be always on some input MDR's. In order for the THG to grow significantly, $E(\omega_{THG})$ must also be on a MDR, commonly referred to as an output MDR. This latter requirement could be problematic, because the linewidth of $E(\omega_{THG})$, $\Delta\omega_{THG}$ is at most the sum of the three times the linewidth of the input MDR's, e.g., $\Delta\omega_{THG} \approx 3\omega_{js}/Q$. The probability that there is a MDR within the narrow $\Delta\omega_{THG}$ is low. The fluctuations in the TSFG signal is probably associated with the droplet radius fluctuating slightly for each laser shots. Sometimes, the radius is just right for an appropriate MDR to be within $\Delta\omega_{THG}$.

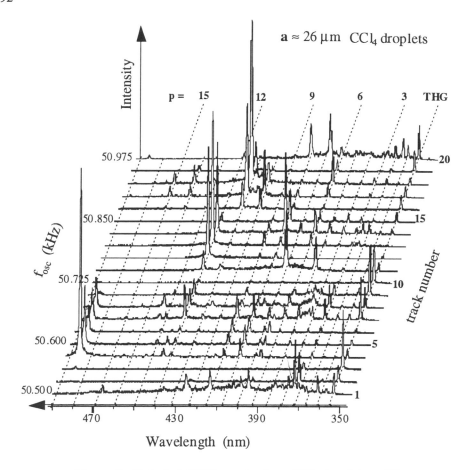

Figure 5. Detected TSFG spectra from CCl_4 droplets at 20 consecutive f_{osc} (shown on the vertical abscissa) starting at $f_{osc} = 50.5$ kHz and ending at $f_{osc} = 50.975$ kHz in 25-Hz increments. The horizontal abscissa indicates wavelength decreasing toward $\lambda_L/3$, and the series of dotted lines correspond to $\omega_{TSFG} = 3\omega_L - p\omega_{vib}$, where ω_{vib}, = 459 cm^{-1} is the frequency of the $\nu 1$ vibrational mode of CCl_4 and p is an integer. The incident laser wavelength is $\lambda_L = 1.064$ μm, and the detector is a position-sensitive resistive-anode device.

The degree of phase matching for THG in droplets [52] involves the amount of spatial overlap among the P^{NLS} and $E(\omega_{THG})$, where P^{NLS} is induced by the product of three input E's. The THG in droplets involve the product of three input fields describable by vector spherical harmonics of MDR's and the generation of one output field that is also describable by a MDR. In contrast to optical cell case which requires

1-d integration of four plane waves (along dz), the droplet case requires 3-d integration ($d\phi$, $\sin\theta d\theta$, and $r^2 dr$) to calculate the spatial overlap for the four MDR's.

The $d\phi$ integration requires the $m_{THG} = 3m_L$ or $m_{THG} = 3m_{js}$, where the subscript specify the frequency of the m-MDR's. Some physical insight can be provided for these m-MDR's requirements. With edge illumination at the equator only those $m \approx n$ modes are involved for $E(\omega_L)$, $E(\omega_{js})$, and $E(\omega_{THG})$. The angular velocity for a wave with $\exp(\pm im\phi - i\omega t)$ is $d\phi/dt = \omega/m$.. The phase velocity on the droplet surface (at $\mathbf{r} = \mathbf{a}$) is $v_f = \mathbf{a}(d\phi/dt)$. The phase matching requires that the v_f's of P^{NLS} and $E(\omega_{THG})$ be matched on the droplet surface, and, hence, $m_{THG} = 3m_L$ or $m_{THG} = 3m_{js}$. That is, the angular momentum in the z direction of the MDR's for the $E(\omega_{THG})$ must be equal to three times those for the $E(\omega_L)$ or $E(\omega_{js})$.

The $\sin\theta d\theta$ and $r^2 dr$ integration were performed for each combination of MDR's, with the input within a size parameter $125 < X_1 < 135$ and the output within $|3X_1 - X_3| < 0.05$. The normal dispersion of $m(\omega)$ is assumed for the liquid, with $m(3\omega) > m(\omega)$. The largest angular-overlap integral occurs when the mode numbers with n-MDR's have $n_3 \approx 3n_1$. That is, the angular momentum of the MDR's for the $E(\omega_{THG})$ must be equal to three times those for the $E(\omega_L)$ or $E(\omega_{js})$. The radial overlap is largest when the mode numbers with n-MDR's have a ratio of $3 < n_3/n_1 < 3.2$ and drops rapidly when n_3/n_1 decreases less than 3. When the mode orders with ℓ-MDR's are considered, the largest radial overlap occurs when $\ell_3/\ell_1 \approx 3$. In the future, analytical solutions involving addition theorem for spherical harmonics may provide rapidly calculable Clebsch-Gordon coefficients needed for phase-matching or spatial overlap considerations.

The result of 3-dimensional integration indicates that the probability is extremely low for finding an output MDR at ω_{THG} that have good spatial overlaps as well as the frequency coincidences, given the fact that all three generating E's are on input MDR's. Even if the shape deformation split the m-MDR degeneracy, the requirement that $m_{THG} = 3m_L$ or $m_{THG} = 3m_{js}$, does not increase much the probability of finding a degeneracy-lifted output MDR that is close to ω_{THG}. TSFG is rarely observed as the time average droplet size is changed in discrete steps. In addition, it is no wonder that experimental study reported TSFG is 10^{-4} - 10^{-6} times the SRS intensity. However, TSFG does provide a testing ground to develop our understanding of phase-matching concepts.

CONCLUSIONS

The subfield of dye lasing and nonlinear optical effects in microdroplets have been selectively reviewed. The emphases were on using these well known nonlinear optical processes to reveal some of the unique features of microcavity. The possible applications of some nonlinear optical techniques are presented in this chapter. We have shown that the dye lasing spectra are highly sensitive to size changes, shape distortions, and temperature changes. In addition, we have shown that the SRS spectra has the potential of being able to provide chemical species information within multicomponent liquid droplet [53]. There exists some hope of increasing the detectivity of minority species in droplets with internal and external seeding, resonance Raman effect, and extra gain provided by the laser-dye gain. Third-order sum frequency generation is reviewed as an example of how the microcavity alters the details related to phase matching. Undoubtedly, many new basic science "surprises" and novel applications to the diagnostics of sprays will continue to emerge from the collective investigation on physics and nonlinear optics of microcavities.

ACKNOWLEDGMENT

We gratefully acknowledge the partial support of this research by the U.S. Air Force Office of Scientific Research (Grant No. F49620-94-1-0135) and by the U.S. Army Research Office (Grant No. DAAH 04-94-G-0031). We acknowledge the collaborations and helpful discussions with Prof. Kenneth Young and Dr. Steven C. Hill, both of whom have taught us and contributed much.

REFERENCES

1. Bohren, C. F., and Huffman, D. R., (1983) *Absorption and Scattering of Light by Small Particles,* Wiley, New York.

2. Barber, P. W., and Chang, R. K., eds. (1988) *Optical Effects Associated with Small Particles*, World Scientific, Singapore.

3. Barber, P. W., and Hill, S. C., (1990) *Light Scattering by Particles: Computational Methods,* World Scientific, Singapore.

4. Johnson, B. R., (1992) Theory of Morphology-Dependent Resonances: Shape Resonances and Width Formulas, J. Opt. Soc. Am. A10, 343.

5. Lock, J. A., (1995) Improved Gaussian Beam Scattering Algorithm, Appl. Opt. 34, 559.

6. Gousbet, G., Maheu, B., and Grehan, G., (1988) Light Scattering From a Sphere Arbitrarily Located in a Gaussian Beam, Using a Bromwich Formulation, J. Opt. Soc. Am. A 5, 1427.

7. Barton, J. P., Alexander, D. R., and Schaub, S. A., (1988) Internal and Near-surface Electromagnetic Fields for a Spherical Particle Irradiated by a Focused Laser Beam, J. Appl. Phys. **64**, 1632.

8. Khaled, E. E. M., Hill, S. C., Barber, P. W., and Chowdhury, D. Q., (1992) Near-resonance Excitation of Dielectric Spheres with Plane Waves and Off-Axis Gaussian Beams, Appl. Opt. **31**, 1166.

9. Lai, H. M., Leung, P. T., Young, K., Hill, S. C., and Barber, P. W., (1990) Time-independent Perturbation for Leaking Electromagnetic Modes in Open Systems with Application to Resonances in Microdroplets, Phys. Rev. A **41**, 5187.

10. Tzeng, H. -M., Wall, K. F., Long, M. B., and Chang, R. K., (1984) Laser Emission from Individual Droplets at Wavelengths Corresponding to Morphology-Dependent Resonances, Opt. Lett. **9**, 499.

11. Knight, J. C., Driver, H. S. T., Hutcheon, R. J., and Robertson, G. N., (1992) Core-Resonance Capillary-Fiber Whispering-Gallery-Mode Laser, Opt. Lett. **17**, 1280.

12. Pinnick, R. G., Fernández, G., Xie, J. -G., Ruekgauer, T., Gu, J., and Armstrong, R. L., (1992) Stimulated Raman Scattering and Lasing in Micrometer-Sized Cylindrical Liquid Jets: Time and Spectral Dependence, J. Opt. Soc. Am. B **9**, 865.

13. Kwok, A. S., Serpengüzel, A., Hsieh, W. -F., Chang, R. K., and Gillespie, J. B., (1992) Two-photon Pumped Lasing in Microdroplets, Opt. Lett. **17**, 1435.

14. Campillo, A. J., Eversole, J. D., and Lin, H. -B., (1991) Cavity Quantum Electrodynamics Enhancement of Stimulated Emission in Microdroplets, Phys. Rev. Lett. **67**, 437.

15. Ching, S. C., Lai, H. M., and Young, K., (1987) Dielectric Microspheres as Optical Cavities: Thermal Spectrum and Density of States, J. Opt. Soc. Am. B **4**, 2004.

16. Chỹlek, P., Lin, H. -B., Eversole, J. D., and Campillo, A. J., (1991) Absorption Effects on Microdroplet Resonant Emission Structure, Opt. Lett. **16**, 1723.

17. Lin, H. -B., Huston, A. L., Eversole, J. D., and Campillo, A. J., (1992) Internal Scattering Effects on Microdroplet Resonant Emission Structure, Opt. Lett. **17**, 970.

18. Armstrong, R. L., Xie, J. -G., Ruekgauer, T. E., Gu, J., and Pinnick, R. G., (1993) Effects of Submicrometer-Sized Particles on Microdroplet Lasing, Opt. Lett. **18,** 119.

19. Eversole, J. D., Lin, H. -B., and Campillo, A. J., (1992) Cavity-Mode Identification of Fluorescence and Lasing in Dye-Doped Microdroplets, Appl. Opt. **31,** 1982.

20. Biswas, A., Latifi, H., Armstrong, R. L., and Pinnick, R. G. (1989) Double-Resonance Stimulated Raman Scattering from Optically Levitated Glycerol Droplets, Phys. Rev. A **40,** 7413.

21. Lin, H. -B., Eversole, J. D., and Campillo, A. J., (1990) Vibrating Orifice Droplet Generator for Precision Optical Studies, Rev. Sci. Instruments. **61,** 1018.

22. Tzeng, H. -M., Wall, K. F., Long, M. B., and Chang, R. K., (1984) Evaporation and Condensation Rates of Liquid Droplets Deduced from Structure Resonances in the Fluorescence Spectra, Opt. Lett. **9,** 273.

23. Chen, G., Serpengüzel, A., Chang, R. K., Acker, W. P., (1993) Relative Evaporation Rates of Droplets in a Segmented Stream Determined by Cavity Droplet Fluorescence Peak Shifts, in *Proceedings of the SPIE Conference on Laser Applications in Combustion and Combustion Diagnostics* **1862,** SPIE, Bellingham, Washington, pp. 200.

24. Tzeng, H. -M., Long, M. B., Chang, R. K., and Barber, P. W., (1985) Laser-Induced Shape Distortions of Flowing Droplets Deduced from Morphology-Dependent Resonances in Fluorescence Spectra, Opt. Lett. **10,** 209.

25. Chen, G., Mazumder, Md. M., Chemla, Y. R., Serpengüzel, A., Chang, R. K., and Hill, S. C., (1993) Wavelength Variation of Laser Emission Along the Entire Rim of Slightly Deformed Droplets, Opt. Lett. **18,** 1993.

26. Mazumder, Md. M., Chen, G., Kindlmann, P. J., Chang, R. K., and Gillespie, J. B., (1995) Temperature-Dependent Wavelength Shifts of Dye Lasing in Microdroplets with a Thermochromic Additive, submitted to Opt. Lett.

27. Scheggi, A. M., Bacci, M., Brenci, M., Conforti, G., Falciai, R., and Mignani, A. G., (1987) Thermometry by Optical Fibers and a Thermochromic Transducer, Opt. Eng. **26,** 534

28. Lin, H. -B., Huston, A. L., Eversole, J. D., and Campillo, A. J., (1990) Double-Resonance Stimulated Raman Scattering in Micrometer-Sized Droplets, J. Opt. Soc. Am. B **7,** 2079.

29. Lin, H. -B., Eversole, J. D., and Campillo, A. J., (1992) Continuous-Wave Stimulated Raman Scattering in Microdroplets, Opt. Lett. **17**, 828.

30. Thurn, R., and Kiefer, W., (1985) Structural Resonances Observed in the Raman Spectra of Optically Levitated Liquid Droplets, Appl. Opt. **24**, 1515.

31. Lettieri, T. R., and Preston, R. L., (1985) Observation of Sharp Resonances in the Spontaneous Raman Spectrum of a Single Optically Levitated Microdroplet, Opt. Comm. **54**, 349.

32. Snow, J. B., Qian, S. -X., and Chang, R. K., (1985) Stimulated Raman Scattering from Individual Water and Ethanol Droplets at Morphology-Dependent Resonances, Opt. Lett **10**, 37.

33. Lin, H. -B., Eversole, J. D., and Campillo, A. J., (1990) Identification of Morphology-Dependent Resonances in Stimulated Raman Scattering from Microdroplets, Opt. Comm. **77**, 407.

34. Lin, H. -B., Eversole, J. D., and Campillo, A. J., (1990) Frequency Pulling of Stimulated Raman Scattering in Microdroplets, Opt. Lett **15**, 387.

35. Qian, S. -X., and Chang, R. K. (1986) Multiorder Stokes Emission from Micrometer-Size Droplets, Phys. Rev. Lett. **56**, 926.

36. Leach, D. H., Chang, R. K., Acker, W. P. and Hill, S. C., (1993) Third Order Sum Frequency Generation in Droplets: Experimental Results, J. Opt. Soc. Am. B **10**, 34.

37. Hsieh W. -F., Zheng, J. -B., and Chang, R. K., (1988) Time Dependence of Multiorder Stimulated Raman Scattering from Single Droplets, Opt. Lett **13**, 497.

38. Pinnick, R. G., Biswas, A., Chỹlek, P., Armstrong, R. L., Latifi, H., Creegan, E., Srivastava, V., Jarzembski, M., and Fernández, G., (1988) Stimulated Raman Scattering in Micrometer-Sized Droplets: Time-Resolved Measurements, Opt. Lett. **13**, 494.

39. Zhang, J. -Z., Chen, G., and Chang, R. K., (1990) Pumping of Stimulated Raman Scattering by Stimulated Brillouin Scattering within a Single Droplet: Input Laser Linewidth Effects, J. Opt. Soc. Am. B **7**, 108.

40. Mazumder, Md. M., Schaschek, K., Chang, R. K., and Gillespie, J. B., (1995) Efficient Pumping of Minority Species Simulated Raman Scattering (SRS) by Majority Species SRS in a Microdroplet of a Binary Mixture, Chem. Phys. Lett., to be published.

41. Zhang, J. -Z., Leach, D. H., and Chang, R. K., (1988) Photon Lifetime within a Droplet: Temporal Determination of Elastic and Stimulated Raman Scattering, Opt. Lett. **13**, 270.

42. Zheng, J. -B., Hsieh, W. -F., Chen, S. -C., and Chang, R. K., (1989) Growth, Decay, and Quenching of Stimulated Raman Scattering in Transparent Liquid Droplets, in Wang, Z. and Zhang, Z., (eds.), *Laser Materials and Laser Spectroscopy*, World Scientific, Singapore, pp. 259.

43. Swindal, J. C., Leach, D. H., Chang, R. K., and Young, K. (1993) Precession of Morphology-Dependent Resonances in Non-Spherical Droplets, Opt. Lett. **18**, 191.

44. Chen, G., Chang, R. K., Hill, S. C., and Barber, P. W., (1991) Frequency Splitting of Degenerate Spherical Cavity Mode: Stimulated Raman Scattering Spectrum of Deformed Droplets, Opt. Lett **16**, 1269.

45. Chen, G., Swindal, J. C., and Chang, R. K., (1992) Frequency Splitting and Precession of Cavity Modes of a Droplet Deformed by Inertial Forces in *Proceedings of Shanghai International Symposium on Quantum Optics,* Vol. 1726 SPIE, Bellingham, Washington, pp. 292.

46. Chen, G., Acker, W. P., Chang, R. K., and Hill, S. C., (1991) Fine Structures in the Angular Distribution of Stimulated Raman Scattering from Single Droplets, Opt. Lett **16**, 117.

47 Hill, S. C., and Chang, R. K., (1995) Nonlinear Optics in Droplets in O. Keller, (ed) *Studies in Classical and Quantum Nonlinear Optics,* Nova Science Publishers, New York, pp. 171.

48. Serpengüzel, A., Chen, G., Chang, R. K., and W. -F., Hsieh, (1992) Heuristic Model for the Growth and Coupling of Nonlinear Processes in Droplets, J. Opt. Soc. Am. B **9**, 871.

49. Kwok, A. S., and Chang, R. K., (1992) Fluorescence Seeding of Weaker-Gain Raman Modes in Microdroplets: Enhancement of Stimulated Raman Scattering, Opt. Lett **17**, 1334.

50. Kwok, A. S., and Chang, R. K., (1993) Stimulated Resonance Raman Scattering of Rhodamine 6G, Opt. Lett. **18**, 1703.

51. Kwok, A. S., and Chang, R. K., (1993) Suppression of Lasing by Stimulated Raman Scattering in Microdroplets, Opt. Lett. **18,** 1597.

52. Hill, S. C., Leach, D. H., and Chang, R. K. (1993) Third Order Sum Frequency Generation in Droplets: Model with Numerical Results for Third Harmonic Generation, J. Opt. Soc. Am B **10**, 16.

53. Acker, W. P., Serpengüzel, A., Chang, R. K., and Hill, S. C., (1990) Stimulated Raman Scattering of Fuel Droplets: Chemical Concentration and Size Determination, Appl. Phys B **51**, 9.

PHOTONIC BAND STRUCTURES

J.W. HAUS

Rensselaer Polytechnic Institute, Troy, NY 12180-3590, USA

1 Introduction

The zero point energy associated with each mode of the electromagnetic spectrum is the source of quantum fluctuations that have physically measurable effects; this field of research has been called quantum electrodynamics. The origins of quantum electrodynamics is traced to the early contributions of Planck and Einstein on Blackbody radiation, but modern developments have clarified the role that vacuum fluctuations play in the atomic and molecular phenomena[1, 2]. The topic of this school attests to the significance of the field and its continued growth and evolution. It has not remained only a laboratory curiosity, but like its historic beginnings, it continues to find technological applications.

The subfield of cavity quantum electrodynamics has emerged as an important and active topic of research, whose origin is traced to a brief note published in 1946 by Purcell[3]. In this paper he shows that spontaneous emission rates are not immutable, because depend upon the density of the electromagnetic modes; a quantity that can be designed in the laboratory. This work was followed up by predictions made by Casimir and Polder on the force of attraction between two conducting plates and experiments to confirm the predictions.

The use of periodic dielectric structures was suggested[4, 5] as a practical means of altering the electromagnetic field modes to observe novel effects similar to those seen in cavity QED. The name coined for these structures is *photonic band structures* (PBS's). Despite the brief history of this subtopic, the topic has been the subject of reviews[6], special issues[7] and a book[8]. The objective of most recent research on these structures has been the creation of a complete bandgap that persists in all directions and over a band of frequencies. It was conjectured that, by a suitable choice of low-loss dielectric materials (such as germanium with a refractive index of $n = 4$) and geometry, one could design a PBS in which the periodicity of the medium prohibits the propagation of electromagnetic modes over a range of frequencies.

It is now established by experiment and theory that a variety of three-dimensionally, periodic dielectric structures possess stop bands (often called photonic band gaps). The ratio of the refractive indices required for a complete bandgap to open is found to be large; typically, the ratio is 3-4 depending on the geometry of the structure.

The periodic structures are a particular example of inhomogeneous media. Maxwell and Rayleigh made early studies of inhomogeneous media, but focused mainly on the long wavelength regime and the effective long wavelength dielectric constant was calculated using various analytical approximations. The revived interest in these materials is due to a variety of potential applications which require the prohibition or suppression of spontaneous emission through a certain channel. For instance, a laser has emission through the sides of the cavity that result in a additional losses and laser linewidth. By eliminating the electromagnetic modes in the lateral direction, the laser becomes a more efficient source.

2 One Dimension

The simplest case to consider is a multilayered structure[9, 10]. It exhibits interesting phenomena, such as, high reflectivity over a frequency range, a so-called forbidden band or *stop band*. The theory has a strong correspondence with the quantum theory of periodic lattice,

M. Ducloy and D. Bloch (eds.), Quantum Optics of Confined Systems, 101-141.
© *1996 Kluwer Academic Publishers. Printed in the Netherlands.*

so we draw heavily on physical concepts developed in solid state physics.

Figure 1: Schematic of a one-dimensional periodic layered structure. The wave propagates in the direction that is perpendicular to the interfaces. The period of the lattice is a.

The simple one dimensional structures are modeled with a lattice constant, a, in which a dielectric material with material coefficients, (ϵ_1, μ_1), has a depth d_1 and a second material with coefficients (ϵ_2, μ_2) has a depth d_2 ($a = d_1 + d_2$).

The electromagnetic plane waves are assumed to propagate perpendicular to the layers. The vector potential, which is perpendicular to the propagation direction satisfies the scalar wave equation for a plane wave of frequency ω propagating along the z-axis. For definiteness, let $\mathbf{A} = \hat{\mathbf{e}}_x A \exp{(-i\omega t)}$, then

$$\frac{d}{dz}\frac{1}{\mu(z)}\frac{dA}{dz} + \left(\frac{\omega}{c}\right)^2 \epsilon(z)A = 0. \tag{1}$$

This equation has two independent solutions, which we denote as $F_1(z)$ and $F_2(z)$. The Wronskian, defined by the determinate of the solutions and their first derivative,

$$W(z) = \begin{vmatrix} F_1(z) & F_2(z) \\ F_1'(z) & F_2'(z) \end{vmatrix}; \tag{2}$$

is non-zero when the solutions are independent. Moreover, it is constant over a region in which the coefficients ϵ and μ are kept constant and the value of the Wronskian is equal in the layers with identical physical properties. The solutions translated by a lattice constant can be written as a linear combination of the original solutions

$$\begin{aligned} F_1(z+a) &= M_{11}F_1(z) + M_{12}F_2(z); \\ F_2(z+a) &= M_{21}F_1(z) + M_{22}F_2(z); \end{aligned} \tag{3}$$

The Wronskian of the translated solutions is identical, i.e. $W(z+a) = W(z)$.

$$W(z+a) = \begin{vmatrix} F_1(z+a) & F_2(z+a) \\ F_1'(z+a) & F_2'(z+a) \end{vmatrix} = \begin{vmatrix} F_1(z) & F_2(z) \\ F_1'(z) & F_2'(z) \end{vmatrix} \cdot \begin{vmatrix} M_{11} & M_{21} \\ M_{12} & M_{22} \end{vmatrix} \tag{4}$$

A result that can be shown by multiplying the matrices on the right hand side before taking the determinants and comparing both sides with Eq. (3). It follows from the constancy of the Wronskian that the determinant of the matrix $\mathbf{M}$ must be equal to unity:

$$M_{11}M_{22} - M_{12}M_{21} = 1. \tag{5}$$

Eigenvalues are sought that have the form, where $\alpha = 1, 2$,

$$F_\alpha(z + a) = \lambda F_\alpha(z); \tag{6}$$

When this is inserted into Eq. (3), the following determinant vanishes for nonzero solutions

$$\begin{vmatrix} M_{11} - \lambda & M_{21} \\ M_{12} & M_{22} - \lambda \end{vmatrix} = 0. \tag{7}$$

In other words the equation for the eigenvalues is

$$\lambda^2 - (M_{11} + M_{22})\lambda + 1 = 0. \tag{8}$$

Eq. (5) was used to simplify the determinant. From this we deduce that the two eigenvalues $\lambda_\pm$ have the property that

$$\lambda_+ \lambda_- = 1;$$

and

$$\lambda_+ + \lambda_- = M_{11} + M_{22}. \tag{9}$$

It is common to define

$$\lambda_+ = e^{ika}, \tag{10}$$

where k is the wavenumber; then the eigenvalue equation is a manifestation of the Floquet–Bloch theorem

$$F_\alpha(z + a) = e^{ika} F_\alpha(z); \tag{11}$$

The functions $F_\alpha(z)$ are called the Floquet–Bloch functions[1], which are strictly periodic in z[9].

2.1   Kronig-Penney Model

In this section the above discussion will be applied to the case of a simple, but realistic model of layered, periodic media[10–12]. The dielectric permittivity and the magnetic permeability are constant in each material, i.e., (ϵ_1, μ_1) in a film of thickness d_1 and (ϵ_2, μ_2) in a film of thickness d_2; we consider two layers per period. The wavenumber is $k_\alpha = \omega \sqrt{\epsilon_\alpha \mu_\alpha}/c$, where $\alpha = 1, 2$. The vector potential is related to the fields by the usual relations in electrodynamics

---

[1]Floquet's name is added to the one-dimensional version of the theory to honor the mathematician who made important contributions to the solution of Mathieu's equation

and the boundary conditions for this case demand the continuity of the electric field and of the H-field. The general solution in each region is

$$A_1 = C_1 e^{ik_1 z} + D_1 e^{-ik_1 z}; \tag{12}$$

$$A_2 = C_2 e^{ik_2(z-d_1)} + D_2 e^{-ik_2(z-d_1)}. \tag{13}$$

The solution A_1 is valid in the region $(0, d_1)$ and all other regions displaced from this by a lattice constant translation; the wavenumber is related to the frequency by

$$k_1 = n_1 \frac{\omega}{c} = \sqrt{\epsilon_1 \mu_1} \frac{\omega}{c}. \tag{14}$$

A_2 is valid in the region (d_1, a) and all its translations by a and the wavenumber for this region is similarly defined

$$k_2 = n_2 \frac{\omega}{c} = \sqrt{\epsilon_2 \mu_2} \frac{\omega}{c}. \tag{15}$$

n_1 and n_2 are the refractive indices of each medium.

According to the Floquet–Bloch theorem, which imposes a periodicity on the solutions, the vector potential in corresponding regions is connected by

$$A_{\alpha k}(z + a) = e^{ika} A_{\alpha k}(z). \tag{16}$$

where the additional subscript, k, has been added to remind the reader that the solutions depend on the wavenumber. The wavenumber is normally restricted to the first Brillouin zone (BZ), since values outside this zone are redundant. To solve the eigenvalue problem and determine the connection between the wavenumber k and the frequency ω, the boundary conditions are applied. The continuity of the electric field and of the H–field can be written in a compact form

$$\begin{pmatrix} 1 & 1 \\ 1 & -1 \end{pmatrix} \begin{pmatrix} C_2 \\ D_2 \end{pmatrix} = \begin{pmatrix} e^{ik_1 d_1} & e^{-ik_1 d_1} \\ Z_1 e^{ik_1 d_1} & -Z_1 e^{-ik_1 d_1} \end{pmatrix} \begin{pmatrix} C_1 \\ D_1 \end{pmatrix}; \tag{17}$$

and

$$\lambda \begin{pmatrix} 1 & 1 \\ 1 & -1 \end{pmatrix} \begin{pmatrix} C_1 \\ D_1 \end{pmatrix} = \begin{pmatrix} e^{ik_2 d_2} & e^{-ik_2 d_2} \\ \frac{1}{Z_1} e^{ik_2 d_2} & -\frac{1}{Z_1} e^{-ik_2 d_2} \end{pmatrix} \begin{pmatrix} C_2 \\ D_2 \end{pmatrix}; \tag{18}$$

where $Z_1 = \sqrt{\epsilon_1 \mu_2 / \epsilon_2 \mu_1}$ is the ratio of impedances. The factor λ is introduced as a manifestation of the Floquet–Bloch theorem.

Combining the above matrix equations, the eigenvalue equations has the from

$$\lambda \begin{pmatrix} C_1 \\ D_1 \end{pmatrix} = \frac{1}{4} \begin{pmatrix} (1 + \frac{1}{Z_1}) e^{ik_2 d_2} & (1 + \frac{1}{Z_1}) e^{-ik_2 d_2} \\ (1 - \frac{1}{Z_1}) e^{ik_2 d_2} & (1 - \frac{1}{Z_1}) e^{-ik_2 d_2} \end{pmatrix} \tag{19}$$

$$\times \begin{pmatrix} (1 + Z_1) e^{ik_1 d_1} & (1 + Z_1) e^{-ik_1 d_1} \\ (1 - Z_1) e^{ik_1 d_1} & (1 - Z_1) e^{-ik_1 d_1} \end{pmatrix} \begin{pmatrix} C_1 \\ D_1 \end{pmatrix} \tag{20}$$

The eigenvalues are solutions of the transcendental equation, Eq. (9), where the properties of the eigenvalues derived in the previous subsection have been used. The two matrix coefficients are

$$M_{11} = M_{22}^* = \frac{(1 + Z_1)^2}{4Z_1} e^{i(k_1 d_1 + k_2 d_2)} - \frac{(Z_1 - 1)^2}{4Z_1} e^{i(k_2 d_2 - k_1 d_1)}. \tag{21}$$

The dispersion equation is

$$\cos ka = \frac{(Z_1 + 1)^2}{4Z_1} \cos(k_1 d_1 + k_2 d_2) - \frac{(Z_1 - 1)^2}{4Z_1} \cos(k_1 d_1 - k_2 d_2) \tag{22}$$

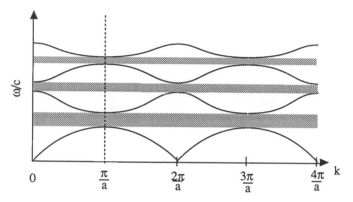

Figure 2: Schematic depicting the dispersion curve of a one dimensional lattice. The vertical dotted line is one edge of the first Brillouin zone (BZ) and the bands are numbered consecutively starting from the lowest frequency band.

The band structure for the 1-dimensional lattice always possesses a stop band, no matter how close Z_1 is to unity. A schematic is plotted in Fig. 2 displays the strong distortion of the dispersion relation from a straight line. The eigenfrequencies are ordered and labeled by their band index, n. The eigenvectors are the Floquet–Bloch functions, which reveal the variations of the field intensity in the structure. There are resonances, especially near the stop band edges where standing wave patterns are found. The center of the first stop band corresponds to layers that are a quarter wavelength thick. This is the choice for distributed Bragg reflectors. Also, this resonance can be applied to enhancing the nonlinear response and will be discussed later. The dispersion relation is periodic in the wavenumber with a period $2\pi/a$. To determine the band structure we need only consider a zone of this width. The first BZ is defined to be symmetric about the origin. In Fig. 2 the vertical dotted line denotes one boundary of the first BZ.

There are several properties of this transcendental equation that should be noted:

- In the long wavelength limit the wavenumber and the frequency are linearly related. This can be used to define an effective index of refraction by $k = \bar{n}\omega/c$

$$\bar{n}^2 = \frac{4n_1 n_2 d_1 d_2 (1 + Z_1^2) + 4Z_1 (n_1^2 d_1^2 + n_2^2 d_2^2)}{4Z_1 a^2}. \tag{23}$$

- The left side of the transcendental equation, Eq. (22), is restricted to the range $(1, -1)$, when the wavenumber is real. On the right hand side the function can lie outside that range. When this occurs, the wavenumber is complex and the electromagnetic wave does not propagate; there is, however, an evanescent wave in this frequency region and tunneling is possible with finite-sized structures. This is the stop gap region of the structure.

- Vertical cavity semiconductor lasers are designed with a stack of dielectrics that are chosen to be a quarter wavelength thick, i.e. $k_1 d_1 = k_2 d_2 = \frac{\pi}{2}$. In this case the last term in Eq. (22) is a constant independent of ω. The dispersion equation for this case simplifies to

$$\cos ka = \frac{(Z_1 + 1)^2}{4Z_1} \cos(K) - \frac{(Z_1 - 1)^2}{4Z_1}. \tag{24}$$

The variable $K = k_1 d_1 + k_2 d_2$ is a related to the frequency of the wave.

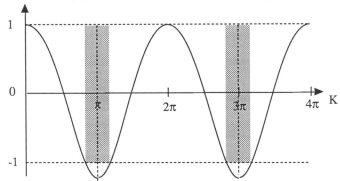

Figure 3: The allowed band lie between $(-1, 1)$ on the ordinate. The curve is the right hand side of Eq. (22) plotted versus K for the case of a quarter–wave stack. The shaded area is the stop band, the center of the stop band and the width of the stop band are discussed in the text.

- This structure always possesses a stop band, no matter how close Z_1 is to unity.

- To find the band structure, the wavenumber is restricted to the first BZ and the eigenfrequencies are ordered and labeled by their band index, n; the numbers are consecutively given beginning at the lowest frequency band. The eigenvectors are the Floquet–Bloch functions, which reveal the variations of the field intensity in the structure.

- The phase velocity is given by $v_p = \omega/k$ and the values are simply read off the band structure graph, eg. Fig. 2. The group velocity is defined by $v_g = \frac{d\omega}{dk}$; for the special case of Eq. (24), the function is:

$$v_g = \frac{4n_1 n_2 ac \sin ka}{(n_1 + n_2)^2 (n_1 d_1 + n_2 d_2) \sin K}; \tag{25}$$

which vanishes at the band edges, as shown in Fig. 4. The density of states, which is

$$\rho(\omega) = \left|\frac{dk}{d\omega}\right|, \tag{26}$$

in one dimension, diverges at the band edge for an infinite lattice.

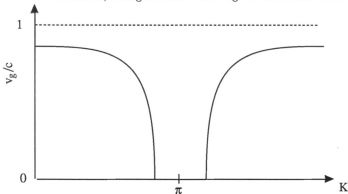

Figure 4: The group velocity close by and away from the stop band. The inverse of the group velocity is the density of states in one dimension.

- There are resonances, especially near the stop band edges where standing wave patterns are found.

2.2 Small index differences

Under the condition that the index difference is tiny compared to the indices, i.e., $|n_1 - n_2| \ll n_{1,2}$, the band edges can be determined by using a parabolic approximation to the right hand side of the dispersion equation. For simplicity, we consider the special case of a quarter–wave stack, given by Eq. (24), and expand the variable K about the minimum on the right hand side at $K = \pi$. The edge is reached when the right hand side meet the line at -1.

$$(K - \pi) = \pm 2\frac{|1 - Z_1|}{1 + Z_1}. \tag{27}$$

For the usual dielectric material $\mu_{1,2} = 1$, the edge, expressed in terms of the stop gap center frequency, $\omega_0 = \pi c/(n_1 d_1 + n_2 d_2)$, is

$$\omega - \omega_0 = \pm\frac{2\omega_0}{\pi}\frac{|n_1 - n_2|}{n_1 + n_2}. \tag{28}$$

The width of the stop gap is a small fraction of the central frequency in this limit. Ultrashort pulses whose band width is comparable in size to the central frequency would have only a small band of its spectrum reflected. To contain the spectrum of an ultrashort pulse within the stop band, the difference between the indices must be large.

2.3   Fourier Series

The wave equation Eq. (1) can be numerically solved by introducing a Fourier series representation. The Fourier series for the dielectric function is

$$\epsilon(z) = \sum_{\nu=-\infty}^{\infty} \epsilon_\nu e^{-iG_\nu z} \tag{29}$$

where the variable G_ν represents the reciprocal lattice vectors

$$G_\nu = \nu \frac{2\pi}{a} \tag{30}$$

and as indicated, the integers ν are both positive and negative.

The vector potential amplitude is expanded according to the Floquet–Bloch theorem

$$A(z) = e^{ikz} \sum_\nu A_\nu e^{iG_\nu z}. \tag{31}$$

The wave equation $(\mu(z) = 1)$ after Fourier transformation takes the form:

$$- (k + G_\nu)^2 A_\nu + \frac{\omega^2}{c^2} \sum_{\nu'} \epsilon_{\nu-\nu'} A_{\nu'} = 0. \tag{32}$$

In general, many terms need to be retained in the series to obtain accurate representations of the eigenvalues and the eigenfunctions. For the one–dimensional case this is an easy task and the results show a rapid convergence. Later we will discuss analytical methods developed around the Floquet–Bloch solutions of the wave equation in one dimension.

2.4   Coupled-mode Analysis

Suppose that k is near the edge of the first BZ and we are interested in the region near the first stop band. The Fourier series solution can be truncated to obtain approximate results for the gap parameters in this region. Consider the series representation in Eq. (32), the component A_ν is written

$$A_\nu = \frac{\left(\frac{\omega}{c}\right)^2 \sum_{\nu' \neq \nu} \epsilon_{\nu-\nu'} A_{\nu'}}{(k + G_\nu)^2 - \epsilon_0 \left(\frac{\omega}{c}\right)^2}. \tag{33}$$

For specific values of the wavenumber k, the denominator will be almost resonant in few selected terms. For instance, when k is near the first Brillouin zone boundary, i.e. $|k| \approx |k + G_1| \approx \epsilon_0 \frac{\omega}{c}$, the amplitudes (A_0, A_1) have near–resonant denominators and therefore, the coupling between these modes is retained, while the other modes are neglected:

$$A_0 = \frac{\left(\frac{\omega}{c}\right)^2 \epsilon_1 A_1}{k^2 - \epsilon_0 \left(\frac{\omega}{c}\right)^2}; \tag{34}$$

and

$$A_1 = \frac{\left(\frac{\omega}{c}\right)^2 \epsilon_{-1} A_0}{(k + 2\pi/a)^2 - \epsilon_0 \left(\frac{\omega}{c}\right)^2}. \tag{35}$$

The amplitude A_0 describes a plane-wave propagating in the direction $k\hat{e}_z$ and A_1 is a plane-wave propagating in the opposite direction with wave vector $(k + \frac{2\pi}{a})\hat{e}_z$. The Fourier component $\epsilon_{-1} = \epsilon_1^*$ when the dielectric constant is real. The eigenvalue equation is a quadratic function of ω^2

$$(\epsilon_0^2 - |\epsilon_1|^2)\left(\frac{\omega}{c}\right)^4 - \epsilon_0(k^2 + (k + \frac{2\pi}{a})^2)\left(\frac{\omega}{c}\right)^2 + k^2\left(k + \frac{2\pi}{a}\right)^2 = 0. \tag{36}$$

At the zone boundary $k = -\pi/a$, the roots are

$$\omega_{\pm}^{(1)} = \pm\frac{|k|}{c\sqrt{\epsilon_0 \pm |\epsilon_1|}}. \tag{37}$$

These frequencies lie at the upper and lower band edges of the first gap and their difference is the first stop gap width between band 1 and band 2 in Figure 2; the superscript 1 denotes the first stop gap. The corresponding eigenfunctions are orthogonal, i.e., they have different parity. The eigenfunction solution for ω_- is constructed, from Eq. (31), assuming ϵ_1 is real, to be

$$A_-^{(1)} = \sin\frac{\pi z}{a}, \tag{38}$$

and the solution for ω_+ is

$$A_+^{(1)} = \cos\frac{\pi z}{a}. \tag{39}$$

At the band edges standing wave solutions are found. The phases at the bottom, ω_+, and top, ω_-, of the band are in a quadrature relation. The amplitude at the bottom of the band corresponds to an intensity pattern that peaks in the high dielectric region; whereas, the intensity at the band edge ω_- peaks in the low dielectric region. As the frequency is tuned across the band gap, the phase difference between the two amplitudes continuously changes by a factor of of π and the wavenumber k becomes complex in this region.

The gap of the higher bands is similarly obtained by taking $|k| \approx |k + G_\nu| \approx \epsilon_0\frac{\omega}{c}$. The general result for the νth stop gap is

$$\omega_{\pm}^{(\nu)} = \pm\frac{|k|}{c\sqrt{\epsilon_0 \pm |\epsilon_\nu|}}. \tag{40}$$

Again, standing–wave patterns are found at the band edges with a $\pi/2$–phase difference between the top and the bottom. In the gap the phase between the amplitudes undergoes a continuous change.

This method is a type of slowly varying envelope method in which particular Fourier components of the field are chosen as the dominant contribution. Near the stop gaps it can be useful, as long as the Fourier components of the dielectric function are small; away from the gaps, there is no dominant contribution and many terms in the series are required. Further discussion can be found in Ref. [10].

3 Two Dimensions

The scalar wave equation has limited application in photonic band structures. Two-dimensional periodic structures, such as, rods in a lattice arrangement, with the field polarized parallel to the rod axis satisfy scalar wave equations [13–18]. In this section, it is relevant to consider for a moment the analysis of the simpler scalar problem and contrast the results with those of the vector equations in Section 4.

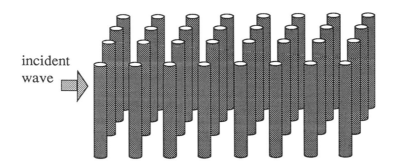

incident wave

Figure 5: A lattice of two–dimensional rods. The incident light lies in the plane perpendicular to the rods.

The derivation of the appropriate scalar wave equation depends upon the chosen field polarization. For the case where the electric field is along the symmetry axis, the equation is:

$$\nabla^2 E - \frac{1}{c^2}\epsilon(\mathbf{x})\frac{\partial^2 E}{\partial t^2} = 0. \tag{41}$$

This forms a generalized eigenvalue problem because the frequency eigenvalues, obtained by Fourier transformation in the time coordinate, are multiplied by the periodic dielectric function. For this reason, it is not precisely equivalent to the Schrödinger equation, since the eigenvalue multiplies the 'potential'. In one-dimensional periodic materials with the field propagating perpendicular to the surfaces, this is equivalent to the Kronig-Penney model[11, 12].

For the case with the magnetic field along the symmetry axis, the reduction of Maxwell's equations to a wave equation takes the form

$$\nabla \cdot \eta(\mathbf{x})\nabla H - \frac{1}{c^2}\frac{\partial^2 H}{\partial t^2} = 0; \tag{42}$$

for convenience the definition $\eta(\mathbf{x}) = 1/\epsilon(\mathbf{x})$ has been made. This equation forms an ordinary eigenvalue problem for the frequency that is equivalent to the Schrödinger equation with a periodically varied mass.

As discussed above, there are several methods that can be applied to solving these equations. The simplest, most widely used and most flexible one is derived from the Bloch wave analysis of periodic structures[19]. The eigenvalues and eigenvectors can be found by introducing the Bloch functions

$$\phi_{n\mathbf{k}}(\mathbf{x}, t) = e^{i(\mathbf{k}\cdot\mathbf{x}-\omega_{n\mathbf{k}}t)} \sum_{\mathbf{G}} \phi_{\mathbf{k}}(\mathbf{G}) e^{i\mathbf{G}\cdot\mathbf{x}}. \tag{43}$$

where $\mathbf{G}$ is the reciprocal lattice vector for the chosen lattice and the wave vector $\mathbf{k}$ lies within the first BZ. The index n labels the band for a particular wave vector $\mathbf{k}$.

A triangular lattice and its reciprocal lattice are depicted in Fig. 6. A lattice is decomposed into unit cells that repeat and tile the space. Each lattice point can be reached from the origin by a linear combination of two primitive basis vectors, which we denote by $(\mathbf{b}_1, \mathbf{b}_2)$. The corresponding reciprocal lattice vectors are defined by the following relations:

$$\mathbf{G}_i \cdot \mathbf{b}_j = 2\pi \delta_{ij}. \tag{44}$$

For example, the triangular lattice in Fig. 6 has the basis vectors: $\mathbf{b}_1 = a\hat{\mathbf{e}}_x$ and $\mathbf{b}_2 = \frac{a}{2}\hat{\mathbf{e}}_x + \frac{\sqrt{3}a}{2}\hat{\mathbf{e}}_y$. The corresponding reciprocal lattice vectors are: $\mathbf{G}_1 = \frac{2\pi}{a}(\hat{\mathbf{e}}_x - \frac{1}{\sqrt{3}}\hat{\mathbf{e}}_y)$, $\mathbf{G}_2 = \frac{4\pi}{\sqrt{3}a}\hat{\mathbf{e}}_y$, as depicted in the figure. The reciprocal lattice is constructed by summing combinations of the two basis reciprocal lattice vectors. Also shown is the construction for the first Brillouin zone (BZ); perpendicular bisectors of the reciprocal lattice points are drawn and where they intersect, the edge of the first BZ is formed. The first BZ is a hexagon; the point at the center of the zone is called the Γ–point; two other important points are shown, the X-point at $\frac{2\pi}{\sqrt{3}a}\hat{\mathbf{e}}_y$ and the J-point at $\frac{2\pi}{\sqrt{3}a}(\frac{1}{\sqrt{3}}\hat{\mathbf{e}}_x + \hat{\mathbf{e}}_y)$. These points are relevant symmetry points in the lattice that will appear again when the band structure is presented.

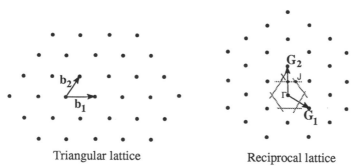

Triangular lattice   Reciprocal lattice

Figure 6: A triangular lattice constructed from basis vectors $\mathbf{b}_1$ and $\mathbf{b}_2$. The reciprocal lattice is also shown with the corresponding reciprocal lattice basis vectors $\mathbf{G}_1$ and $\mathbf{G}_2$ and the boundary of the first BZ. The Γ–point is at the center of the zone and the X- and J-points lie on the boundary.

The equations are transformed to matrix equations when this expansion is inserted into the wave equations, along with the transform of the dielectric function

$$\epsilon(\mathbf{x}) = \sum_{\mathbf{G}} \epsilon(\mathbf{G}) e^{i\mathbf{G}\cdot\mathbf{x}}, \tag{45}$$

or its inverse $\eta(\mathbf{x})$.

The Fourier transform of Eq. (41) is

$$-|\mathbf{k} + \mathbf{G}|^2 E_{\mathbf{G}} + \left(\frac{\omega}{c}\right)^2 \sum_{\mathbf{G}'} \epsilon(\mathbf{G} - \mathbf{G}')E_{\mathbf{G}}' = 0; \qquad (46)$$

and similarly for Eq. (42) the Fourier transform is

$$-(\mathbf{k} + \mathbf{G}) \cdot \sum_{\mathbf{G}'}(\mathbf{G} - \mathbf{G}')\eta(\mathbf{G} - \mathbf{G}')H_{\mathbf{G}'} + \left(\frac{\omega}{c}\right)^2 H_{\mathbf{G}} = 0. \qquad (47)$$

Determination of the eigenvalues again proceeds by truncating the series of amplitudes and diagonalizing the matrices. Truncation has its pitfalls and extra care must be exercised to obtain accurate and reliable results [20]. The dielectric function has large discontinuities, which means that there are important Fourier components at large reciprocal lattice vectors. The convergence of the Fourier amplitudes proceeds slowly and the number of terms required for a specified accuracy increases as the power of the lattice dimensionality. When only the basis reciprocal lattice vectors are included, the dielectric constant varies as a sinusoidal function of position. Several authors have used this assumption to examine the band structure, but it is not even in qualitative agreement with results generated from the sharp-boundary discontinuity model.

3.1   Dielectric Fluctuations

The dispersion curves are a result of the spatial variation of the dielectric constant. The homogeneous medium has a linear relation between the wavenumber and the frequency; but in a inhomogeneous medium, the waves are scattered and interfere with one another and quantitative results require numerical computations. The resulting dispersion curve is complicated. Nevertheless, we can crudely learn about the size of the perturbation by considering the fluctuations of the dielectric constant.

The relative fluctuation of $\epsilon(\mathbf{x})$ from its spatial average provides a measure of the spectrum's deviation from the homogeneous medium. Express the dielectric constant as

$$\epsilon(\mathbf{x}) = \bar{\epsilon}\left[1 + \epsilon_r(\mathbf{x})\right], \qquad (48)$$

where $\bar{\epsilon}$ is the spatial average of the dielectric constant

$$\bar{\epsilon} = \int_{cell} \epsilon(\mathbf{x})d^D x. \qquad (49)$$

The superscript D denotes the dimension of the lattice. Note that $\bar{\epsilon}$ is an overall scaling factor and therefore, $\epsilon_r(\mathbf{x})$ is the relative ripple of the dielectric constant.

The perturbation parameter is related to the variance of the ripple

$$<\epsilon_r^2> \equiv <\epsilon_{fluc}^2>/\bar{\epsilon}^2;$$

the average is taken with respect to the unit cell volume. For $D > 1$ we find that when $< \epsilon_r^2 > \sim 1$ or larger, significant deviations from the linear dispersion relation is found and nonperturbative effects, such as band gaps for both polarizations, begin to appear.

For any two-component medium where $\epsilon(\mathbf{x})$ can assume only two values ϵ_a and ϵ_b, with the a-type medium occupying a volume fraction β of space, one has

$$< \epsilon_r^2 > = \frac{< \epsilon^2 >}{\bar{\epsilon}^2} - 1 = \frac{\beta \epsilon_a^2 + (1 - \beta) \epsilon_b^2}{[\beta \epsilon_a + (1 - \beta) \epsilon_b]^2} - 1. \tag{50}$$

Note that for fixed β the ripple saturates in value as either dielectric constant goes to infinity. Given ϵ_a and ϵ_b, the value of β that maximizes $< \epsilon_r^2 >$ is

$$\beta_{max} = \frac{\epsilon_b}{\epsilon_a + \epsilon_b}, \tag{51}$$

The corresponding maximum variance of the ripple is

$$< \epsilon_r^2 > = \frac{(\epsilon_a - \epsilon_b)^2}{4 \epsilon_a \epsilon_b} \tag{52}$$

In Fig. 7, contours of $< \epsilon_r^2 >$ are plotted against (ϵ_a, β) for $\epsilon_b = 1$. The dashed line is the optimal value of β given by Eq. (51). The computed relative gap $\Delta\omega/\omega$ as a function of ϵ_a and/or β peaks roughly where $< \epsilon_r^2 >$ does. For scalar waves in three dimensions, the quantitative agreement is excellent [21–23]. For vector waves, the agreement is only approximate due to competition from the effects of connectivity.

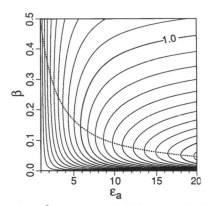

Figure 7: Contours of $< \epsilon_r^2 >$ vs ϵ_a and β for $\epsilon_b = 1$. The dashed line is the value of β that maximizes $< \epsilon_r^2 >$ for a given ϵ_a. The line labeled by 1 is the contour where the ripple variance is unity. (From Ref. [20])

The line labeled by 1 in Fig. 7 delineates the boundary between the regions of the ripple variances smaller and larger than unity. The ratio of the dielectric constants must be large, about 6, for large enough perturbations to open a gap. Also the corresponding volume fraction is mostly low dielectric material, eg. $\epsilon_b = 6$, $\beta_{max} \approx 0.86$ when the ripple variance is unity. These numbers provide a guide to finding where to expect band gaps. *This simple analysis is in accord with the observations that a band gap for both polarizations only exists when the volume is predominantly low dielectric material and the contrast between the two dielectric constants is large; also, increasing the ratio of dielectric constants eventually leads to a saturation of the band structure features, such as, the band gap size.*

3.2   Band Structure

The band structure is computed by a straightforward matrix procedure. The Fourier transform of the dielectric function for the case of non-overlapping rods of radius r and dielectric constant ϵ_a in a background dielectric ϵ_b is

$$\epsilon(\mathbf{G}) = \epsilon_b \delta_{\mathbf{G},0} + 2\beta(\epsilon_a - \epsilon_b)\frac{J_1(Gr)}{Gr}; \tag{53}$$

where $J_1(Gr)$ is the first-order Bessel function. For $\mathbf{G} = 0$, the average dielectric constant is $\epsilon(0) = \beta\epsilon_a + (1 - \beta)\epsilon_b$.

Truncating the Fourier series results in a rounding-off of the dielectric function at the interfaces. This occurs because a sharp interface represents a structure with very high Fourier components and the convergence there is not uniform. There is also an overshoot phenomenon, called Gibb's phenomenon, which persists even in the limit of an infinite number of plane waves. Convergence can be a difficult problem indeed for the plane–wave expansion. This problem is magnified in three dimensions and will be discussed again later. A representation of the dielectric function is shown in Fig. 8. With 271 plane waves, there are ripples in evidence; the convergence for this case is very good.

The band structure for the polarization of the H-field parallel to the rods is shown in Fig. 9. There is a large gap that opens up between the 2-3 bands. The volume fraction for this case is $\beta = .906$ and the ratio of dielectric constants is 5. The itinerary in the BZ can be understood by reference to Fig. 6. The point labeled 0 is the J-point at the zone boundary, the path leads directly to the center of the zone until at 20 the Γ-point is reached. Next a path is taken directly to the X-point (40) and from there the path is closed back on the J-point (60).

The band structure for the electric field parallel to the rods is displayed in Fig. 10. The volume fraction of $\beta = 0.906$ corresponds to the case of close–packed rods. The same itinerary is chosen, as in Fig. (9). For several values of ϵ_a, the bands transform from straight lines without dispersion in the homogeneous limit, $\epsilon_a = \epsilon_b = 1$, to a strong fluctuation limit in which large band gaps are in evidence. In the case where the electric field is parallel to the rods, the gap opens up between the 2-3 bands and it opens for a value of the dielectric function around 10. A complete gap for both polarizations is only found for very large dielectric ratios. Contrast this result with the simple one-dimensional structures, where a gap is present no matter how small the difference between dielectric constants.

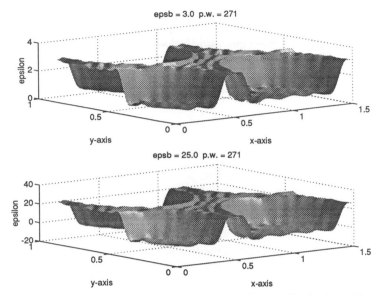

Figure 8: The dielectric function for a two–dimensional triangular lattice with two separate ratios for the dielectric constants: 3 and 25. The convergence in both cases is about the same.

3.3   Infrared and Visible Band gap Materials

For two–dimensional structures, there are both theoretical analyses and experimental results in the microwave regime, where the lattice constant can be machined on the order of millimeters [15, 24–27]. The author participated in a Japan-US collaboration in which two dimensional hollow-rod structures in glass were fabricated with a lattice constant of $1.17\mu m$, i.e. suitable for the near infrared regime[28]. This fabrication technique opens up the way for future studies of Raman active processes to resonantly enhance the nonlinear optical properties and improve the Kerr coefficient at resonant frequencies by using constructive interference effects.

The group of Prof. K. Inoue in Sapporo, Japan had two-dimensional triangular lattice samples fabricated by Hamamatsu. The samples are a specially fabricated design using the technology for producing micro-channel plates. A top view of the structure is shown in Figure 11; it is a periodic, parallel array of cylinders with a dielectric constant of 1 (called air-rods hereafter) each having a circular cross-section of diameter 0.90 μm in a background (clad) material of Pb-O glass with a dielectric constant of 2.62. The lattice constant of the periodic lattice is 1.17 μm. The arrays were pulled to a hexagonal shape, whose parallel edges were separated by 64 μm. The sample was fabricated by placing 466 of the hexagonal shaped arrays together to form a quasi-circular bundle of hexagons; the diameter of the bundle is about 1.5 mm. Outside the bundle, the lattice structure is supported by a similar Pb-O glass with a refractive index almost identical with the cladding. The length of the structure

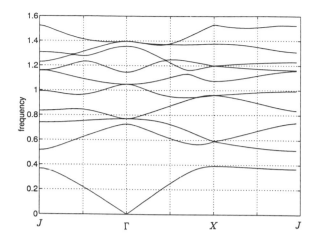

Figure 9: Band structure for the case of the H-field parallel to the rod axis. The rods are just touching, which corresponds to a volume fraction of $\beta = 0.906$.

is 1.0 mm. The core glass was dissolved by using a HCl acid etch. Direct inspection of the sample by an optical microscope reveals a regular array of holes, as shown in Figure 11.

The two-dimensional Brillouin zone of the triangular lattice possesses three relevant points, see Fig. 6. Propagation parallel to these directions was studied; the samples were made by polishing two samples whose front and back surfaces of the support glass were flat and perpendicular to the X-point or the J-point [28].

In Fig. 12 results are shown for the transmittance spectra for the p- and s-polarizations, i.e. E polarized perpendicular (p-polarization) and E polarized parallel (s-polarization) to the rod axis. In the leftmost figure the light is propagated along the Γ- X direction and on the right are the experimental results for propagation along the Γ-J direction.

Consider first the spectra for the X-point direction. An opaque region exists from $3900 \mathrm{cm}^{-1}$ to $4600 \mathrm{cm}^{-1}$ for the p-polarization and from $3800 \mathrm{cm}^{-1}$ to $4300 \mathrm{cm}^{-1}$ for the s-polarization. This is the inherent photonic band gap of the structure and is supported by calculations. Beyond $4600 \mathrm{cm}^{-1}$ no complete reflections were observed. For the s-polarization, on the other hand, no opaque region is recognized over the entire observed range.

With the lattice spacing and hole diameter for our sample the filling factor of air-rods is 0.537. Calculations to verify the nature of the observed transmittance were made using the plane-wave analysis; 271 plane waves for each polarization determined the dispersion spectra. When the H-field is parallel to the rod axis (see Fig. 13), the band structure has a large gap for all directions in the 2D BZ, even though the dielectric constant is not very large. For the s-polarization though, the degeneracy at the J-point persists. This degeneracy was broken by using two rods per unit cell[29], which has produced small gaps in the dispersion curves between the 1-2 bands.

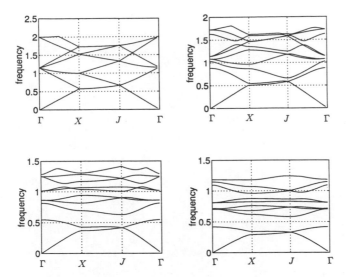

Figure 10: Band structure when the rods are just touching; the volume fraction of rod material is $\beta = 0.906$. The dielectric constants are: 1,3,10 and 18. A gap has formed between the 2-3 bands in the third frame; this is close to the expectation for band gap formation given by the dielectric fluctuation argument

4 Three Dimensions

The three-dimensional scalar wave equation, analogous to Eq. (41), has been studied in detail by Datta et al. in Ref. [30]; all the cubic structures have been investigated with spheres around each lattice point: simple cubic (SC), body centered cubic (BCC) and FCC, as well as the diamond structure consisting of two spheres per unit cell. When the spheres are low dielectric material embedded in a matrix of high dielectric material, we denote this case as the *air-sphere* material. For the scalar wave equation, band gaps open up for all cases between the first and second bands and again between the fourth and fifth bands. The cases of dielectric spheres and air spheres are both interesting and display common features; the gaps can be as large as about 30 % and the size of the gap saturates as the ratio of dielectric constants becomes large. This is consistent with the saturation of the dielectric fluctuations that were discussed in Section 3.1.

The volume fraction at which the level gap widest is closely given by the maximum of dielectric constant fluctuations, Eq. (51). The gaps begin to appear roughly when the magnitude of the dielectric fluctuations is near unity. The plane-wave method has the advantage of allowing a simple check on the convergence; namely, the actual dielectric function can be compared with the truncated one [20]. The results for the optimal volume filling fraction [22, 23, 30] are in good quantitative agreement with the estimates from Eq. (51).

Unfortunately, the scalar wave equation does not provide predictive power in guiding the design of experimental structures. This brings up another convergence problem for the

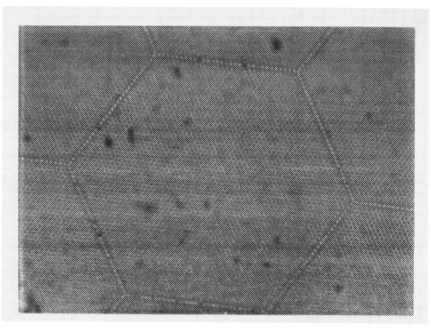

Figure 11: An optical microscope view of channel glass from the top of the sample and magnified by 1000 times. The lattice constant is 1.17 μm and the diameter of the holes is 0.9 μm. The structure is a triangular lattice of air-rods.

vector field case, which is discussed below. There are two, equivalent plane–wave methods for solving the vector wave equation; their rates of convergence depends on several factors: the dielectric constants, the volume fractions of each material and the geometry or topology of the structure. We then summarize our findings for the FCC, diamond, SC and the layered lattices lattices, called woodpile structures. Already a large body of literature exists on vector-wave band structure calculations in three dimensions[31–37]. A number of experiments have been designed to prove the basic principle, to learn about the structures and to investigate defects[38, 39]. Possible quantum electrodynamic effects, including Casimir-Polder forces, have also been proposed[42, 43].

4.1 Vector Wave Equations

Our starting point is Maxwell's equations for $\mathbf{E}$ and $\mathbf{H}$ in a dielectric medium

$$\nabla \times \nabla \times \mathbf{E}(\mathbf{x}, t) + \frac{1}{c^2} \frac{\partial^2}{\partial t^2} \epsilon(\mathbf{x}) \mathbf{E}(\mathbf{x}, t) = 0, \tag{54}$$

$$\nabla \times \eta(\mathbf{x}) \nabla \times \mathbf{H}(\mathbf{x}, t) + \frac{1}{c^2} \frac{\partial^2}{\partial t^2} \mathbf{H}(\mathbf{x}, t) = 0. \tag{55}$$

where $\eta(\mathbf{x}) \equiv 1/\epsilon(\mathbf{x})$ and $\epsilon(\mathbf{x})$ is linear, locally isotropic, positive-definite and periodic with lattice vectors $\mathbf{R}$

$$\epsilon(\mathbf{x}) = \epsilon_b \Theta_b(\mathbf{x}) + \epsilon_a \Theta_a(\mathbf{x}); \tag{56}$$

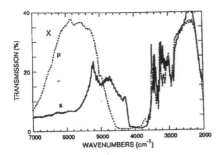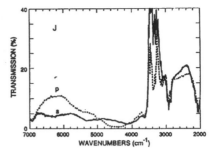

Figure 12: Transmittance spectra for p- and s-polarizations corrected for the absorbance of the intrinsic glass below 3600 cm^{-1} wavenumber s. (left) the absorbance along the Γ-X direction; (right) the absorbance along the Γ-J direction. After Ref. [28]

where ϵ_b is the dielectric constant of the background material and ϵ_a is the inclusion material. The Heaviside functions $\Theta_\alpha(\mathbf{x})$ are unity in material α and vanish otherwise. For non-overlapping spheres of radius r, its Fourier transform is

$$\epsilon(\mathbf{G}) = \frac{1}{V_{\text{cell}}} \int_{\text{WS cell}} d^3x \, e^{-i\mathbf{G}\cdot\mathbf{x}} \, \epsilon(\mathbf{x}) = \epsilon_b \delta_{\mathbf{G},0} + 3\beta(\epsilon_a - \epsilon_b)\frac{j_1(Gr)}{Gr}. \tag{57}$$

The function $j_1(Gr)$ is the spherical Bessel function of order 1. The eigenfunctions of Eqs. (54) and (55) are Bloch functions of the form

$$\mathbf{E}_{n\mathbf{k}}(\mathbf{x}, t) = \exp\left[i(\mathbf{k}\cdot\mathbf{x} - \omega_{n\mathbf{k}}t)\right] \sum_{\mathbf{G}} \mathbf{E}_{n\mathbf{k}}(\mathbf{G}) \exp\left[i(\mathbf{G}\cdot\mathbf{x})\right]; \tag{58}$$

$$\mathbf{H}_{n\mathbf{k}}(\mathbf{x}, t) = \exp\left[i(\mathbf{k}\cdot\mathbf{x} - \omega_{n\mathbf{k}}t)\right] \sum_{\mathbf{G}} \mathbf{H}_{n\mathbf{k}}(\mathbf{G}) \exp\left[i(\mathbf{G}\cdot\mathbf{x})\right]; \tag{59}$$

where $\mathbf{G}$ is a reciprocal lattice vector, $\mathbf{k}$ is the reduced wave vector in the first BZ and n is the band index including the polarization. The Fourier coefficients $\mathbf{E}_{\mathbf{G}} \equiv \mathbf{E}_{n\mathbf{k}}(\mathbf{G})$ and $\mathbf{H}_{\mathbf{G}} \equiv \mathbf{H}_{n\mathbf{k}}(\mathbf{G})$ satisfy, respectively, the infinite-dimensional matrix equations

$$(\mathbf{k} + \mathbf{G}) \times [(\mathbf{k} + \mathbf{G}) \times \mathbf{E}_{\mathbf{G}}] + \frac{\omega^2}{c^2} \sum_{\mathbf{G}'} \epsilon_{\mathbf{G}\mathbf{G}'} \mathbf{E}_{\mathbf{G}'} = 0, \tag{60}$$

$$(\mathbf{k} + \mathbf{G}) \times \left[\sum_{\mathbf{G}'} \eta_{\mathbf{G}\mathbf{G}'}(\mathbf{k} + \mathbf{G}') \times \mathbf{H}_{\mathbf{G}'}\right] + \frac{\omega^2}{c^2} \mathbf{H}_{\mathbf{G}} = 0, \tag{61}$$

with $\epsilon_{\mathbf{G}\mathbf{G}'} \equiv \epsilon(\mathbf{G} - \mathbf{G}')$ and $\eta_{\mathbf{G}\mathbf{G}'} \equiv \eta(\mathbf{G} - \mathbf{G}')$. We call these the E and the H method respectively. The choice of other fields to express the wave equation is redundant to these, at

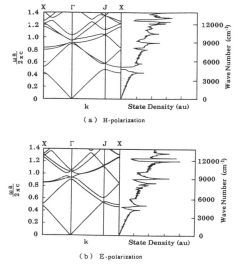

(a) H-polarization

(b) E-polarization

Figure 13: The band structure for the two-dimensional triangular lattice, as described in the text. Also shown is the two-dimensional density of states. After Ref. [28]

least for the case when $\mu(\mathbf{x}) = 1$. For instance, Zhang and Satpathy[35] used the displacement field, which has the wave equation

$$\nabla \times \nabla \times \eta(\mathbf{x})\,\mathbf{D}(\mathbf{x},t) + \frac{1}{c^2}\frac{\partial^2}{\partial t^2}\,\mathbf{D}(\mathbf{x},t) = 0. \qquad (62)$$

However, multiplication by $\nabla \times \eta(\mathbf{x})$ and the definition: $\mathbf{Q} = \nabla \times \eta(\mathbf{x})\mathbf{D}(\mathbf{x})$, reveals that this is identical to the H method [20].

The Bloch functions satisfy the following ortho-normality relations

$$\int_{\text{all }\mathbf{x}} d^3x\, e^{-i(\mathbf{k}-\mathbf{k}')\cdot\mathbf{x}}\, \epsilon(\mathbf{x})\, \mathbf{E}_{n'\mathbf{k}'}^*(\mathbf{x}) \cdot \mathbf{E}_{n\mathbf{k}}(\mathbf{x}) = \delta_{nn'}\delta(\mathbf{k}-\mathbf{k}'), \qquad (63)$$

$$\int_{\text{all }\mathbf{r}} d^3x\, e^{-i(\mathbf{k}-\mathbf{k}')\cdot\mathbf{r}}\, \mathbf{H}_{n'\mathbf{k}'}^*(\mathbf{x}) \cdot \mathbf{H}_{n\mathbf{k}}(\mathbf{x}) = \delta_{nn'}\delta(\mathbf{k}-\mathbf{k}'), \qquad (64)$$

It is important to note that the **E** field is normalized with $\epsilon(\mathbf{x})$ as the weight function and **H** on the other hand does not include this weight factor.

The solution proceeds by truncation of the reciprocal lattice vectors. N lattice vectors produces a matrix of size $3N \times 3N$. Using $\nabla \cdot \nabla \times \mathbf{E} = 0$ and $\nabla \cdot \mathbf{H} = 0$, the matrix equations Equations (60) and (61) can be reduced to $2N \times 2N$ equations.

Although the two methods yield the same spectrum when an infinite number of plane waves are included, their truncated forms yield, in general, very different spectra even when as many as a few thousand plane waves are used. We note that

$$\epsilon(\mathbf{x})\eta(\mathbf{x}) = 1 \quad \Rightarrow \quad \sum_{\mathbf{G}''} \epsilon_{\mathbf{G}\mathbf{G}''}\eta_{\mathbf{G}''\mathbf{G}'} = \delta_{\mathbf{G}\mathbf{G}'} \qquad (65)$$

that is, the $\infty \times \infty$ matrix $\eta_{\mathbf{G}\mathbf{G}'}$ is the inverse of the $\infty \times \infty$ matrix $\epsilon_{\mathbf{G}\mathbf{G}'}$. However, when the matrix $\epsilon_{\mathbf{G}\mathbf{G}'}$ is first truncated at $N \times N$, and then inverted, the matrix obtained, $\tilde{\epsilon}_{\mathbf{G}\mathbf{G}'}^{-1}$ is quite different from $\eta_{GG'}$. Thus the finite forms of Eqs. (60) and (61) in general yield different spectra. In the following subsections, some earlier results are highlighted. All lengths are in units of $a/2\pi$ and the wave vectors $\mathbf{k}$, $\mathbf{G}$ are in units of $2\pi/a$ where a is the side of the real space conventional cubic unit cell for the relevant Bravais lattice.

4.2   The Face Centered Cubic Lattice

The earliest treatment of three-dimensionally periodic lattices was the face centered cubic (FCC) structure. The first experiments preceeded theoretical calculations and were guided by intuition and the so-called 'Edisonian' trial-and-error approach. The choice of an FCC lattice was based upon the idea that Brillouin zone without protruding edges were more likely to form full band gaps. The BZ for the FCC structure is the roundest shape; it's shape is depicted by the inset in Figure 14.

In treating the FCC lattice, there are two situations to examine: dielectric spheres embedded in a host with the dielectric constant of air (dielectric spheres) and spherical voids in a dielectric background (air spheres). Experiments examined the case of air spheres. The frequency regime was investigated in a range where the second and third bands lie and no complete gap was found; however, a pseudo–gap was identified. The speudogap is evident at the W-point in Figure 14 as a point of a degeneracy between the second and third bands. The density of states is reduced in this region making an appearance similar to a full band gap. The air spheres were overlapping on the lattice, so the volume fraction of air is very high ($\approx 72\%$).

Theoretical results were performed soon after the experiments were published. For the air–sphere case, the calculations revealed a degeneracy at the W-point, as mentioned already, that prevented a full gap from opening in that frequency regime. However, a full gap was identified for the air–sphere FCC lattice between the eighth and ninth bands. This gap lies above the frequency regime region reported by the experiments and is it missing in the earliest papers due to poor convergence of the plane–wave expansion. Figure 14 is a plot of the band structure for $\epsilon_b=16$, $\epsilon_a=1$ and $\beta=0.74$. a gap was found using the E method, the asymptotic size of the 8-9 gap extrapolated to a value around 7.5 %. The convergence of the H method is much worse, as would be expected.

In an attempt to break the degeneracy at the W-point of the FCC BZ, the band structure for spheroidal atoms periodically arranged in an FCC lattice was calculated[37]. After trying spheroids of various eccentricities and orientations, the largest gaps were found for prolate air-spheroids embedded in a high dielectric background. For dielectric spheroids no gaps were found. The 2-3 gap peaks to a value of $\sim$12% around $\beta \sim 0.75$. Moreover, a second gap opens between the 8th and 9th bands and peaks to $\sim$6% around $\beta \sim 0.81$. This structure remains a theoretical curiosity because ellipsoidal holes can't be easily fabricated.

4.3   Diamond Structure

There is an alternative possibility to break the 2-3 band degeneracy of the FCC lattice at the W-point. In the diamond lattice, a second air sphere is added to the unit cell displaced

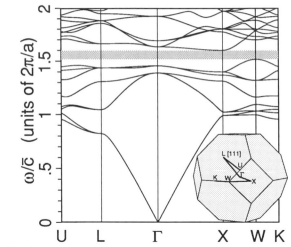

Figure 14: Band structure for EM waves for close-packed air spheres on a FCC lattice. $\bar{c} \equiv c/\sqrt{\bar{\epsilon}}$. $\epsilon_b = 13$, $\epsilon_a = 1$, $\beta = 0.74$, $\|\epsilon_r\| = 1.34$, $N = 749$. Inset shows the path in the BZ. (From Ref. [20])

along the (111) direction by a quarter of the lattice constant. Band gaps for structures with spheres at each diamond lattice site was first reported by Ho *et al.*[32] who used the E method. The H method works much better for these structures[41]. With either method, when the dielectric spheres are even slightly separated, *i.e.* for $\beta < 0.34$, the gap between the second and third bands closes as the number of plane waves N is increased. At the close-packed fraction of $\beta = .34$, our best estimate of the gap is $\Delta_{23} < 3.5\%$ for $\epsilon_a = 13$. On the other hand, a gap that was not reported in Ref. [32] opens between the 8th and the 9th bands. This has been independently confirmed[34] and serves as a reminder that convergence should be thoroughly investigated. With air spheres in the diamond structure, the E method converges much faster, and one observes that the 2-3 gap opens just as the spheres touch. When the spheres "overlap" there are indeed large gaps. This structure is also difficult to fabricate in the laboratory and has never been experimentally verified. The diamond lattice showed that structures with a full band gap were indeed possible, even though it did not fulfill the requirement that the structure also should be easy to fabricate (on a sub-micron scale).

4.4 The Simple Cubic Lattice

The study of photonic bands with the periodicity of the SC lattice was undertaken because of the simplicity of the geometry which could possibly translate into easier fabrication. The SC lattice also provides a framework in which structures with different topologies can be investigated, since the computational results obtained for the FCC structures indicate a strong relationship between connectivity and photonic band gaps. The band structures of a variety of geometries including spheres, overlapping spheres and the topologically equivalent structures with rods of circular and square cross section along the three Cartesian axes were reported in[33].

The simplest "square-rod" structure is when the faces of the rods are oriented parallel to those of the unit cell(Fig. 15). When the volume fraction of the dielectric is 50%, the geometry of both types of regions become identical. There are gaps for these structures. Although the fabrication of any micron-sized structure remains an experimental challenge.

The band structure corresponding to the square square rod simple cubic lattice is shown in Figure 16. Due to the sharp edges of the rods, convergence is slow for this case. The inset shows the Brillouin zone (BZ) for the simple cubic lattice. It is the squarest of the Bravais lattice type and provides a different understanding of the conditions for a complete gap . The presence of a gap between the fifth and the sixth bands is another piece of evidence that the shape of the BZ does not play an important role in the appearance of complete gaps. It is also noted that there is no degeneracy between the second and third bands.

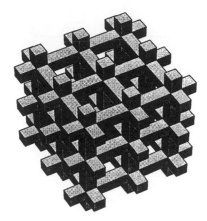

Figure 15: The Square-Rod Structure.(From Ref. [33])

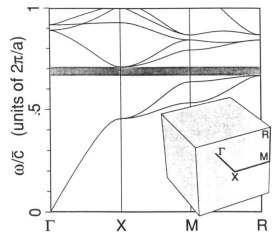

Figure 16: The photonic bands for the square-rod structure in the sc lattice. $\epsilon_b = 13$, $\epsilon_a = 1$ and $\beta = 0.82$.
(From Ref. [33])

4.5 Woodpile Structures

A number of different geometries have been shown to possess band gaps, including the face centered cubic lattice [20], the diamond lattice[32], the simple cubic lattice[33] and intersecting rod geometries[39]. However, the structures are difficult to fabricate on the scale of centimeters and scaling them to the submicron regime has not yet been successful. Instead, a new strategy had to be formulated.

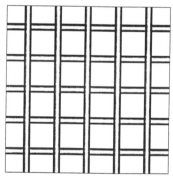

Figure 17: A top view of two layers in a woodpile structure. The dielectric rods have a two layer geometry that is repeated. Ref.[45]

A remarkably simple structure can be designed by extending a geometry originally proposed by Pendry [44]. The lattice is constructed by placing parallel rods in a plane; the rods can have a rectangular or circular cross-section. A second layer is placed on top of the first again by using parallel rods, but now rotated with respect to the row underneath. The stacking process is continued to build the lattice to macroscopic dimensions[45, 46]. The number of layers in a unit cell is one more design parameter. This class of lattices has been called *woodpile structures* in Ref. [45].

To illustrate the success of this approach, Sozuer and Dowling[45] considered several different structures. Only one will be shown here. Consider their rectangular rod structure with a volume fraction of 0.5 for each type of material and a two layer simple cubic symmetry structure. The dielectric constants of each material is 1 and 16. There is no degeneracy between the second and third bands; the direct band gap, i.e. the vertical frequency range between points in each band, is enormous and the overall size of the gap is 2-3 %. This size increases to 7-10 % when the dielectric is increased to a value of 25.

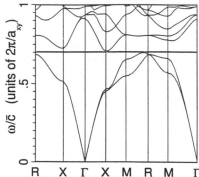

Figure 18: The photonic band structure for a rectangular rod structure with dielectric constants 1 and 25 and a volume fraction of 0.5. After Ref.[45]

Other versions of these structures have been micro-machined and their optical properties were measured[47]. These structures can be fabricated with band gaps sub-millimeter wavelength regime and work is now proceeding to push the technology toward the infrared. These structures provide the best pathway toward providing structures suitable for the visible regime. There are many variants of the simple structure in Figure 17. The lattice symmetry and the number of rods per unit cell are design parameters that leave a lot of room for exploration.

5 A Novel Numerical Approach

The plane–wave method discussed above is suitable for finding eigenvalues and eigenvectors of an infinite lattice with a lossless, linear dielectric. These methods are hampered by convergence problems, that can be overcome by using several different methods that hopefully converge to one another in the limit of an infinite number of plane waves or by using softer edges of the dielectric function that remove the high spatial-frequency components. However, other approaches [44, 48] are developed to handle more complicated and physically meaningful situations. New methods have improved the flexibility of the modeling.

Scalora and Crenshaw [48] developed a generalization of the beam propagation algorithm that has found wide application in nonlinear optics. This method has many advantages over the plane-wave method; it can be used for nonlinear media and is useful to describe pulse propagation. This method is the most powerful numerical procedure yet to be developed for studies of photonic band structures.

We have applied this new method to two different problems: 1. the propagation of a pulse through a photonic bandgap structure and 2. the emission rate of a dipole embedded in a photonic band structure.

The first problem is important because it addresses the nature of the so-called super-luminal tunneling times of a pulse[50]. After examining the energy flow through the lattice, we found that the apparent super-luminal tunneling is due to a reshaping of the pulse in passage through the lattice. The local momentum and energy density in the structure reveal that the speed of the pulse never exceed the speed of light in vacuum.

In the second problem the power radiated by classical dipole oscillators is studied as a function of frequency. The power radiated peaks at an order of magnitude above the free space radiation rate near the band edge and the spontaneous emission rate was suppressed by three orders of magnitude with respect to free space in the bandgap region.

We discuss Scalora's method for one dimensional lattices. The wave is incident perpendicular to the interface and satisfies the wave equation:

$$\frac{\partial^2 E}{\partial z^2} - \left(\frac{1}{c}\right)^2 \frac{\partial^2 E}{\partial t^2} = \frac{4\pi}{c^2} \frac{\partial^2 P}{\partial t^2}; \tag{66}$$

where P is the polarization of the medium. This can be as simple as an expression proportional to the electric field or as complicated as a contribution that is a nonlinear function of the electric field.

The equation is approximated by a slowly varying envelope expansion, but only the rapid time variable is approximated. Let, the field be represented by

$$E(z,t) = \hat{E}e^{-i\omega t} + c.c.; \tag{67}$$

where $\hat{E}$ is an envelope function and the polarization is also decomposed into a rapid and slow varying contribution. Expressing the polarization envelope as a linear susceptibility

$$\hat{P} = \chi\hat{E} \tag{68}$$

The wave equation is approximated as

$$\frac{\partial \hat{E}}{\partial \tau} - D\hat{E} = i\chi\hat{E}. \tag{69}$$

The spatial variable has been scaled to the wavelength $\xi = z/\lambda$; time has been scaled by the oscillation frequency $\tau = \nu t$. The operator D is

$$D = \frac{i}{4\pi} \frac{\partial^2}{\partial \xi^2}. \tag{70}$$

The solution of Eq. (69) is formally written as

$$\hat{E}(\xi, \tau) = \mathcal{T}e^{\int_0^\tau (D+\chi)d\tau} \hat{E}(\xi, 0). \tag{71}$$

The factor $\mathcal{T}$ is the time-ordering operator. The differential operator D can be diagonalized by Fourier transformation, but the susceptibility ξ may be a complicated function of the field and the spatial and temporal variables. Therefor, this equation is solve by a spectral method called the split-step propagation algorithm or beam propagation method. A second-order version of this algorithm is

$$\hat{E}(\xi, \tau) = e^{D\tau/2} e^{\int_0^\tau (\chi) d\tau} e^{D\tau/2} \hat{E}(\xi, 0). \tag{72}$$

It is solved by applying fast Fourier transforms to diagonalize the operator D and then solving for the susceptibility in the original space.

5.1   Superluminal Communication?

There is a recent controversy surrounding tunneling times in quantum physics associated with the time it takes for an evanescent wave to penetrate a barrier. Chiao's experiments[49] were designed to examine this question; they propagated an ultrshort pulse of light that was attenuated so that each pulse contained only a single photon. The attenuated pulse was then sent through a one-dimensional periodic structure with the center frequency of the pulse tuned to the center of the gap. After detecting many photons, the maximum of the envelope appeared at a time prior to the maximum of the pulse that propagated through free space. This is obviously not super-luminal communication, but are the results dependent on having a single quantum of light or can the results be interpreted by classical waves.

Scalora et al.[50] using the above numerical procedure set out to address the classical physics of pulse propagation in a periodic dielectric structure with the pulse carrier frequency tuned to the center of the gap. The structure is a quarter-wave stack with 22 periods. In this case the susceptibility is

$$\chi = n^2(z) - 1. \tag{73}$$

Figure 19 is a plot of the electric and magnetic fields super-imposed on the dielectric lattice for the case where the center frequency is tunde to the transmission maximum at the band edge. Note that the electric and magnetic fields are π out of phase with one another and therefore, the energy velocity is small. The fields are spatially anti-correlated; this is an indication that the energy velocity is small and therefore, the pulse travels slowly through the structure.

The inset in Fig. 20 is a comparison between the transmitted pulse maximum, when the center frequency is tuned to the center of the stop band, with that of a pulse that propagates through free space. The maxima of each pulse have been scaled to unity. The location of the pulse that 'tunneled' through the structure is ahead of the peak that when through free space; this result is similar to the experiments of Chiao. This behavior can be classically understood by looking at Fig. 20, the intensity of the evanescently transmitted pulse is actually far below the intensity of the free space pulse. When the same amplitude for free-space signal is compared with the maximum of the 'tunneled' signal, then there is no doubt that the free-space signal arrives first; what has occured in the tuneeling region is a reshaping of the pulse. In other words, the initial edge of the pulse propagates through the structure without interference effects, but of course, it is attenuated by reflections at

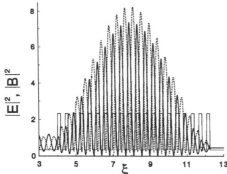

Figure 19: Shown are the electric field squared(solid line) and the magnetic field squared(dotted line) when the pulse carrier frequency is tuned to the lower edge of the first stop band in a 1-dimensional structure (see Fig. 21). The dielectric constant of the high dielectric layer is 2.

each surface; as the pulse has time to reflect back and forth in the structure the destructive interference pattern builds up and reflects the remainder of the pulse.

What is significant in pulse propagation is not the speed of the pulse maximum, but rather the leading edge of the pulse. The initial information is carried by the first photons that can be detected. In the free space propagating signal, it is clearly the front edge that has greater intensity and arrives first at the dectector.

5.2 Dipole Radiation

Since Purcell[3] predicted that the emission rate of an atom could be affected by cavity effect, there have been a number of experiments on micro-cavities to demonstrate this effect[51]. This is simply understood by reference to the Fermi golden rule result for the emission rate:

$$\gamma = \frac{2\pi}{\hbar} |< i|\mathbf{d} \cdot \mathbf{E}|f >|^2 \rho(\omega). \tag{74}$$

The interaction energy coupled the atomic dipole operator, $\mathbf{d}$ and the electric field operator $\mathbf{E}$; $\hbar\omega$ is the energy of the atomic transition. The matrix element between the initial, $|i >$, and final, $|f >$ states of the atom-field system determine part of the emission, but the electromagnetic density of states $\rho(\omega)$ is also needed. In other words, the excited atom requires a bath of electromagnetic modes into which the energy is released.

Dipole emission in a one-dimensional periodic structure was studied in Ref. [52]. The same lattice was considered as in the previous Subsection with the polarization consisting of a linear dielectric term, as before, and an additional term due to a dipole sheet placed in

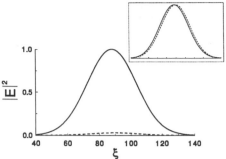

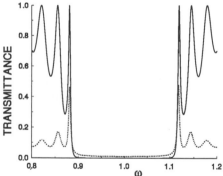

Figure 20: The pulse intensities of a free-space propagating pulse (solid line) and a pulse 'tunneling' through the periodic structure (dashed line). The frequency is tuned to the mid-gap. In the same time, the pulse maximum from the periodic structure is ahead of the free space maximum. Note the scaled intensities on the inset.

the lattice. The additional dipole sheet satisfies the equation of motion

$$\frac{\partial \hat{P}}{\partial \tau} = i\delta \hat{P} + i\frac{\omega_p^2}{8\pi\omega}\hat{E} \tag{75}$$

Figure 21: The transmittance curve (solid line) and the density of states (dotted line) for a 22 layer periodic structure. The parameters are identical to those in Fig. 20.

The constant ω_p is the plasma frequency, which expresses the dipole density, and δ is the detuning of the dipole from resonance with the electric field. In these studies it was assumed that $\omega_p/\omega_{cg} = 10^{-4}$. The dipole sheet was initially excited and no field was present, so it oscillated at its natural frequency; the coupling between the field and the dipole produced electromagnetic radiation, whose power could be simply measured. For the

22 period photonic band structure, the transmittance and the density of state $\rho(\omega)$ is shown in Fig. 21. From this result, we observe a reduced density of states in the gap and at the band edges, the density is enhanced; thus, we expect that the spontaneous emission rate is suppressed in the gap and enhanced near the band edges. To obtain enhanced or suppressed emission, the dipole frequency should be appropriately tuned.

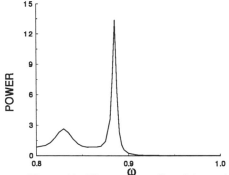

Figure 22: The power radiated from the structure shown in Fig. 21

Fig. 22 clearly shows the sensitive dependence of the radiated power on the oscillation frequency of the dipole. At the band edge the power is about 14 times larger than its free space value, while in the gap the radiated power is reduced by about three orders of magnitude. The maxima in the radiated power occurs at the frequencies where the density of states in Fig. 21 have maxima. The first maximum corresponds to the maximum closest to the band edge, but the second maximum corresponds to the third peak in the density of states. This is due to a near-zero value of the field mode at the dipole position.

A recent publication by Suzuki and Yu[53] calculated the emission from a point dipole in a close–packed FCC lattice; a much more complicated situation than the one considered here. In the FCC lattice, recall that there is a gap between the 8-9 bands, see Subsection 4.2. They studied the dependence of the emission on dipole orientation and position in the structure. The emission was indeed enhanced near the band edges by as much as 15 times and suppressed in the gap region; very similar to the results above. Note that the density of states doesn't give the whole picture because the matrix elements in Eq. (74) also depend upon the Bloch functions in the lattice. In three dimensions, these functions are indeed complicated and the dipole may be orthogonal to the field and not radiate, in spite of a generous density of states.

6 Nonlinear Electrodynamics in Photonic Band Structures

The nonlinear dielectric response of the material can be enhanced through the electromagnetic resonances at stop–band edges in the medium. The use of periodic dielectric materials can controllably enhance the local field using interference effects.

Periodic structures that incorporate a nonlinear response of the medium exhibit inter-esting phenomena. Chen and Mills[54] considered a periodic structure with an intensity–dependent polarizability. The equation of evolution in their case is

$$\frac{d^2 E}{dz^2} + \left(\frac{\omega}{c}\right)^2 (\epsilon(z) + \chi^{(3)}(z)|E|^2)E = 0. \tag{76}$$

$\chi^{(3)}(z)$ is the Kerr nonlinearity, which is periodic and a small perturbation. When the light frequency is tuned inside the stop gap and has a low intensity, the transmissivity is low, as expected. As the intensity is increased though, the transmissivity begins to increase until the sample is transparent. There is a bistable behavior of the transmissivity very similar to the behavior of a Kerr nonlinearity in a Fabry-Perot cavity[55].

This phenomenon can be physically explained as a shift of the dielectric values that moves the edge of the stop band toward the excitation frequency. As it moves out of the stop band, the transmission dramatically changes.

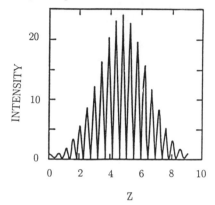

Figure 23: The spatial profile of the field intensity for the input power at which the trans-mission becomes unity. The oscillations are the Floquet-Bloch function and the envelope is a hyperbolic–secant squared.

Chen and Mills also discovered that the field-envelope inside the structure has a hy-perbolic secant shape, which is a well-known form of soliton solutions, see Fig. 23. Thus, they named the field structure a gap soliton. As with many such numerical discoveries, the simplicity of the field envelope suggested that it would be possible to find an appropriate analytical theory to describe the nonlinear phenomena in periodic structures.

6.1 Multiple Scales Method

The method of multiple scales(MMS), is a systematic perturbation method [56]; it has been applied to a number of interesting nonlinear problems and provides simple analytic results that complement the numerical simulations. MMS has also been applied to analytic studies of gap solitons [57–60].

All of these approaches assume that the periodic dielectric variations are weak; they can be generalized to include large variations when the Floquet–Bloch functions replace the plane-wave functions of the fields, as outlined above. A $\mathbf{k} \cdot \mathbf{p}$ formulation has been used with success for electron waves [61]; input is provided from electron band structure calculations and expansions around extremal points in the Brillouin zone provide useful quantities, eg. effective mass and dielectric constant, for further analytic calculations. deSterke et al. [58–60] developed a formalism for one-dimensional structures that is equivalent to this electronic technique. In the same manner, Johnson et al. [62] have proposed using an extension to the scalar electromagnetic problem. This method cannot eliminate the need for numerical calculations of the band structure; the method can be applied to problems with disorder and nonlinear response to examine the wave propagation in a perturbed medium. In this sense the band structure calculations provide the lowest order solution of the problem.

The regime of many scatterers, whose size and separations are on the order of a wavelength, is a difficult regime for which only a few analytical techniques have been developed. The method of the slowly varying envelope approximation (SVEA) and the method of multiple scales (MMS) [56, 57] are two perturbative techniques that can be mentioned in this regard. The MMS method to the multiple–fields problem of periodic dielectric structures in order to simplify the analysis of more complicated situations that cannot be numerically analyzed. Only a brief description is possible here, more details are provided in the literature, see Ref.([63]) which applies this method to the study of short pulse propagation in materials with quadratic nonlinearities, i.e. $\chi^{(2)}$ materials. A multi–field MMS method is required to calculate the effective equations of motion including diffraction effects and to study $\chi^{(2)}$ and $\chi^{(3)}$ materials. Numerical simulations have been underway and provide important clues for the theoretical development. At the band edges, there are standing wave interference patterns that show field enhancements of several orders of magnitude. Four orders of magnitude enhancement of a second–harmonic signal have already been seen in numerical simulations by Scalora[64].

Consider the nonlinear wave equation with periodic coefficients $\epsilon(z)$ and $\chi^{(3)}(z)$. The nonlinear wave equation has the form

$$\mathcal{L}E = c^2 \frac{\partial^2 E}{\partial z^2} - \epsilon(z) \frac{\partial^2 E}{\partial t^2} = 4\pi \chi^{(3)}(z) \frac{\partial^2 E^3}{\partial t^2} \tag{77}$$

We begin with an expansion of the electric field in powers of a perturbation parameter λ, the scalar fields are used here

$$E(z, t) = \lambda E_1(z, t) + \lambda^2 E_2(z, t) + \lambda^3 E_3(z, t) + \cdots. \tag{78}$$

This expansion is applied to the vector wave equation including the nonlinear polarization term. The independent variables of the time, t, and propagation direction, z, are likewise expanded in powers of the parameter λ

$$t_n = \lambda^n t, \tag{79}$$
$$z_n = \lambda^n z, \tag{80}$$
$$\tag{81}$$

t_0 and z_0 represent the shortest time and length scales, resp.; the variables t_n and z_n are progressively slower scales as the index n increases. The the nonlinearity is weak and does not contribute to the first order in λ; the equation to this order captures the field's amplitude at the fastest scales and the equation is identical with Eq. (1). This equation must be numerically solved and the eigenfunction solutions are required input for the higher order perturbation results. The order in with the nonlinearity appears is dependent on the model for the material; or a nonlinearity that is cubic in the field, the contribution is to order λ^3 [57].

Accordingly, the operators are also expanded in powers of the perturbation parameter

$$\frac{\partial}{\partial z} = \frac{\partial}{\partial z_0} + \lambda \frac{\partial}{\partial z_1} + \lambda^2 \frac{\partial}{\partial z_2} + \cdots; \tag{82}$$

and

$$\frac{\partial}{\partial t} = \frac{\partial}{\partial t_0} + \lambda \frac{\partial}{\partial t_1} + \lambda^2 \frac{\partial}{\partial t_2} + \cdots; \tag{83}$$

The dielectric function varies on the shortest length scale: $\epsilon(z) = \epsilon(z_0)$. Next expand the wave operator in powers of the perturbation

$$\mathcal{L} = \mathcal{L}_0 + \lambda \mathcal{L}_1 + \lambda^2 \mathcal{L}_2; \tag{84}$$

where

$$\mathcal{L}_0 = c^2 \frac{\partial^2}{\partial z_0^2} - \epsilon(z_0) \frac{\partial^2}{\partial t_0^2}; \tag{85}$$

$$\mathcal{L}_1 = 2(c^2 \frac{\partial^2}{\partial z_0 \partial z_1} - \epsilon(z_0) \frac{\partial^2}{\partial t_0 \partial t_1}); \tag{86}$$

$$\mathcal{L}_2 = c^2 \frac{\partial^2}{\partial z_1^2} - \epsilon(z_0) \frac{\partial^2}{\partial t_1^2} + 2(c^2 \frac{\partial^2}{\partial z_0 \partial z_2} - \epsilon(z_0) \frac{\partial^2}{\partial t_0 \partial t_2}). \tag{87}$$

The field E_1 is decomposed into a function of short spatial and time variations and an envelope function that depends on the longer scales.

$$E_1 = a(z_1, z_2, \cdots; t_1, t_2, \cdots)\phi_{kn}(z_0)e^{-i\omega_n t_0} + c.c.. \tag{88}$$

The function $\phi_{kn}(z_0)$ is the Floquet–Bloch function, which satisfy

$$\mathcal{L}_0 \phi_{kn}(z_0)e^{-i\omega_n t_0} = 0; \tag{89}$$

or equivalently,

$$\phi_{kn}''(z_0) + \omega_n^2 \epsilon(z_0)\phi_{kn} = 0. \tag{90}$$

The amplitude a depends on the higher-order scales and therefore, it commutes with $\mathcal{L}_0$. Now proceed to order λ^2 and define

$$E_2 = \sum_{n'} b_{n'}(z_1, z_2, \cdots; t_0, t_1, t_2, \cdots)\phi_{kn'}(z_0). \tag{91}$$

Note that $b_{n'}$ depends on the fastest time scale. The equation at second order is

$$\mathcal{L}_0 E_2 = -\mathcal{L}_1 E_1. \tag{92}$$

Since $\phi_{kn'}$ is the only function depending on the z_0 scale, we can eliminate derivative with respect to z_0 on the left hand side by substitution from Eq. (90)

$$\mathcal{L}_0 E_2 = -\sum_{n'} \left(\omega_{n'}^2 b_{n'} + \frac{\partial^2 b_{n'}}{\partial t_0^2} \right) \epsilon(z_0) \phi_{kn'}(z_0). \tag{93}$$

The orthogonality properties of the Floquet–Bloch functions reduce the equation to

$$\frac{d^2}{dt_0^2} b_{n'} + \omega_{n'}^2 b_{n'} = (2c \frac{\partial a}{\partial z_1} < n' | c \frac{\partial}{\partial z_0} | n > + 2i\omega_{n'} \frac{\partial a}{\partial t_1} < n' | \epsilon(z_0) | n >) e^{-i\omega_n t_0} + c.c.; \tag{94}$$

where the bracket notation has been introduced as a notational convenience:

$$< n' | A(z_0) | n > = \int \phi_{kn'}^*(z_0) A(z_0) \phi_{kn} dz_0. \tag{95}$$

Secular terms in a differential equation are those contributions with a resonance frequency. For example, a harmonic oscillator driven by a force

$$\frac{\partial^2}{\partial t_0^2} b_n + \omega_n^2 b_n = F e^{-i\omega_n t_0}. \tag{96}$$

The solution is a sum of the homogeneous equation's solution and the inhomogeneous equations solution. Since, the frequency of the forcing term is resonant with the natural frequency of the oscillator, the amplitude of b_n, (i.e. $n' = n$) grows without limit. To prevent this catastrophe in the solution, the secular term in Eq. (94) must vanish. It is not difficult to identify that this contribution is

$$\frac{\partial a}{\partial z_1} + \frac{1}{v_g} \frac{\partial a}{\partial t_1} = 0; \tag{97}$$

where v_g is the group velocity of the envelope

$$v_g = \frac{c < n | - ic \frac{\partial}{\partial z_0} | n >}{\omega_n}. \tag{98}$$

The solution a moves along characteristic curves

$$\xi = z_1 - v_g t_1. \tag{99}$$

The field at second order is

$$E_2 = \sum_{n' \neq n} \left(2ic \frac{\partial a}{\partial \xi} \phi_{kn'}(z_0) \frac{< n' | - ic \frac{\partial}{\partial z_0} | n >}{\omega_{n'}^2 - \omega_n^2} e^{-i\omega_n t_0} + c.c. \right) \tag{100}$$

Now proceed to third order in the perturbation and let

$$E_3 = \sum_{n'} d_{n'}(z_1, z_2, \cdots; t_0, t_1, t_2, \cdots) \phi_{kn'}(z_0). \tag{101}$$

The evolution equation is

$$\mathcal{L}_0 E_3 = -\mathcal{L}_1 E_2 - \mathcal{L}_2 E_1 + 4\pi\chi^{(3)}(z_0)\frac{\partial^2}{\partial t_0^2}E_1^3.$$ (102)

This is the first order that the nonlinearity has made its appearance. Following the identical procedure as before, the equation can be expressed as

$$\frac{\partial^2}{\partial t_0^2}d_n + \omega_n^2 d_n = \sum_{n'} <n|\mathcal{F}|n'> .$$ (103)

At this order we are only interested in eliminating the secular terms, i.e $n = n'$. By construction, the first term on the right hand side has no secular contribution when $n' \neq n$). The first two terms yields a secular term

$$<n|\mathcal{L}_1 E_2|n> + <n|\mathcal{L}_2 E_1|n> = 2i\omega_n\left(v_g\frac{\partial a}{\partial z_2} + \frac{\partial a}{\partial t_2}\right) + \omega_n''\omega_n\frac{\partial^2 a}{\partial \xi_1^2};$$ (104)

where

$$\omega_n''\omega_n = c^2 <n|n> -\omega'^2 + 4c^2 \sum_{n'\neq n}\frac{|<n'| - ic\frac{\partial}{\partial z_0}|n>|^2}{\omega_{n'}^2 - \omega_n^2}.$$ (105)

ω_n'' is related to the group velocity dispersion. The final term is found by expanding Eq. (88) and retaining only the secular contribution, which is

$$- 12\pi\omega_n^2\bar{\chi}^{(3)}|a|^2 a.$$ (106)

The nonlinear coefficient is

$$\bar{\chi}^{(3)} = \int_0^L \chi^{(3)}(z_0)|\phi_{kn}(z_0)|^4 dz_0.$$ (107)

Using the characteristic time $\tau_2 = t_2 - z_2/v_g$ for the first contribution, we find the amplitude is a solution of the nonlinear Schrödinger equation

$$i\frac{\partial a}{\partial \tau_2} + \frac{\omega_n''}{2}\frac{\partial^2 a}{\partial \xi_1^2} + 6\pi\omega_n\bar{\chi}^{(3)}|a|^2 a = 0.$$ (108)

This equation is known to have soliton solutions that are obtainable by applying an inverse scattering procedure. It has extensive applications, for instance, optical fiber communications systems may be based on ultrashort optical pulses that are solitons.

Covering the inverse scattering procedure would take us too far afield; we will concern ourselves here with the simple case of a fundamental soliton solution

$$a(\xi,\tau) = \psi(\xi)e^{i\delta\tau};$$ (109)

the nonlinear Schrödinger equation reduces to an ordinary differential equation (ψ real)

$$- \delta\psi + \frac{\omega_n''}{2}\frac{d^2\psi}{d\xi_1^2} + \gamma_n\psi^3 = 0;$$ (110)

where $\gamma_n = 6\pi\omega_n\bar{\chi}^{(3)}$ and δ is the detuning of th soliton from the band edge frequency ω_m. The first integral of this equation is

$$-\delta\psi^2 + \frac{\omega_n''}{2}\left(\frac{d\psi}{d\xi_1}\right)^2 + \frac{\gamma_n}{4}\psi^4 = 0; \qquad (111)$$

where the integration constant has been set to zero to satisfy the asymptotic conditions that the amplitude and its derivative vanish as $\xi_1 \rightarrow \pm\infty$. Soliton solutions of this equation are found when the dispersion is positive, i.e. $\omega_n'' > 0$,

$$\sqrt{\gamma_n}\psi(\xi) = \sqrt{2\delta}\, \text{sech}\left(\sqrt{2\delta/\omega_n''}\xi\right). \qquad (112)$$

The parameter δ is related to the amplitude and the width of the wave; this means that the amplitude is not arbitrary and the width and amplitude are determined together.

The method has been detailed by deSterke and Sipe[59], who exemplify solutions for finite systems and reproduce the quantitative features found in the numerical work of Chen and Mills. The method needs the infinite period Floquet-Bloch functions as input, but thereafter the task of solving the propagation properties is reduced to finding the amplitude a at the boundaries and solving for its propagation in the structure. This is far simpler than numerically integrating the equations.

This method is analogous to the envelope function methods for determining electronic properties, called $\mathbf{k} \cdot \mathbf{p}$ methods, already familiar in semiconductor physics [61] and is also related to the coupled–mode expansion in Section 2.4. In our case the numerical solutions of the band structure at the lowest order of the perturbation theory is used to generate the coefficients of the differential equations at the next order of the expansion; this procedure has been successfully used in the study of gap solitons [57] and a scalar version has been used for a three dimensional structure [62].

6.2 Gap Soliton Experiments

Two types of experiments have reported verification of the gap–soliton predictions. Sankey et al.[65] use a corrugated silicon-on-insulator waveguide geometry. The measurements were made using a Nd:YAG laser at a wavelength of 1.06 μm. The pulse duration was around 30 ns and the pulse energies were in the μJ range. The carrier dynamics was complicated, but was also the source of interesting unstable dynamical evolution.

The experiments of Herbert et al.[66] used a dye-doped colloidal crystal as a distributed Bragg reflector. They use a cw dye laser to tune across the stop gap and ramp the intensity while holding the frequency constant. The lattice spacing was around 215 nm and the index modulation is weak; there were about 400 periods of the unit cell. Fig. 24 is a brief summary of their results. Depending on the tuning position within the gap, the transmissivity showed a monotonic behavior, bistable or multi-stable operation.

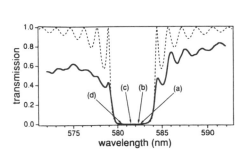

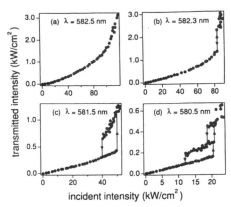

Figure 24: Left: the transmission through a colloidal crystal showing the stop gap; experiment (solid line) and theory (dashed line)are in good agreement for the stop band position. The points a-d are the tuning positions for the transmission versus intensity curves; the corresponding transmission curves are shown to the right.

7 Summary

One–dimensional PBS's have been available for many years, as multiple thin film structures. These materials are used, for instance, as high reflectivity, low loss mirrors in semiconductor laser cavities. One dimensional structures are also fabricated in a fiber geometry by using interference of ultraviolet light in the fiber that creates a periodic defect in the glass. In one dimension a gap is formed for any index difference; in the case of optical fibers, $\Delta n \approx 10^{-2}$; as a result, the gap is quite small.

Despite the appearance of a small gap in many one–dimensional lattice systems, some concepts can be examined for this case. In particular, the change of the spontaneous emission rate for a dipole sheet sandwiched between two PBS structures has been studied and compared to theoretical results. Other interesting features have been predicted for the nonlinear optical response near the band edge.

In two dimensions, the appearance of a gap is not guaranteed. The index difference must be sufficiently large between the two media to create a band gap; also, the polarization behave quite differently. A full band gap for each polarization is typically found for any lattice symmetry, but they are often not overlapping. A complete gap for both polarizations was found for the triangular lattice and this case is one of experimental relevance, as well.

Experiments in the near IR or visible regime are now a reality in triangular-lattice, 2-dimensional structures with the channel glass technology. This opens up the possibility of using lasers to study optical properties of the new structures and this means a number of new experiments can be designed to study diffraction during propagation, nonlinear multiple–wave mixing and short–pulse propagation and dispersion. The gaps are large enough so that an ultrashort optical pulse, whose carrier frequency lies in the center of the gap, will have its bandwidth contained within the gap. For many applications, a full band gap is not required; for instance, enhancing nonlinear response or gain near a band edge by taking advantage of the wave interference is a strategy that applies to all dimensions.

Three–dimensional structures have not yet been fabricated in the visible regime, but there are good candidate structures that have band gaps and may be constructed by etching and overgrowth methods. The simple cubic lattice and the woodpile structures have the advantage of simplicity. Three-dimensional PBS's have only been reported in the microwave regime. From theoretical and experimental studies, there are several design features that must be taken into account

- The structure should be mostly air, i.e. low dielectric material.

- The low and high dielectric regions should be overlapping.

- The ratio of the dielectric constants must be large (around 10).

Acknowledgment

The author is especially indebted to H. Sami Sozuer during the collaboration on three-dimensional structures and to Mike Scalora for discussions and collaboration on one-dimensional structures. The contributions of Dawne Yarne and Henry Schneider on two–dimensional lattices is also gratefully acknowledged.

References

[1] P. MILONNI, *The Quantum Vacuum:An Introduction to Quantum Electrodynamics*, (Academic Press, Boston, 1994).

[2] P. R. BERMAN, ED., *Cavity Quantum Electrodynamics*, (Academic Press, Boston, 1994).

[3] E. M. PURCELL, *Phys. Rev.* **69**, 681 (1946).

[4] S. JOHN, *Phys. Rev. Lett.* **58**, 2486 (1987).

[5] E. YABLONOVITCH, *Phys. Rev. Lett.* **58**, 2059 (1987).

[6] E. YABLONOVITCH, *J. Opt. Soc. Am.* B10, *283 (1993); J. Mod. Opt.* **41**, *185 (1994);* J. W. HAUS, *J. Mod. Opt.* **41**, *195 (1994);* P. ST. J. RUSSELL, *Physics World, p. 37 (August, 1992).*

[7] G. KURITZKI AND J. W. HAUS, EDS., *J. Mod. Opt.*41, *Issue No. 2 (1994);* C.M. BOWDEN, J. D. DOWLING AND H. EVERITT *J. Opt. Soc. Am.* B10, *issue No. 2 (1993).*

[8] J. D. JOANNOPOULOSI, R. D. MEADE AND J. N. WINN, *Photonic Crystals (Princeton Univ. Press, NJ, 1995).*

[9] L. BRILLOUIN, *Wave Propagation in Periodic Structures (Wiley, NY, 1946).*

[10] A. YARIV AND P. YEH, *Optical Waves in Layered Media (Wiley, NY, 1988)*.

[11] J. DOWLING AND C.M. BOWDEN, *Phys. Rev. A* **46**, *612 (1992)*.

[12] J. DOWLING AND C.M. BOWDEN, *J. Opt. Soc. Am. B* **10**, *353 (1993)*.

[13] M. PLIHAL AND A.A. MARADUDIN *Phys. Rev. B* **44**, *8565 (1991)*.

[14] M. PLIHAL, A. SHAMBROOK, A.A. MARADUDIN AND P. SHENG *Opt. Comm.* **80**, *199 (1991)*.

[15] S. L. MCCALL, P. M. PLATZMAN, R. DALICHAOUCH, D. SMITH AND S. SCHULTZ *Phy. Rev. Lett.* **67**, *2017 (1991)*.

[16] R.D. MEADE, K. D. BROMMER, A. M. RAPPE AND J. D. JOANNOPOULOS *Appl. Phys. Lett.* **61**, *495 (1992)*.

[17] A. A. MARADUDIN AND A.R. MCGURN *J. Opt. Soc. Am. B* **10**, *307 (1993)*.

[18] T. K. GAYLORD, G. N. HENDERSON AND E. N. GLYTSIS *J. Opt. Soc. Am. B* **10**, *333 (1993)*.

[19] N. W. ASHCROFT AND N.D. MERMIN, *Solid State Physics, (Holt Rinehart and Winston, 1976), Ch. 11*.

[20] H. S. SOZUER, J. W. HAUS AND R. INGUVA, *Phys. Rev. B* **45**, *13962 (1992)*.

[21] E. N. ECONOMOU AND A. ZDETSIS, *Phys. Rev. B* **40**, *1334 (1989)*.

[22] S. SATPATHY, Z. ZHANG AND M. R. SALEHPOUR, *Phys. Rev. Lett.* **64**, *1239 (1990)*.

[23] K. M. LEUNG AND Y. F. LIU, *Phys. Rev. B* **41**, *10188 (1990)*.

[24] W. ROBERTSON, G. ARJAVALINGHAM, R. D. MEADE, K. D. BROMMER, A. M. RAPPE AND J. D. JOANNOPOULOS, *Phys. Rev. Lett.* **68**, *2023 (1992)*.

[25] W. ROBERTSON, G. ARJAVALINGHAM, R. D. MEADE, K. D. BROMMER, A. M. RAPPE AND J. D. JOANNOPOULOS, *J. Opt. Soc. Am. B* **10**, *322 (1993)*.

[26] R. D. MEADE, K. D. BROMMER, A. M. RAPPE AND J. D. JOANNOPOULOS, *Appl. Phys. Lett.* **61**, *495 (1992)*.

[27] R. D. MEADE *et al.*, *Phys. Rev. Lett.*, **67**, *3380 (1991)*; R. D. MEADE *et al.*, *Phys. Rev. B*, **44**, *10961 (1991)*.

[28] K. INOUE, M. WADA, K. SAKODA, A. YAMANAKA, M. HAYASHI AND J. W. HAUS, *Japanese Journal of Applied Physics* **33**, *L 1463-L 1465 (1994)*.

[29] D. YARNE, *Rensselaer Polytechnic Institute, unpublished results (1995)*.

[30] S. DATTA, C.T. CHAN, K.M. HO AND C.M. SOUKOULIS, *Phys. Rev. B* **46**, *10,650 (1992)*.

140

[31] K. M. LEUNG AND Y. F. LIN, *Phys. Rev. Lett.* **65**, *2646 (1990)*.

[32] K. M. HO, C. T. CHAN AND C. M. SOUKOULIS, *Phys. Rev. Lett.* **65**, *3152 (1990)*.

[33] H. S. SÖZÜER AND J. W. HAUS, *J. Opt. Soc. Am. B* **10**, *296 (1993)*.

[34] I. H. H. ZABEL AND D. STROUD, *Phys. Rev. B* **48**, *13962 (1992)*.

[35] Z. ZHANG AND S. SATPATHY, *Phys. Rev. Lett.* **65**, *2650 (1990)*.

[36] C. T. CHAN, K. M. HO AND C. M. SOUKOULIS, *Europhys. Lett.*, **16** *(6), pp. 563-568 (1991)*.

[37] J. W. HAUS, H. S. SOZUER AND R. INGUVA, *J. Mod. Optics* **39**, *1991 (1992)*.

[38] E. YABLONOVITCH AND T. J. GMITTER, *Phys. Rev. Lett.* **63**, *1950 (1989)*.

[39] E. YABLONOVITCH, T. J. GMITTER AND K. M. LEUNG, *Phys. Rev. Lett.* **67**, *2295 (1991)*.

[40] E. YABLONOVITCH, T. J. GMITTER, R. D. MEADE, A. M. RAPPE, K. D. BROMMER AND J. D. JOANNOPOULOS, *Phys. Rev. Lett.* **67**, *3380 (1991)*.

[41] H. S. SOZUER, *private communication*.

[42] C. L. ADLER AND N. M. LAWANDY, *Phys. Rev. Lett.* **66**, *2617 (1991); erratum, ibid.* **66**, *3321 (1991)*.

[43] J. MARTORELL AND N. M. LAWANDY, *Phys. Rev. Lett.* **65**, *1877 (1990)*.

[44] J. B. PENDRY, *J. Mod. Opt.* **41**, *208 (1994)*; J. B. PENDRY AND A. MACKINNON, *Phys. Rev. Lett.* **69**, *2772 (1992)*.

[45] H. S. SOZUER AND J. D. DOWLING, *J. Mod. Opt.* **41**, *231 (1994)*.

[46] K. M. HO, C. T. CHAN, C. M. SOUKOULIS, R. BISWAS AND M. SIGALAS *Sol. St. Commun.***89**, *413 (1994)*.

[47] E. OZBAY, E. MICHEL, G. TUTTLE, R. BISWAS, M. SIGALAS AND K. M. HO *Appl. Phys. Lett.*, *(1994)*.

[48] M. SCALORA AND M. CRENSHAW, *Opt. Commun.* **108**, *191 (1994)*.

[49] R. CHIAO, P. G. KWIAT AND A. M. STEINBERG, *Physica B***175**, *257 (1991)*; A. M. STEINBERG, P. G. KWIAT AND R. Y. CHAIO, *Phys. Rev. Lett.* **71**, *708 (1993)*.

[50] M. SCALORA, J. D. DOWLING, A. S. MANKA, C. M. BOWDEN AND J. W. HAUS, *Pulse Propagation near Highly Reflecting Surfaces: Applications to Photonic Bandgap Structures and the Question of Superluminal Tunneling Times, Phys. Rev. A, in press (1995)*.

[51] D. KLEPPNER, *Phy. Rev. Lett.* **47**, *233 (1981);* R.G. HULET, E. S. HILFER AND D. KLEPPNER, *Phys. Rev. Lett.* **20**, *2137 (1985);* S. HAROCHE AND D. KLEPPNER, *Physics Today* **42**, *24 (1989);* F. DE MARTINI, G. INNOCENTI, G. R. JACOBOVITZ AND P. MATALONI, *Phys. Rev. Lett.* **59**, *2955 (1987);* G. BJÖRK ET AL., *Phys. Rev.* *A* **44**, *669 (1991);* H. *Yokoyama and S.D. Brorson, J. Appl. Phys.* **66**, *4801 (1989);* J. MARTORELL AND N. M. LAWANDY, *Phys. Rev. Lett.* **65**, *1877 (1990).*

[52] M. SCALORA, M. TOCCI, M. J. BLOEMER, C. M. BOWDEN AND J. W. HAUS, *Appl. Phys. B* **60** , *S57 (1995).*

[53] T. SUZUKI AND P. K. L. YU, *J. Opt. Soc. Am. B* **12**, *576 (1995).*

[54] W. CHEN AND D. L. MILLS, *Phys. Rev. Lett.* **58**, *160 (1987).*

[55] H. GIBBS, *"Optical Bistability" (Academic Press, New York, 1985).*

[56] J. KEVORKIAN AND J. D. COLE, *"Perturbation Methods in Applied Mathematics," (Springer Verlag, New York, 1981).*

[57] J. E. SIPE AND H. G. WINFUL, *Opt. Lett.* **13**, *132 (1988).*

[58] C. M. DESTERKE AND J. E. SIPE, *Opt. Lett.* **14**, *754 (1989).*

[59] C. M. DESTERKE AND J. E. SIPE, *Phys. Rev. A* **38**, *5149 (1988).*

[60] C. M. DESTERKE AND J. E. SIPE, *Phys. Rev. A* **39**, *5163 (1989).*

[61] G. BASTARD, J. A. BRUM AND R. FERREIRA, *Phys. Rev. B* **38**, *10101 (1988). "Solid State Physics,", Vol. 44, H. Ehrenreich and D. Turnbull, eds., Academic Press, Boston, 1991, p. 229.*

[62] N.F. JOHNSON AND P.M. HUI, *J. of Physics - Cond. Matter* **5**, *L355 (1993).*

[63] A. G. KALOCSAI AND J. W. HAUS, *Opt. Commun.* **97**, *239 (1993); Phys. Rev. A* **49**, *574 (1994).*

[64] M. SCALORA, *Private Communication (1995).*

[65] S. D. SANKEY, D. F. PRELEWITZ AND T. G. BROWN, *Appl. Phys. Lett.* **60**, *1427 (1992);* S. D. SANKEY, D. F. PRELEWITZ, T. G. BROWN AND R. C. TIBERIO, *J. Appl. Phys.* **73**, *7111 (1993);* D. F. *Prelewitz and T. G. Brown, J. Opt.. Soc. Am. B* **11**, *304 (1994).*

[66] C. J. HERBERT, W. S. CAPINSKI AND M. S. MALCUIT, *Opt. Lett.* **15**, *1037 (1992).* C. J. HERBERT AND M. S. MALCUIT, *Opt. Lett.* **18**, *1783 (1993).*

SOLID STATE PHYSICS: BASIC AND OPTICAL PROPERTIES

A. QUATTROPANI, V. SAVONA AND F. TASSONE
Institut de Physique Théorique, Ecole Polytechnique Fédérale,
CH-1015 Lausanne, Switzerland

AND

P. SCHWENDIMANN
Defense Procurement and Technology Agency, System Analysis
Division,
CH-3003 Bern, Switzerland

1. Introduction

The scope of this lecture is to give an overview of the basic solid states physics which is used in describing solid state confined system and to discuss the fundamentals of the optical properties of excitons and of polaritons in bulk and in confined system. Therefore in the first two sections the definitions, approximations and equations which are used to describe electron states in the context of this lecture are introduced. The last two sections are devoted to the discussion of the excitons in bulk and in confined systems and to their optical properties and of polaritons both in bulk and in confined systems and in particular in quantum wells embebbed in microcavities.

2. Electronic band structures of crystals

2.1. THE BASIC APPROXIMATIONS

A crystal is constituted by a very large number of atoms., which are arranged in a regular lattice. The lattice structure of crystals implies that their properties are invariant under translations in space. In fact the lattice of any crystal can be built out from a fundamental cell, called unit cell, by spatial translations along the three fundamental axes which we call x_1, x_2,

143

M. Ducloy and D. Bloch (eds.), Quantum Optics of Confined Systems, 143-180.
© 1996 *Kluwer Academic Publishers. Printed in the Netherlands.*

$\mathbf{x}_3$ [1]. The translation vector is written as

$$\mathbf{r}_n = n_1\mathbf{x}_1 + n_2\mathbf{x}_2 + n_3\mathbf{x}_3 \tag{1}$$

where n_1, n_2, n_3 are integers. It is also useful for later purposes to introduce a reciprocal lattice whose vectors are defined through the relation $\exp(\mathbf{k}_n \cdot \mathbf{r}_n) = 1$ [1] and which are written in terms of the fundamental axes of the unit cell of the reciprocal lattice as

$$\mathbf{k}_n = m_1\mathbf{k}_1 + m_2\mathbf{k}_2 + m_3\mathbf{k}_3 \tag{2}$$

where m_1, m_2, m_3 are integers The set of all plane waves whose wave vector $\mathbf{k}$ has the periodicity of the lattice belong to the reciprocal lattice.

In order to obtain the energy levels and the states of a solid a number of approximations have to be introduced [2]. The first approximation is the adiabatic approximation which allows to split the eigenvalue problem into two separate, though correlated, problems for the nuclei and the electrons. This approximation relies on the fact that there is a huge difference between the masses of the electrons and the nuclei and that the velocity of the motion of the nuclei around their equilibrium positions in the lattice is several orders of magnitude smaller than that of the electrons. This means that the electrons may be considered to follow adiabatically the slower motion of the nuclei. The wave function of the crystal is then represented as the product of a wave function for the nuclei and of a wave function for the electrons . These wave functions obey coupled Schroedinger equations. The adiabatic approximation allows to solve these two equations separately. The Schroedinger equation for the electrons is solved in the potential of the nuclei which are frozen in their equilibrium positions. The equation for the nuclei then is approximated by neglecting terms corresponding to interactions between electrons and phonons i.e. lattice oscillations and contains the contribution of the electrons only through a global potential term . As we are ultimately interested in transitions between electronic states let us discuss the equation for the electronic part of the wave function. The solution of this equation is a very difficult many electron problem. Therefore a second approximation is introduced: the one-electron approximation. This approximation is obtained as follows: the wave function of N electrons is written as the antisymmetrized product of N one electron wave functions. We formally write

$$\psi(\mathbf{r}_1, \mathbf{r}_2, \ldots, \mathbf{r}_n) = \mathcal{A}(\psi(\mathbf{r}_1)\psi(\mathbf{r}_2)\ldots\psi(\mathbf{r}_n)) , \tag{3}$$

where $\mathcal{A}$ is the operator which antisymmetrizes the product of the one electron functions. This wave function is usually written in the form of a determinant (Slater determinant).

This approximation implies that electron correlations are neglected at this stage. Finally we have to set up a calculation scheme for the energies and states of the electrons in the crystal. We do it by a variational method. We evaluate the expectation value of the energy using the Slater determinants as a basis. thus obtaining a functional whose minimum has to be evaluated. The corresponding Euler equations have the form of a non linear Schroedinger equation with an effective hamiltonian given by

$$H_{HF}\psi_n = E_n\psi_n \tag{4}$$

with

$$H_{HF} = \frac{p^2}{2m} - V_{eL} + V_{coul} + V_{exch} , \tag{5}$$

Here the electron-nuclei interaction is

$$V_{eL} = \sum_I \frac{Z_I e^2}{|\mathbf{r} - \mathbf{R}_I|} , \tag{6}$$

where $\mathbf{R}_I$ are the positions of the nuclei, the Coulomb interaction between electrons is

$$V_{coul} = e^2 \sum_j \int \frac{|\psi_j(\mathbf{r}_1)|^2}{|\mathbf{r} - \mathbf{r}_1|} \, d\mathbf{r}_1 , \tag{7}$$

where $\mathbf{r}_1$ are the positions of the electrons, and the exchange interaction is

$$V_{exch} = -e^2 \sum_{j,n} \int \frac{\psi_j^*(\mathbf{r}_1)\psi_n(\mathbf{r}_1)}{|\mathbf{r} - \mathbf{r}_1|} \, d\mathbf{r}_1 . \tag{8}$$

A particularity of H_{HF} compared to one-electron Hamiltonians is the dependence on its own eigenfunctions through the electron density in the Coulomb interaction between electrons and through the non-local exchange potential V_{exch}. In order to reduce the mathematical difficulties, V_{exch} is often approximate by a local potential depending on the electron density [2]. The Hartree-Fock equation (4) is a complicated non-linear problem. It is solved e.g. by an iterative procedure. One chooses a set of approximate one electron wave functions $\psi_1^{(1)}, \psi_2^{(1)}, \ldots, \psi_n^{(1)}$ and calculates the approximate expression $V^{(1)}$ for the non-linear part of the potential. The Hartree-Fock equation is then solved and its solution are used in order to find a new approximate expression for the non-linear potential. The procedure is repeated until the difference between the $n - 1^{th}$ approximation and the nth approximation of the non linear potential can be neglected in the desired approximation. In particular by this procedure we can calculate the ground state of the crystal. An example of the variational calculation for

146

a one electron excited state in a semiconductor will be given in the section on excitons.

In the following we will discuss optical transitions between electronic states. Therefore we are interested in the energies of excited electronic states. We quote here two important results which allow to exploit the Hartree-Fock approximation in order to calculate the energy of the excited states of the crystal from the Hartree-Fock energies. It can be shown that the energy which is required to remove an electron from the system without influencing the other electron states, is equal to the Hartree-Fock energy of that electron (Koopman's theorem). Now consider the ground state, with energy E_0 in the Hartree-Fock approximation and a state in which only one electron is excited, with energy E_i. This state is written as:

$$\Phi_{\mathbf{k}_n,\mathbf{k}'_n} = \mathcal{A}\{\psi_{v,\mathbf{k}_1}, \cdots, \psi_{c,\mathbf{k}'_n}, \cdots, \psi_{v,\mathbf{k}_N}\}. \tag{9}$$

where $\psi_{c,\mathbf{k}'_n}$ is a state of the first unoccupied band of the crystal (conduction band). It can be shown that the difference of the Hartree-Fock energies $E_i - E_0$ is a good approximation to the energy difference between the true ground state and the one electron excited state of the crystal (Koopman's approximation). These approximations shows that although the Hartree-Fock energies may be poor approximations of the true energies of the crystal, their difference will give a quite good result when compared with the experimental energy differences between crystal states.

As we have seen the solution of self-consistent Hartree-Fock equations is obtained by iteration using an arbitrary set of one-particle wave functions. This is still a very difficult and cumbersome approach. Therefore in order to describe the electron states in the crystal one often adopts the band approximation. In this case the basic equation is

$$\left[\frac{p^2}{2m} + V(\mathbf{r})\right] \psi_n(\mathbf{k},\mathbf{r}) = E_n \psi_n(\mathbf{k},\mathbf{r}) , \tag{10}$$

where $V(\mathbf{r})$ is a periodic potential, is invariant under all symmetry operation of the crystal and is the same for all states. The periodic potential $V(\mathbf{r})$ is expressed in terms of the Hartree-Fock potentials as follows

$$V(\mathbf{r}) = -V_{eL} + V_{coul} + V_{exch} + V_{correlation} , \tag{11}$$

In particular the Coulomb and exchange components of $V(\mathbf{r})$ are calculated using the charge distribution of the ground state also for the evaluation of the excited states. The last term describes an approximation to the correlation energy which is in general assumed to be proportional to a non integer power of the charge density. In the section on excitons we discuss the limitation of such an approximation.

Using translation invariance and the fact that translations in space form a group we obtain an important result concerning the one-particle wave functions $\psi(\mathbf{r})$ of the crystal [2]. Let us call $T(\mathbf{r})$ the translation operator, then

$$T(\mathbf{r}_n)\psi(\mathbf{r}) = \psi(\mathbf{r} - \mathbf{r}_n) = \psi(\mathbf{r})e^{i\mathbf{k}\cdot\mathbf{r}_n} \tag{12}$$

Here $\mathbf{k}$ is a wave vector which has to be considered as a label for the eigenstates of the translation operator $T(\mathbf{r})$. In order to classify these states we have to consider only the vectors $\mathbf{k}$ which differ in magnitude by less than the magnitude of a vector of the reciprocal lattice. It is useful to define a volume of $\mathbf{k}$ space inside which this condition is satisfied. A particularly convenient volume is the so called first Brillouin zone. In order to preserve the translation invariance in a finite crystal, periodic boundary conditions at the crystal boundary have to be introduced. It can be shown that the function $\psi(\mathbf{r})$ on the r.h.s. of (12) is a periodic function on the lattice when periodic boundary conditions are considered. The functions thus obtained are called Bloch functions and are written as $\psi(\mathbf{k}, \mathbf{r}) = \exp(i\mathbf{k} \cdot \mathbf{r})u(\mathbf{k}, \mathbf{r})$ where $u(\mathbf{k}, \mathbf{r}) = u(\mathbf{k}, \mathbf{r} + \mathbf{r}_n)$ is the periodic part which has the period of the lattice [2]. The solutions of (10) with periodic boundary conditions will be functions of the Bloch type and describe the energy bands of the crystal. In a semiconductor we will deal with valence and conduction bands which are well separated in energy by the so called energy gap E_g. Typically the energy gap in a semiconductor varies between 0.5 eV and 2 eV at room temperature.

2.2. THE $\mathbf{K} \cdot \mathbf{P}$ APPROXIMATION

The calculation of the band structure of the crystal is in general a heavy task. There are different procedures for obtaining crystal states within the band approximation. We quote here as a few examples the tight binding method, the othogonalized plane wave method, the pseudopotential method, the cellular method and the Green's function method. For details on these approaches we refer the reader to the literature [2]. However these global descriptions of the crystal states rarely provide explicit expressions for the quantities of common use in semiconductor physics like effective masses wave functions etc. In order to obtain these quantities a local approach which refers to a small range of values of $\mathbf{k}$ around the band extrema is sufficient. We consider here an interpolation approach which is very useful when the eigenstates of the crystal for some value $\mathbf{k}_0$ of $\mathbf{k}$ are known [2, 3]. The eigenstates of equation (10) when solved with periodical boundary verify Bloch theorem $\psi_n(\mathbf{k}, \mathbf{r}) = \exp(i\mathbf{k} \cdot \mathbf{r})u_n(\mathbf{k}, \mathbf{r})$. The periodic part

148

of the Bloch function verifies the eigenvalue equation

$$\left[\frac{(\mathbf{p}+\hbar\mathbf{k})^2}{2m} + V(\mathbf{r})\right] u_n(\mathbf{k},\mathbf{r}) = E_n(\mathbf{k})u_n(\mathbf{k},\mathbf{r}) \ . \tag{13}$$

For each fixed wave vector $\mathbf{k}_0$ the set $\{u_n(\mathbf{k}_0,\mathbf{r})\}$ is complete and $u_l(\mathbf{k},\mathbf{r})$ can be expanded in terms of $\{u_n(\mathbf{k}_0,\mathbf{r})\}$. By replacing this expression into (12) we obtain a linear system of equations whose solution determines the states completely. Useful information is already obtained by using a perturbation approach. We treat the term $\mathbf{k}\cdot\mathbf{p}$ in (12) as a perturbation. For $\mathbf{k}$ close to $\mathbf{k}_0$, one obtains to first order in the perturbation for the wave function and to second order in the energy

$$u_l(\mathbf{k},\mathbf{r}) = u_l(\mathbf{k}_0,\mathbf{r}) + \frac{\hbar}{m}\sum_{n\neq l} \frac{(\mathbf{k}-\mathbf{k}_0)\cdot\mathbf{M}_{nl}(\mathbf{k}_0)}{E_l(\mathbf{k}_0) - E_n(\mathbf{k}_0)} u_n(\mathbf{k}_0,\mathbf{r}) \ , \tag{14}$$

with

$$\mathbf{M}_{nl}(\mathbf{k}_0) = \langle \psi_n(\mathbf{k}_0)|\mathbf{p}|\psi_l(\mathbf{k}_0)\rangle \tag{15}$$

and

$$\begin{aligned}
E_l(\mathbf{k}) &= E_l(\mathbf{k}_0) + \frac{\hbar^2}{2m}(\mathbf{k}-\mathbf{k}_0)^2 + \frac{\hbar}{m}(\mathbf{k}-\mathbf{k}_0)\cdot\mathbf{M}_{ll}(\mathbf{k}_0) \\
&+ \frac{\hbar^2}{m^2}\sum_{n\neq l}\frac{|(\mathbf{k}-\mathbf{k}_0)\cdot\mathbf{M}_{nl}(\mathbf{k}_0)|^2}{E_l(\mathbf{k}_0) - E_n(\mathbf{k}_0)} u_n(\mathbf{k}_0,\mathbf{r}) \ .
\end{aligned} \tag{16}$$

When the energy band $E_l(\mathbf{k})$ has an extremum at $\mathbf{k}=\mathbf{k}_0$, $\mathbf{M}_{ll}(\mathbf{k}_0) = 0$. In this case (16) is rewritten as

$$E_i(\mathbf{k}) = E_i(\mathbf{k}_0) + \hbar\frac{(\mathbf{k}-\mathbf{k}_0)^2}{m_c} \ , \tag{17}$$

where we have introduced the effective mass

$$\frac{1}{m_c} = \frac{1}{m} + \frac{2}{m^2}\sum_{n\neq l}\frac{|(\mathbf{s}\cdot\mathbf{M}_{nl}(\mathbf{k}_0)|^2}{E_l(\mathbf{k}_0) - E_n(\mathbf{k}_0)} \ . \tag{18}$$

Here the vector $\mathbf{s}$ indicates a principal direction. From this definition we see that a correction to the electron mass appears in consequence of the interaction with the bands. The interaction with bands of higher energy will lower the mass whereas the interaction with bands of lower energy will increase the mass.

Once the band structure of the crystal is known, we can calculate the transition probabilities for excitation of electrons by a radiation field and thus study the optical properties of the crystal. In a semiconductor the optical transitions happen between the valence and the conduction band and the transition probabilities are calculated using the Fermi golden rule. Examples of these calculations will be given in the following sections.

3. The effective mass equation (The envelope function method)

3.1. IMPURITY STATES (DONORS)

The theory developed so far has considered the states of a perfect crystal. However there are situations in which this picture must be refined. We will deal with two examples of such a situation: the impurity case in which some atoms in the crystal lattice are replaced by "foreign" atoms and the exciton case. This appears when an electron excited from the valence into the conduction band interacts with the positive excess charge left in the valence band after the transition (hole). In the following we discuss the impurity case in order to introduce the envelope function method The exciton will be discussed in detail in the next sections.

The envelope function method is illustrated in the case where a foreign atom which can release an electron in the conduction band of the host crystal (donor) is introduced into the lattice. The transitional periodicity of the lattice is broken and an additional potential originating from the presence of the impurity atom, is introduced. In the one-electron approximation the Schrödinger equation can be written as

$$\left[\frac{p^2}{2m} + V(\mathbf{r}) + U(\mathbf{r}) \right] \Phi(\mathbf{r}) = E\Phi(\mathbf{r}) , \qquad (19)$$

where $V(\mathbf{r})$ is as usual the potential of the perfect lattice and $U(\mathbf{r})$ is the additional potential due to the impurity. We consider the case very common in semiconductor physics, where the electronic states are very weakly bound to the impurity as it is the case of donor impurities like arsenic or phosphorus in germanium. The eigenvalue problem (19) is solved under the following assumptions:

  i The unperturbed conduction band $E_c(\mathbf{k})$ is non degenerate, has an extremum at $\mathbf{k} = \mathbf{k}_0$ and is well separated from the other bands. The corresponding unperturbed Bloch state is denoted by $\psi_c(\mathbf{k}, \mathbf{r})$.
  ii The perturbation potential $U(\mathbf{r})$ is weak and its interband matrix element can be disregarded.
  iii $U(\mathbf{r})$ is a slowly varying function over the crystal cell, i.e. for the Fourier transform of the potential the following inequality holds: $\tilde{U}(\mathbf{k}) >> \tilde{U}(\mathbf{k} + \mathbf{G})$, where $\mathbf{G}$ is a reciprocal lattice vector and $\mathbf{k}$ belongs to the first Brillouin zone.

iv The $\mathbf{k} \cdot \mathbf{p}$ approximation holds.

These assumptions show clearly also the limits of validity of this approach.
We first notice that usual perturbation theory cannot be used because the unperturbed eigenvalues $E_n(\mathbf{k})$ depend continuously on the wave vector $\mathbf{k}$. The solution of (19) is expanded in terms of unprturbed Bloch functions

$$\Phi(\mathbf{r}) = \sum_n \int \phi(\mathbf{k}) \psi_n(\mathbf{k}, \mathbf{r}) dk . \tag{20}$$

Substituting (20) into (19) and projecting the obtained equation on the unperturbed state $\psi_c(\mathbf{k}, \mathbf{r})$ one obtains

$$[E_c(\mathbf{k}) - E]\phi(\mathbf{k}) + \int d\mathbf{k}' \langle \psi_c(\mathbf{k})|U|\psi_c(\mathbf{k}')\rangle \phi(\mathbf{k}') = 0 . \tag{21}$$

We now substitute in the matrix element of (21) the $\mathbf{k} \cdot \mathbf{p}$ expansion of $\psi_c(\mathbf{k}', \mathbf{r}) = \exp(i\mathbf{k}' \cdot \mathbf{r}) u_c(\mathbf{k}', \mathbf{r})$ for $\mathbf{k}' = \mathbf{k}$ and we obtain as a consequence of the assumptions introduced above $\langle \psi_c(\mathbf{k})|U|\psi_c(\mathbf{k}')\rangle = \tilde{U}(\mathbf{k} - \mathbf{k}')$, namely the Fourier transform of the perturbing potential $U(\mathbf{r})$. For an isotropic and parabolic conduction band $E_c(\mathbf{k})$ with an effective mass m_c the Fourier transform of (21) takes the form

$$\left[-\frac{\hbar^2 k^2}{2m_c} + U(\mathbf{r}) \right] F(\mathbf{r}) = EF(\mathbf{r}) , \tag{22}$$

where $F(\mathbf{r})$ is the Fourier transform of the envelope function $\phi(\mathbf{k})$. For the special case of a screened Coulomb potential $U(\mathbf{r}) = -e^2/(\epsilon_0 r)$, $F(\mathbf{r})$ is a hydrogenic function and the binding energy in the ground state is the effective Rydberg $Ry^* = Ry \times m_c/(m\epsilon_0^2)$; for typical semiconductors $Ry^* \sim 10 \ meV$. The expression of the impurity wave function in terms of the envelope function is given by (20); after expanding the Bloch function around the extremum $\mathbf{k}_0$ and performing the integration over $\mathbf{k}$, we obtain in first order

$$\Phi(\mathbf{r}) = F(\mathbf{r})\psi_c(\mathbf{k}_0, \mathbf{r}) . \tag{23}$$

This result represents a useful introduction to the exciton problem which is discussed in detail in the next section. In fact the exciton which results from the interaction of an electron in the conduction band and of a hole in the valence band is described through an extension of the approach outlined here.

4. Optical properties of excitons

4.1. EXCITONS IN BULK SYSTEMS

In Sec. 2.1 we showed how the ground and excited states of the crystal are described in the Hartree-Fock approximation: the wavefunction of the N electrons of the crystal is taken as a Slater determinant of single particle wavefunctions which obey the Hartree-Fock equations. Given the ground state wavefunction, and therefore N single particle wavefunctions of the occupied states, it is also possible to find the orbitals of the unoccupied states. In the case of the semiconductor, the lowest band of these states form the conduction band, while the highest occupied band forms the valence band. One-electron excited states of the crystal have been formed by 'raising' an electron from the valence to the conduction band, i.e. considering a Slater determinant as in (9). We showed that this state describes well the true excited state since the charge distribution of the ground state is barely modified in the limit $N \to \infty$. However, this is only approximatively correct, since it can be shown that the modification of the charge distribution of the ground state in the excited crystal is similar to the presence of a positive charge, a hole, in it. Even if the Coulomb energy of a delocalized electron and hole is negligible, it is well understood that if the promoted electron in the conduction band correlates to this hole in a 'localized' way, a finite amount of this energy is found. Experimentally, the existence of such correlated electron-hole states has been observed as the presence of excitonic absorption peaks below the gap-energy, as described by Delalande in his lectures.

Let's show how to introduce the electron-hole correlation, using the excited one-electron many-body wavefunctions of (9). The translational quantum number is a good quantum number also for the excited states of the crystal. In order to describe the excited state with translational quantum number $\mathbf{k}$, we consider a combination of states of the type (9), all having this translational quantum number $\mathbf{k}$. Since it's easy to show that the state (9) transforms as a Bloch state with k-vector $\mathbf{k}'_n - \mathbf{k}_n$. we obtain for this excited state:

$$\Phi_{exc}(\mathbf{k}) = \sum_{\mathbf{k}'} A_{\mathbf{k}}(\mathbf{k}')\Phi_{\mathbf{k}'-\frac{m_h}{M}\mathbf{k},\mathbf{k}'+\frac{m_e}{M}\mathbf{k}} \tag{24}$$

In this expression we introduced two parameters which are the hole and the electron effective masses m_h and m_e, and $M = m_e + m_h$. Although it is strictly not necessary to do so, this form simplifies the following calculations. In particular, it allows a direct interpretation of the coefficient $A_{\mathbf{k}}(\mathbf{k}')$ as the Fourier transform of a wavefunction of the excited state in real space. The normalization of (24) gives:

$$\sum_{\mathbf{k}'} |A_{\mathbf{k}}(\mathbf{k}')|^2 = 1 \tag{25}$$

The coefficients $A_{\mathbf{k}}(\mathbf{k}')$ are chosen by minimizing the expectation value of the *total* hamiltonian $\mathcal{H}_0$. We start by calculating the matrix elements of the hamiltonian on the states (9).

$$\langle \Phi_{\mathbf{k}_h,\mathbf{k}_e}|\mathcal{H}_0|\Phi_{\mathbf{k}_h',\mathbf{k}_e'}\rangle = \delta_{\mathbf{k}_h,\mathbf{k}_h'}\delta_{\mathbf{k}_e,\mathbf{k}_e'}[E_g + E_c(\mathbf{k}_e) - E_v(\mathbf{k}_h)] \tag{26}$$
$$+ \ \delta_{\mathbf{k}_e-\mathbf{k}_h,\mathbf{k}_e'-\mathbf{k}_h'}[\langle \psi_{c,\mathbf{k}_e}\psi_{v,\mathbf{k}_h'}|V|\psi_{v,\mathbf{k}_h}\psi_{c,\mathbf{k}_e'}\rangle - \langle \psi_{c,\mathbf{k}_e}\psi_{v,\mathbf{k}_h'}|V|\psi_{c,\mathbf{k}_e'}\psi_{v,\mathbf{k}_h}\rangle]$$

where E_g is the ground state energy, as obtained in the Hartree-Fock treatment. The terms $E_c(\mathbf{k}_e)$ and $E_v(\mathbf{k}_h)$ are the energies of conduction and valence band if we neglect some small corrections due to the change in the charge distribution of the excited state with respect to the ground state. The two remaining terms, where

$$\langle \psi_{\mathbf{k}_1}\psi_{\mathbf{k}_2}|V|\psi_{\mathbf{k}_1'}\psi_{\mathbf{k}_2'}\rangle = \int \psi_{\mathbf{k}_1}^*(\mathbf{r}_1)\psi_{\mathbf{k}_2}^*(\mathbf{r}_2)\frac{e^2}{|\mathbf{r}_1-\mathbf{r}_2|}\psi_{\mathbf{k}_1'}(\mathbf{r}_1)\psi_{\mathbf{k}_2'}(\mathbf{r}_2)d\mathbf{r}_1 d\mathbf{r}_2,$$
$$\tag{27}$$

represent the Coulomb and exchange interaction between the electron and the hole.

We use (26) and we calculate the matrix element of the total hamiltonian on the state (24). In what follows we neglect the first of the two matrix elements of V in (26). This corresponds to neglect the Coulomb interaction between the electrons, which is small in the Wannier exciton case which is the only we are going to consider. We will see that the other matrix element, which is the exchange interaction between all the N-1 electrons in the ground state and the conduction electron, is at the end equivalent to a Coulomb interaction between the electron and the hole.

The matrix element of the exciton function is

$$\langle \Phi_{exc}(\mathbf{k})|\mathcal{H}_0|\Phi_{exc}(\mathbf{k})\rangle = \tag{28}$$
$$\sum_{\mathbf{k}'}|A_{\mathbf{k}}(\mathbf{k}')|^2[E_g + E_c(\mathbf{k}' - \frac{m_h\mathbf{k}}{M}) - E_v(\mathbf{k}' + \frac{m_e\mathbf{k}}{M})]$$
$$+ \ \sum_{\mathbf{k}',\mathbf{k}''}A_{\mathbf{k}}^*(\mathbf{k}')A_{\mathbf{k}}(\mathbf{k}'')\langle \psi_{c,\mathbf{k}'+\frac{m_e}{M}\mathbf{k}}\psi_{v,\mathbf{k}''-\frac{m_h}{M}\mathbf{k}}|V|\psi_{c,\mathbf{k}''+\frac{m_e}{M}\mathbf{k}}\psi_{v,\mathbf{k}'-\frac{m_h}{M}\mathbf{k}}\rangle$$

Minimizing with respect to $A_{\mathbf{k}}^*(\mathbf{k}')$ we get

$$[E_g - E + E_c(\mathbf{k}' - \frac{m_e\mathbf{k}}{M}) - E_v(\mathbf{k}' + \frac{m_h\mathbf{k}}{M})]A_{\mathbf{k}}(\mathbf{k}') \tag{29}$$
$$+ \ \sum_{\mathbf{k}''}A_{\mathbf{k}}(\mathbf{k}'')\int d\mathbf{r}_1 d\mathbf{r}_2 \psi_{c,\mathbf{k}'+\frac{m_e}{M}\mathbf{k}}^*(\mathbf{r}_1)\psi_{v,\mathbf{k}''-\frac{m_h}{M}\mathbf{k}}^*(\mathbf{r}_2)\frac{e^2}{|\mathbf{r}_1-\mathbf{r}_2|}$$
$$\times \ \psi_{c,\mathbf{k}''+\frac{m_e}{M}\mathbf{k}}(\mathbf{r}_1)\psi_{v,\mathbf{k}'-\frac{m_h}{M}\mathbf{k}}(\mathbf{r}_2) = 0$$

We introduce some assumptions. Since we are considering the case of a Wannier or weakly bound exciton, we suppose its 'wavefunction' spreaded in real space compared to the lattice period. We can equivalently assume that the $A_{\mathbf{k}}$ in (24) involves only a small range of $\mathbf{k}'$ around zero. This hypothesis depends of course on the material we are considering and may be justified only a posteriori. We may therefore conveniently use an effective mass description of the energies and Bloch states involved in the exciton wave function, as already introduced in Sec. 2.2.

$$
E_c(\mathbf{k}) = \frac{\hbar^2 k^2}{2m_e}
$$

$$
E_v(\mathbf{k}) = -\frac{\hbar^2 k^2}{2m_h} \tag{30}
$$

and $\psi_{\mathbf{k}}(\mathbf{r})$ where we put

$$
\begin{aligned}
\psi_{\mathbf{k}}(\mathbf{r}) &= u_{\mathbf{k}}(\mathbf{r}) \exp(i\mathbf{k} \cdot \mathbf{r}) \\
&\simeq u_0(\mathbf{r}) \exp(i\mathbf{k} \cdot \mathbf{r})
\end{aligned} \tag{31}
$$

All the terms in the integral in (29), except the periodic part of the Bloch functions, are slowly varying with respect to $\mathbf{r}$. If we consider all the slowly varying parts as constants inside each elementary cell, the only varying part of the integrand is the function $u_0(\mathbf{r})$. Because these functions are normalized on the elementary cell, we can finally write:

$$
[E_g - E + \frac{\hbar^2}{2\mu}k'^2 + \frac{\hbar^2}{2M}k^2]A_{\mathbf{k}}(\mathbf{k}') + \sum_{\mathbf{k}''} A_{\mathbf{k}}(\mathbf{k}'') \frac{e^2}{|\mathbf{k}'' - \mathbf{k}'|^2} = 0 \tag{32}
$$

This integral equation can be Fourier-transformed in $\mathbf{k}'$ in order to obtain a differential equation for $F_{\mathbf{k}}(\mathbf{r})$ which is the Fourier transform of $A_{\mathbf{k}}(\mathbf{k}')$.

$$
-\frac{\hbar^2}{2\mu}\nabla_{\mathbf{r}}^2 F_{\mathbf{k}}(\mathbf{r}) + [E_g - E + \frac{\hbar^2}{2M}k^2 + \frac{e^2}{|\mathbf{r}|}]F_{\mathbf{k}}(\mathbf{r}) = 0 \tag{33}
$$

We remark that the last equation is exactly the Schroedinger equation for the hydrogen atom. In this equation $\mathbf{k}$ represents the wavevector for the center of mass motion and $\mathbf{r}$ is the relative coordinate of the electron and the hole.

In a realistic situation the electric field between the two charges which form the exciton is screened by the presence of all the other electrons in the material. This screening originates from the polarization of the medium around the charges. Responsible for this polarization are all the virtual transitions to the excited states of the system, of which only the ones to the lower conduction band have been included in our model. In principle we should therefore introduce all the electron states belonging to different bands in (24) to obtain the effect of screening. In practice this amounts to the introduction of a constant screening of the Coulomb interaction between the electron and the hole in eq. (33):

$$\frac{e^2}{\epsilon |\mathbf{r}|} \tag{34}$$

Here ϵ should be taken as the high-frequency dielectric constant of the material. In most small gap semiconductors this is a rather large dielectric constant, and is frequently above the value of 10.

We can now write the two parameters which characterize a hydrogenic system, the effective Rydberg and the Bohr radius:

$$Ry^* = \frac{\mu e^4}{2\epsilon^2 \hbar^2}$$
$$a_B^* = \frac{\hbar^2 \epsilon}{\mu e^2} \tag{35}$$

Due to the small effective masses of the conduction and valence bands in most small gap semiconductors, and to the large dielectric screening, the effective Rydberg is typically of the order of few meV only.

4.2. OPTICAL TRANSITIONS

Let us now consider the creation of the excited states introduced in the preceding section by the absorption of electromagnetic radiation. The interaction hamiltonian is obtained replacing $\mathbf{p}_i$ by $\mathbf{p}_i - (e/c)\mathbf{A}_i$ in the total hamiltonian $\mathcal{H}_0$. This gives an interaction Hamiltonian

$$\mathcal{H}_{int} = \frac{e}{mc} \sum_i \mathbf{p}_i \cdot \mathbf{A}(\mathbf{r}_i), \tag{36}$$

where we are neglecting the $\mathbf{A}^2$ term. Unperturbed states of the crystal are not anymore stationary when an electromagnetic field is switched on. As it is well known from quantum mechanics, the ground state starts to evolve into the excited states of the system soon after the perturbation is turned on. In the case of a 'slow' evolution, a transition rate to the excited states may be defined, and is given by the Fermi Golden rule

$$\frac{dP}{dt} = \frac{2\pi}{\hbar} |\langle \Phi_{exc}(\mathbf{k})|H_{int}|0\rangle|^2 \delta(\omega(\mathbf{k}) - \omega), \tag{37}$$

where $\hbar\omega(\mathbf{k})$ is the exciton energy. The matrix element which appears in (37) can be easily calculated using the expression given in the previous section for the exciton wavefunction. It is straightforward to show that [2]:

$$\langle \Phi_{exc}(\mathbf{k})|H_{int}|0\rangle = E_{\mathbf{k}_{ph}} \sqrt{V} F_{\mathbf{k}}(0) \boldsymbol{\mu}_{cv} \cdot \boldsymbol{\epsilon} \delta_{\mathbf{k},\mathbf{k}_{ph}} \frac{\omega(\mathbf{k})}{ck}, \tag{38}$$

where V is the normalization volume, $E_{\mathbf{k}_{ph}}$ is the amplitude of the electric field, ϵ its polarization and $\mathbf{k}_{ph}$ the wavevector of the photon. $\boldsymbol{\mu}_{cv}$ is the dipole matrix element between valence and conduction states, defined as

$$\boldsymbol{\mu}_{cv} = \int_{\Omega} d\mathbf{r} \; u_{c,0}^*(\mathbf{r}) e r u_{v,0}(\mathbf{r}). \tag{39}$$

We remark that the total momentum of the photon plus crystal is conserved in the transition, therefore an exciton with given $\mathbf{k}_{ex}$ couples only to a field with the same wavevector. This conservation gives rise to the divergence of the interaction matrix element with the volume of the crystal, as this volume is let to go to infinity as in the bulk crystal. In other words, the squared matrix element is a *macroscopic* quantity. This can also be seen if one tries to calculate the absorption of the crystal, that is related to the transition probability by the well known relation

$$\alpha(\omega) = \frac{2\pi\hbar\omega}{ncE_{\mathbf{k}_{ph}}^2} \frac{dP}{dt} \tag{40}$$

The absorption coefficient is proportional to the volume of the crystal through the square modulus of the dipole matrix element. This fact clearly shows that the hypothesis on which Fermi golden rule is based, i.e. small transition rates to the excited states, do not hold in this case, and the evolution of the perturbed system has to be described differently.

In practice, we may consider the intensity of the electric field as a perturbation parameter, and calculate the time-dependent wavefunction of the perturbed system to the first order in this field (linear response theory, see e.g. Kubo [4]). With this wavefuntion any interesting linear property of the excited system as a function of time may be easily calculated. For example, we may be interested in the macroscopic polarization developed by the crystal. The knowledge of this polarization allows us to calculate the re-radiated electromagnetic fields by solving Maxwell' s equations.

The linear response of the system is characterized by a response function, which is called susceptibility in the case of the polarization. Without carrying out all the steps of derivation, we give the final expression for this excitonic susceptibility:

$$\chi(\mathbf{k}, \omega) = \frac{4\pi}{\hbar} \sum_n \frac{|\boldsymbol{\mu}_{cv} \cdot \epsilon|^2 |F_{n,\mathbf{k}}(0)|^2}{\omega_n^2(\mathbf{k}) - \omega^2} \tag{41}$$

We remark that it is calculated microscopically and contains information about the true excited states of the semiconductor close to the gap. It contains the strength of the coupling of the crystal with radiation through the $\boldsymbol{\mu}_{cv}$ factor, the strength of electron-hole correlation in $|F(0)|^2$, and shows resonant response at the excitonic energies. The absence of any broadening in the resonances follows from neglecting other interactions in the evolution of the excited state. It shows that a perturbative approach in the transition rate is not suitable in this case, as we have remarked using Fermi golden rule.

4.3. EXCITONS IN QUANTUM WELLS

Nanofabrication technology today allows to grow crystals with specially tailored properties. In particular, in the simplest epitaxial growth, a small gap semiconductor may be sandwiched inside a large gap semiconductor. The translational symmetry is broken along the growth direction, and the lowest unoccupied one particle states, i.e. the conduction states, as well as the highest occupied states, i.e. the valence states, become confined in the small-gap material. We refer to the lectures of Delalande for the description of these states in the effective mass approximation. We try instead to solve the excitonic problem in this case, working in analogy to the theory of Sec. 4.3 for the bulk.

Exciton states of a quantum well structure are always constructed from the one particle states of the structure. Since we are aiming to describe Wannier excitons, we can approximate the one particle states in the effective mass approach. The one particle states for an electron in a conduction or valence subband are:

$$
\begin{aligned}
\psi_{c,\mathbf{k}_\parallel,n}(\boldsymbol{\rho},z) &= \exp(i\mathbf{k}_\parallel \cdot \boldsymbol{\rho})c_n(z)u_{c,0}(\boldsymbol{\rho},z) \\
\psi_{v,\mathbf{k}_\parallel,n}(\boldsymbol{\rho},z) &= \exp(i\mathbf{k}_\parallel \cdot \boldsymbol{\rho})v_n(z)u_{v,0}(\boldsymbol{\rho},z)
\end{aligned}
\tag{42}
$$

As we have remarked before, the excited state has only the wavevector in the plane of the well as a good quantum number, since the translational symmetry is broken along the other direction. The excited state with a given $\mathbf{k}_\parallel$ thus is

$$
\Phi_{exc}(\mathbf{k}_\parallel) = \sum_{n_c,n_v,\mathbf{k}'_\parallel} A_{\mathbf{k}_\parallel}(n_c,n_v,\mathbf{k}'_\parallel)\Phi_{n_v,\mathbf{k}'_\parallel-\frac{m_h}{M}\mathbf{k}_\parallel,n_c,\mathbf{k}'_\parallel+\frac{m_e}{M}\mathbf{k}_\parallel}.
\tag{43}
$$

Here the same notation as before is used for Slater determinants, provided the single-particle states given before in (42) are used in place of the bulk ones.

Once again we calculate the matrix element of the total hamiltonian on the state (43) and minimize with respect to the coefficients $A_{\mathbf{k}_\parallel}(n_c,n_v,\mathbf{k}'_\parallel)$. We obtain the following $\mathbf{k}_\parallel$-space equation for the Wannier excitons in a QW:

$$
[E_g + E_{c,n_c}(\mathbf{k}'_\parallel + \frac{m_e}{M}\mathbf{k}_\parallel) - E_{v,n_v}(\mathbf{k}'_\parallel - \frac{m_h}{M}\mathbf{k}_\parallel)]A_{\mathbf{k}_\parallel}(n_c,n_v,\mathbf{k}'_\parallel)
$$

$$+ \sum_{n_c'',n_v'',\mathbf{k}_\parallel''} A_{\mathbf{k}_\parallel}(n_c'',n_v'',\mathbf{k}_\parallel'')\tilde{V}_{n_c',n_v',n_c'',n_v''}(\mathbf{k}_\parallel'' - \mathbf{k}_\parallel') = E A_{\mathbf{k}_\parallel}(n_c,n_v,\mathbf{k}_\parallel')$$

where the Coulomb matrix element is given by

$$\tilde{V}_{n_c',n_v',n_c'',n_v''}(\mathbf{k}_\parallel'' - \mathbf{k}_\parallel') =$$

$$\int d\mathbf{r}_1 d\mathbf{r}_2 \exp[i(\mathbf{k}_\parallel'' - \mathbf{k}_\parallel')(\rho_1 - \rho_2)]\frac{e^2}{|\mathbf{r}_1 - \mathbf{r}_2|}c_{n_c'}^*(z_1)c_{n_c''}(z_1)v_{n_v''}^*(z_2)v_{n_v'}(z_2)$$

Up to now we have included the contributions from all the subbands of the QW electronic states. While in the bulk case we included only the valence and conduction bands, and considered the effect of the other bands as a constant screening of the Coulomb interaction, in the quantum well case this approach of considering two subbands only does not give good results. The difference is that in a QW the different subbands are not sufficiently separated with respect to the exciton binding energies. Nevertheless it is instructive to consider the two band approximation anyway, because it makes possible to introduce a wave equation for the envelope function of the exciton state in real space, in analogy to the bulk one

$$[E_g + \frac{\hbar^2}{2M}k_\parallel^2 + \frac{\hbar^2}{2\mu}\nabla_\rho^2]F_{\mathbf{k}_\parallel}(\rho) + \tilde{V}(\rho)F_{\mathbf{k}_\parallel}(\rho) = E F_{\mathbf{k}_\parallel}(\rho), \qquad (44)$$

where $F_{\mathbf{k}_\parallel}(\rho)$ and $\tilde{V}(\rho)$ are the Fourier transforms of $A_{\mathbf{k}_\parallel}(n_c,n_v,\mathbf{k}_\parallel')$ and $\tilde{V}_{n_c,n_v,n_c,n_v}(\mathbf{k}_\parallel')$ respectively and

$$\tilde{V}(\rho) = \int dz_1 dz_2 \frac{e^2}{|\mathbf{r}_1 - \mathbf{r}_2|}|c_{n_c}(z_1)|^2|v_{n_v}(z_2)|^2. \qquad (45)$$

The potential (45) is an effective Coulomb potential which takes into account the confinement of the electron and the hole in the QW. In the extreme limit where the confinement length is much shorter than the exciton Bohr radius, we can replace $|c_{n_c}(z_1)|^2$ and $|v_{n_v}(z_2)|^2$ by delta functions and obtain a purely two dimensional hydrogenic potential

$$\tilde{V}(\rho) = \frac{e^2}{|\rho|} \qquad (46)$$

In this case we have the well known two dimensional hydrogenic series

$$\hbar\omega_{exc,n} = E_g - \frac{Ry^*}{(n + 1/2)^2}, \qquad n = 0,1,\ldots \qquad (47)$$

For $n = 0$ the lowest exciton state has four times the binding energy of the corresponding state in the bulk case and twice its exciton Bohr radius.

4.4. OPTICAL TRANSITIONS IN QWS

We calculate the optical transition rates to the exciton states using the Fermi golden rule, as in the bulk case. This is still given by (37), provided we calculate the matrix element of the interaction hamiltonian (36) with the quantum well exciton state and the quantum well ground state. We obtain

$$\langle \Phi_{exc}(\mathbf{k}_\parallel)|\mathcal{H}_{int}|0\rangle =$$

$$E_0\sqrt{S}F_{\mathbf{k}_\parallel}(0)\boldsymbol{\mu}_{cv}\cdot\boldsymbol{\epsilon}\delta_{\mathbf{k}_\parallel,\mathbf{k}_{\parallel ph}}\int dz c^*_{n_c}(z)v_{n_v}(z)\exp(ik_{ph,z}z), \quad (48)$$

where all the quantities have been defined before, except for S which is the normalization surface of the QW. This factors shows us again that this simple perturbative scheme is incorrect and we must resort to linear response theory again. In particular, we are interested in the macroscopic polarization of the well, i.e. in the susceptibility. Notice that in the QW case the total momentum of the exciton plus photon system is not conserved. Only the in-plane component of this momentum is conserved while the exciton can couple to photons with any z-component of the wavevector. This fact apports considerable differences in the optical response of the QW as compared to the bulk case. For example, the polarization produced by an electromagnetic field is confined in the QW region since the exciton also is, and the exciton is again sensitive to the field inside the well only. The quantum well susceptibility therefore is nonlocal along the growth direction:

$$\chi(\mathbf{k}_\parallel,\omega,z,z') = \frac{1}{\hbar}\sum_n \frac{|\boldsymbol{\mu}_{cv}\cdot\boldsymbol{\epsilon}|^2|F_{n,\mathbf{k}_\parallel}(0)|^2\xi(z)\xi(z')}{\omega_n^2(\mathbf{k}_\parallel)-\omega^2} \quad (49)$$

where n runs over all the exciton states and $\xi(z) = c^*(z)v(z)$ is defined as the exciton confinement function. The polarization for a given electric field $\mathbf{E}(\mathbf{z})$ then is given by:

$$\mathbf{P}_{\mathbf{k}_\parallel}(z) = \int dz'\,\chi(\mathbf{k}_\parallel,\omega,z,z')\mathbf{E}_{\mathbf{k}_\parallel}(z') \quad (50)$$

5. Polaritons

5.1. POLARITONS IN BULK SYSTEMS

In the previous section we have studied the dipole matrix element on exciton states, which gives the coupling between the excitons and the radiation field. We remember that it is usually not sufficient to treat the optical response of the excitonic system in perturbation theory, using the Fermi golden rule to calculate the transition probabilities, because the squared matrix element is proportional to the volume of the sample. Therefore, the coupling hamiltonian has to be diagonalized exactly, and the exciton and radiation modes become mixed into new elementary excitations of the solid which are called polaritons.

These mixing of the material excitation with the electromagnetic field can also be seen using a classical picture, i.e. considering the classical polarization calculated with linear response theory, eq 41, and a classical electromagnetic field. The resonance form of the susceptibility shows the importance of the re-radiated electric field close to the exciton resonance; the wave equations for the electric field has to be solved self-consistently, and the resulting electric field describes mixed material-electromagnetic modes propagating in the crystal. We therefore consider a single exciton resonance in (41):

$$\chi(\mathbf{k}, \omega) = \frac{f^2}{\omega^2(\mathbf{k}) - \omega^2} \tag{51}$$

where $f^2 = (4\pi/\hbar)\mu_{cv}^2|F(0)|^2$ is the oscillator strength of the transition. The dielectric function is

$$\epsilon(\mathbf{k}, \omega) = \epsilon_\infty + \frac{4\pi f^2}{\omega^2(\mathbf{k}) - \omega^2}. \tag{52}$$

Here ϵ_∞ is a background dielectric constant which takes into account all the other resonances of the medium. In what follows we will neglect the exciton spatial dispersion and set $\omega(\mathbf{k}) = \omega_0$. Defining ϵ_0 as the low frequency dielectric constant, we can rewrite the last equation as

$$\epsilon(\mathbf{k}, \omega) = \epsilon_\infty + \frac{\epsilon_0 - \epsilon_\infty}{1 - \omega^2/\omega_0^2} \tag{53}$$

The wave equation for the electromagnetic field in the medium reads

$$-\nabla \times \nabla \times \mathbf{E} - \epsilon(\mathbf{k}, \omega)k^2\mathbf{E} = 0 \tag{54}$$

This equation has two distinct kinds of solution. First a longitudinal solution, with $\nabla \times \mathbf{E} = 0$, whose energy is given by

$$\epsilon(\mathbf{k}, \omega) = 0 \tag{55}$$

This equation simply gives us $\omega = \omega_L = \omega_0(\epsilon_0/\epsilon_\infty)^{1/2}$

The other solutions are those for transverse modes, with $\nabla \cdot \mathbf{E} = 0$, obtained when considering a plane wave solution for the e.m. field

$$\epsilon(\mathbf{k}, \omega) = \epsilon_\infty + \frac{\epsilon_0 - \epsilon_\infty}{1 - \omega^2/\omega_0^2} = \frac{k^2}{\omega^2}c^2 \tag{56}$$

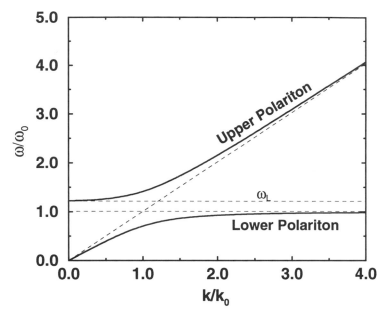

Figure 1. The polariton dispersion in the case of a dispersionless exciton. The non-interacting photon and exciton dispersions are indicated with dashed lines.

For each given $\mathbf{k}$ this is a biquadratic equation which gives the energies of the two transverse modes called upper and lower polariton modes. They are the true propagating modes inside the solid. In particular, The energy of the lower polariton mode approaches the transverse energy ω_T as k goes to infinity. The upper polariton mode starts at an energy ω_L and approaches the photon dispersion for large wavevectors. Therefore, there is a gap in the frequency domain between ω_0 and ω_L in which there are no propagating modes. The width of this gap is usually referred to as longitudinal-transverse splitting, and called ω_{LT}. The existence of such a gap comes from the assumption of no spatial dispersion for the exciton energy. Including the k dependence of the exciton energy, modes in this LT energy region appear at large wavevectors The polariton dispersion, with and without the exciton spatial dispersion, is illustrated in Fig. 1 and 2.

The polariton modes can also be obtained as quantized fields, using a second quantization treatment of excitons and photons and their interaction. We first have to remark that excitons behave as bosons. This can be seen from the form of the exciton state, eq. (24). In fact, the one-electron excitation is spread over many 'fermion' states, and in the limit $N \to \infty$, the occupation per conduction/valence orbital is vanishing. Therefore any finite number of excitations can be piled up in the same state, without sensing the effects of Pauli exclusion principle. We introduce boson creation and destruction operators for excitons and photons respectively $\hat{b}_k^\dagger$ ($\hat{b}_k$) and $\hat{a}_k^\dagger$ (a_k). We consider the interaction term between the transverse Wannier exciton and the radiation field, given in eq. (36), and eventually add the $\mathbf{A}^2$ term also:

$$-\frac{e}{mc}\sum_i \mathbf{A}_i \cdot \mathbf{p}_i + \frac{e^2}{mc^2}\sum_i \mathbf{A}_i^2. \tag{57}$$

We then obtain the exciton-photon or polariton Hamiltonian in second quantization:

$$\mathcal{H} = \sum_k \hbar\omega(\mathbf{k})\hat{b}_k^\dagger \hat{b}_{n,k} + \sum_k \hbar c|\mathbf{k}|\hat{a}_k^\dagger \hat{a}_k +$$

$$+ \sum_k iC_k(\hat{a}_k^\dagger + \hat{a}_{-k})(\hat{b}_k - \hat{b}_{-k}^\dagger) + \sum_k D_k(\hat{a}_k^\dagger + \hat{a}_{-k})(\hat{a}_k^\dagger + \hat{a}_{-k}), \tag{58}$$

with

$$C_k = \frac{1}{c\hbar}\sqrt{\frac{2\pi\hbar c}{kV}}\langle\Phi_k|e\,\mathbf{r}\cdot\epsilon\epsilon^{i\mathbf{k}\cdot\mathbf{r}}|0\rangle\hbar\omega_n(\mathbf{k}) = \sqrt{\frac{2\pi\hbar}{ck}}\boldsymbol{\mu}_{cv}\cdot\epsilon F(0)\omega(\mathbf{k}) \tag{59}$$

and

$$D_k = \frac{C_k^2}{E^{ex}(\mathbf{k})}. \tag{60}$$

The first two terms represent the free hamiltonians of the transverse exciton and photon fields. The other terms represent the exciton-photon interaction, i.e. the two terms of eq. (57) written in second quantization.

The full translational symmetry of the systems allows only the coupling of states which carry the same momentum. For this reason the problem separates in this index, and for each wavevector, we can diagonalize the hamiltonian with a suitable operatorial transformation, called the Hopfield transformation [5], which introduces new boson operators $\hat{\alpha}_k^\dagger$ and $\hat{\alpha}_k$, as linear combinations of the $\hat{a}$ and $\hat{b}$ operators:

$$\hat{\alpha}_k = W(\mathbf{k})\hat{a}_k + Y(\mathbf{k})\hat{a}_{-k}^\dagger + X(\mathbf{k})\hat{b}_k + Z(\mathbf{k})\hat{b}_{-k}^\dagger \tag{61}$$

Here $W(\mathbf{k})$, $Y(\mathbf{k})$, $X(\mathbf{k})$ and $Z(\mathbf{k})$ are complex coefficients, determined by the diagonalization condition

$$\left[\hat{H}, \hat{\alpha}_k^{(i)}\right] = -\hbar\Omega_i(\mathbf{k})\hat{\alpha}_k^{(i)}, \tag{62}$$

and the normalization one:

$$\left[\hat{\alpha}_k^{(i)}, \hat{\alpha}_k^{(j)\dagger}\right] = \delta_{kk'}\delta_{ij}. \tag{63}$$

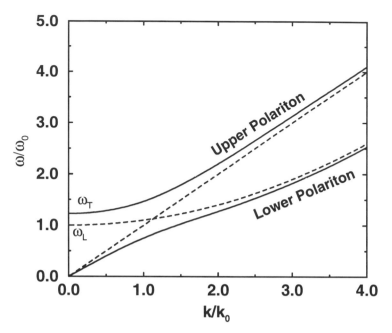

Figure 2. The polariton dispersion in the case of a finite mass exciton. The non-interacting photon and exciton dispersions are indicated with dashed lines.

These conditions give exactly the dispersion of the transverse modes (56) and the following solutions for the coefficients

$$W_1(\mathbf{k}) = \frac{(\omega_0^2 - \Omega_1^2)(ck + \Omega_1)}{2\sqrt{ck\Omega_1}[(\omega_0^2 - \Omega_1^2)^2 + 4\pi f^2 \omega_0^2]^{\frac{1}{2}}} \tag{64a}$$

$$X_1(\mathbf{k}) = -i\frac{(4\pi f^2 \omega_0 ck)^{\frac{1}{2}}}{(\omega_0 - \Omega_1)(ck + \Omega_1)}W_1(\mathbf{k}) \tag{64b}$$

$$Y_1(\mathbf{k}) = \frac{\Omega_1 - ck}{\Omega_1 + ck}W_1(\mathbf{k}) \tag{64c}$$

$$Z_1(\mathbf{k}) = \frac{\omega_0 - \Omega_1}{\omega_0 + \Omega_1}X_1(\mathbf{k}) \tag{64d}$$

The coefficients W_2, Y_2, X_2, Z_2 for the upper polariton mode, are simply obtained by substituting Ω_2 for Ω_1 and multiplying by a phase factor i.

The coefficients here obtained allow us to go from the representation in terms of exciton and photon states to the one in terms of polariton states, which are the diagonal states of the material-radiation system. The problem of the interaction of the semiconductor close to the gap with the radiation field has therefore been solved exactly, and not simply perturbatively as we tried to do in calculating the transition rates. In fact, the polariton state is

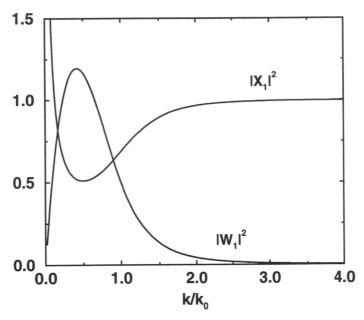

Figure 3. The W and X Hopfield coefficients for the lower branch polariton as functions of k.

a stationary state, and no transition between a photon and an exciton state is described in this picture. Rather, both the excitations are admixed in the true propagating state of the crystal. In an equivalent picture, we may start at time t=0 with a pure photon state, that is not stationary. It gradually evolves into an exciton state, and then backwards to a photon state and so on. This is a fast process, and a characteristic time given by the inverse of the polariton splitting at resonance can be estimated. It is clear from this picture of a periodic oscillation between a photon and an exciton state that the perturbative picture of adiabatic transition between these state is fundamentally incorrect. We remark that this polariton splitting also gives the range of non-negligible exciton-photon mixing in a given polariton state. It is given by $\delta\omega = \sqrt{2\omega_0\omega_{LT}}$, and amounts to several meV in typical small gap semiconductors.

The exciton-photon mixing is represented in Fig. 3, where $|W_1|^2$ and $|Y_1|^2$ are shown as functions of k. We can see that as the polariton starts as a dominantly photon-like mode (with photon-like dispersion too, as seen from Fig. 1), it then becomes exciton-like at larger k or energy. A region of non-neglible mixing of the order of $k_0 = \omega_0/c$ is found. Comparing with Fig. 1, we find that the polariton has non-negligible exciton content still where its dispersion is already photon-like. This fact has further important consequences in the diffusion dynamics of excitons, and in many other optical experiments close to the gap, such as Resonant Raman Scattering, as already observed by Hopfield in 1969.

5.2. POLARITONS IN QWS

In QW's as in the bulk system, polaritons give the correct description of the interaction between the exciton and the electromagnetic field. We have in fact already remarked that also in this case the perturbative approach in the response of the QW system to a macroscopic electric field is inappropriate.

Let us again introduce the polariton hamiltonian in second quantization. The different dimensionality of the QW excitons produces substantial differences with the bulk case. Only the translational quantum number $\mathbf{k}_\parallel$ is conserved in the exciton photon interaction, and therefore an exciton with a given $\mathbf{k}_\parallel$ interacts with a continuum of photons with all the possible k_z. Introducing again the second quantization picture as was done for the bulk, we have in this case the following exciton-photon hamiltonian:

$$\mathcal{H} = \sum_{\mathbf{k}_\parallel} \hbar\omega_{\mathbf{k}_\parallel} b^\dagger_{\mathbf{k}_\parallel} b_{\mathbf{k}_\parallel} + \sum_{\mathbf{k}} \hbar c|\mathbf{k}| a^\dagger_{\mathbf{k}} a_{\mathbf{k}} \tag{65}$$

$$+ i \sum_{\mathbf{k}} C_{\mathbf{k}} (a^\dagger_{\mathbf{k}} + a_{-\mathbf{k}})(b^\dagger_{\mathbf{k}_\parallel} - b_{\mathbf{k}_\parallel})$$

$$+ \sum_{\mathbf{k}_\parallel, k_z, k'_z} D_{\mathbf{k}_\parallel, k_z, k'_z} (a^\dagger_{\mathbf{k}_\parallel, k_z} + a_{-\mathbf{k}_\parallel, -k_z})(a^\dagger_{-\mathbf{k}_\parallel, -k'_z} + a_{\mathbf{k}_\parallel, k'_z}).$$

Here $\hbar\omega_{\mathbf{k}_\parallel}$ is the exciton energy. Again as in the bulk case, the two final terms in the hamiltonian represent the interaction (57) in second quantization for the QW system. The coefficients C and D are simply found from the interaction matrix elements:

$$C_{\mathbf{k}} = \sqrt{\frac{2\pi\hbar}{v|\mathbf{k}|L_z}} \omega_{\mathbf{k}_\parallel} \boldsymbol{\mu}_{cv} \cdot \boldsymbol{\epsilon} F_{QW}(0) \int dz\ \xi(z) e^{ik_z z}, \tag{66}$$

$$D_{\mathbf{k}_\parallel, k_z, k'_z} = \frac{C^*_{\mathbf{k}_\parallel, k_z} C_{\mathbf{k}_\parallel, k'_z}}{\hbar\omega_{\mathbf{k}_\parallel}} \tag{67}$$

where $\xi = c^*(z)v(z)$ has been defined before. We have introduced L_z, the quantization length along z.

It is clear that the eigenmodes of this hamiltonian can be labelled by $\mathbf{k}_\parallel$. For each $\mathbf{k}_\parallel$, we have a discrete exciton state in interaction with a continuum of photon states (the photons at different k_z). We therefore have an example of a Fano-Anderson resonance. However, the photon continuum has a minimum energy given by $\hbar c k_\parallel$. Therefore, those modes having an energy $\Omega < ck_{\parallel}$ remain discrete, while those modes with $\Omega > ck_{\parallel}$ become admixed with the continuum. The problem can again be diagonalized exactly in both regions. In an equivalent picture for the continuum of states region, the discrete level at time t=0 starts to decay into the continuum as soon as the interaction is turned on. The decay time turns out to be the inverse of the spread in energy of the mixing of the discrete state inside the continuum.

To make things clear we report the solution, which can be found in any textbook of many-body physics (see e.g. G. D. Mahan, [6]), without entering into the algebraic details of derivation.

We write the most general polariton operator in analogy to the bulk case, (61), taking into account the freedom in choice of the photon k_z:

$$\hat{\alpha}_{\mathbf{k}_{\parallel}} = \sum_{k_z} \left[W(k_z)\hat{a}_{\mathbf{k}_{\parallel}k_z} + Y(k_z)\hat{a}^{\dagger}_{-\mathbf{k}_{\parallel}-k_z} \right] + \left[X\hat{b}_{\mathbf{k}_{\parallel}} + Z\hat{b}^{\dagger}_{-\mathbf{k}_{\parallel}} \right]. \tag{68}$$

If we impose the diagonalization condition:

$$\left[\hat{H}, \hat{\alpha}_{\mathbf{k}_{\parallel}} \right] = -\hbar\Omega\hat{\alpha}_{\mathbf{k}_{\parallel}}, \tag{69}$$

we then obtain from (69) a system of equations for the coefficients $W(k_z)$, $Y(k_z)$, X and Z

$$\begin{cases} (ck - \Omega)W(k_z) = -C^*_{k_z}(X + Z) \\ (ck + \Omega)Y(k_z) = C^*_{k_z}(X + Z) \\ (\omega - \Omega)Z = \sum_{k_z} C_{k_z}(W(k_z) - Y(k_z)) \\ (\omega + \Omega)Z = \sum_{k_z} C_{k_z}(W(k_z) - Y(k_z)) \end{cases} \tag{70}$$

Here $k = |\mathbf{k}| = \sqrt{\mathbf{k}_{\parallel}^2 + k_z^2}$. For a given $\mathbf{k}_{\parallel}$ two different regions on the energy axis can be distinguished. In the first region, where $\Omega < vk_{\parallel}$, the energy of the polariton does not overlap the continuum of the photon levels since $vk_{\parallel} < v|\mathbf{k}|$. This implies that a solution exists only for one discrete energy value Ω which is found to satisfy the following condition:

$$\Omega^2 - \omega^2 = -4\frac{\omega\Omega^2\mu_{cv}^2 F(0)^2}{\hbar} \int_{-\infty}^{+\infty} dk_z \, dz \, dz' \, \frac{\xi(z)\xi(z')e^{ik_z(z-z')}}{v^2k^2 - \Omega^2}. \tag{71}$$

The integration in k_z is performed on an appropriate contour in the complex plane, and results into the following eigenvalue equation:

$$\Omega^2 - \omega^2 = -4\frac{\omega_T\Omega^2\mu_{cv}^2 F(0)^2}{2\hbar\kappa_z}\frac{2\pi}{v^2} \int dz dz' \xi(z)\xi(z') \exp(-\kappa_z|z - z'|), \tag{72}$$

and $\kappa_z^2 = k_{\parallel}^2 - \Omega^2/v^2$. This equation describes the dispersion of stationary modes which are therefore called stationary QW polaritons. The r.h.s. of the equation is related to the polaritonic energy shift. It turns out that the dispersion is similar to that of the lower branch of the bulk polariton, given in Fig. 1. However recall that the polariton wavefunction is much different, since at any given $\mathbf{k}_\parallel$, all photon modes with any k_z are contained in it, while in the bulk one mode only was.

In the second region, where $\Omega > v k_\parallel$, the polariton levels lie in the continuum of the photon energies, thus polariton modes exist for any frequency. In other words, the system of equations for the coefficients given by (70) has solutions for every value of the Ω. In fact notice that the equation for the coefficient $W(k_z)$ is singular, since on the energy shell, with $k_z = \pm \bar{k}_z = \sqrt{\Omega^2/c^2 - \mathbf{k}_\parallel^2}$, and therefore $\Omega - ck = 0$. This means that a degree of freedom in the choice of $W(\bar{k}_z)$ exists and therefore this function can be adapted in order to find a solution for any given Ω in this continuum region. Notice that in the stationary region instead none of the photon modes is on energy shell, so that all the photons are 'virtual'. In the continuum case the X coefficient (which gives the exciton content in the polariton) has the typical shape of a Lorentzian, centered at the polariton energy. We do not write the polariton energy shift as in the stationary case, because this shift turns out to be very small in real samples. Instead we consider the other parameter which characterizes the Lorentzian, i.e. its width. It is known that it is related to the radiative decay of the discrete level into the continuum of photons, and in fact it is the inverse of the radiative lifetime of the exciton. Its expression is given by:

$$\Gamma_T = \frac{2\pi \mu_{cv}^2 |F(0)|^2}{\epsilon \hbar} \frac{\Omega^2/c^2}{\sqrt{\Omega^2/c^2 - \mathbf{k}_\parallel^2}} \tag{73}$$

We remark here that this value may also be calculated using Fermi Golden Rule for spontaneous decay, i.e. in the case were the external perturbing field is given by the fluctuations of the vacuum. We are going to discuss into this point in the next section, where the emitted photon in the radiative decay is subject to additional confinement by a microcavity. Now, let us just notice the typical divergence as $\bar{k}_z^{-1}$, which reflects the coupling of the exciton to the one-dimensional continuum of photons (i.e. photons with different k_z). Then, $\bar{k}_z^{-1}$ is simply the one-dimensional density of states.

5.3. POLARITONS IN MICROCAVITY-EMBEDDED QWS

We want now to address the problem of exciton polaritons in systems where the electromagnetic field is also partially confined.

The idea of confining the radiation interacting with an excitation of the solid came from atomic physics. In the early eighties experiments were performed in which an atomic beam was directed in a very thin space between two metallic mirrors [7]. Performing spectroscopy measurements on these atoms showed a spectacular modification of the "natural" linewidth of an atomic level in both directions of inhibited and enhanced spontaneous emission, depending on the relative positions of the cavity mode energy and the atomic transition energy. Since spontaneous emission is due to the coupling between the atomic level and the vacuum field fluctuations, the results of these experiments were clearly attributable to the modification of the vacuum field fluctuations caused by the mirrors, an effect which was very well known since the works of Casimir and Purcell [7] in the late forties. Since then the field of cavity quantum electrodynamics (CQED) had a very rapid development. In fact in atom-cavity systems several elementary quantum mechanical effects can be observed. "Schroedinger cat" states can be created and studied. A very important peculiarity of these photon-atom mixed states is that the emission process becomes reversible when the cavity has a very high Q factor because the photon is emitted and reabsorbed several times before exiting the cavity. This regime is known as the "strong coupling" regime, for reasons which will become clear later, and the oscillation implied by this reversible radiation process is called Rabi oscillation. Furthermore, statistical properties of radiation are also modified in cavity systems. Sub-Poissonian fields and squeezed states have been produced by means of radiation confinement. Finally, some experiments can be performed to test the basis of the quantum theory of measurement.

Once the basis of CQED had been verified, the step leading to the application of such knowledge to semiconductor devices has been straightforward. The first experiment to test enhanced spontaneous emission of quantum wells by means of radiation confinement dates 1990 [8] and, since then, the field has received growing attention. The interest in modifying the radiation linewidth of the elementary excitations in solids came from the idea of designing light emitting devices with very high efficiency and semiconductor lasers with very low threshold. Several review works exist on this topic [9] and we refer to them for a detailed description of the state of the art in this field.

The confinement of the electromagnetic field in semiconductor structures is achieved using devices called semiconductor microcavities. In such devices multilayered dielectric structures called distributed Bragg reflectors (DBRs) are used as mirrors to confine radiation. We will describe these structures in some detail later in this section. In what follows we will give a step-by-step description of the physics of microcavity systems, starting from a simple perturbative approach to the QW exciton linewidth in microcavities up to a full polariton scheme. Our aim is twofold. On one side we want to remark some aspects of the coupling between excitations in solids and the electromagnetic field, which constitute an important difference with respect to the atomic counterpart. On the other side we will discriminate between the two regimes of weak and strong coupling, both attainable in semiconductor structures, showing the different physics involved.

Atoms couple to electromagnetic waves traveling in any direction. In other words there is no selection rule for the direction of the wavevector of the radiation coupled to the atom. Thus an atom couples to a three dimensional continuum of electromagnetic modes. In ideal cavities with perfect mirrors the radiation would be totally confined and a set of discrete modes instead of a continuum would exist. Actually, even in the best quality cavities there is some leakage and the modes are not truly discrete modes. For this reason an atom in a cavity is always coupled to a three dimensional continuum of modes. The only difference is that in this case the continuum in "structured", namely the density of states of the electromagnetic field is strongly peaked at energies corresponding to the modes of the cavity. We speak in this case of "quasimodes". In order to obtain quasimodes a cavity has to be designed in such a way that the electromagnetic field is confined along the three directions. If the field is not confined along one direction, the atom will "see" along that direction a flat density of states instead of a strongly peaked one.

The situation in semiconductors is quite different. Because of the in-plane translational simmetry, every excitation in QW's is characterized by an in-plane wavevector. Thus an exciton (or an electron-hole pair) with a given in-plane wavevector is coupled only to radiation with the same in-plane component of its wavevector. This selection rule is very important in microcavity systems because it allows to restrain the radiation confinement to the growth direction only. In fact, inside a planar cavity (a Fabry-Pérot structure) the radiation has full in-plane translational simmetry and each electromagnetic mode is characterized by an in-plane wavevector. The density of radiation modes is strongly peaked only in the growth direction. An excitation with a given in-plane wavevector is thus coupled with all the Fabry-Pérot one dimensional quasimodes with the same in-plane component of the wavevector. This is the exact one dimensional analogous of the coupling between atoms and three dimensional cavity quasimodes.

Actually, because of interface imperfections and thermal noise, any excitation in a QW has a finite coherence length along the plane, and consequently a finite indetermination of the in-plane wavevector. This indetermination in k-space, together with the fact that in usual experiments it is not possible to coherently create an excitation with a given wavevector, usually leads to the modelization of the QW excitation with a dipole, which has no k-space selection rule. This can be a good approximation in most of the cases, but is generally misleading. In what follows we will describe the

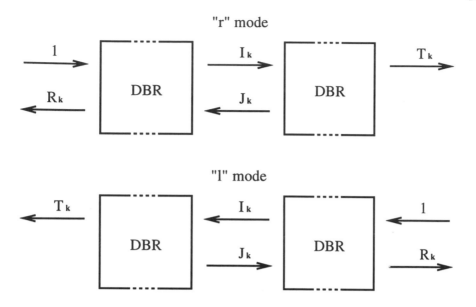

Figure 4. The FP left and right traveling modes. In this picture thick DBRs appear as cavity mirrors. The choice of the boundary conditions is the same for thin mirrors.

coupling between QW excitons and cavity modes in complete analogy to what has been done in the previous sections, thus using a realistic model for the exciton which is necessary in order to develop a polariton formalism.

Before going to the polariton formalism we find it more interesting to consider the exciton-radiation coupling in the framework of the Fermi golden rule. To do this we start by writing a second quantization form of the electric field inside a microcavity.

Let us consider a planar Fabry-Pérot (FP) of thickness L, whose mirrors have a real reflection coefficient r. We suppose that the refraction index inside the FP is n while it is unity outside. An electromagnetic mode with frequency ω and in-plane wavevector $\mathbf{k}_{\|}$ can be written as

$$\mathbf{E}_{\mathbf{k}_{\|},\omega}(\mathbf{x}) = \epsilon_{\mathbf{k}_{\|}} U_{\mathbf{k}_{\|},\omega}(z) e^{i\mathbf{k}_{\|}\cdot\boldsymbol{\rho}} \ . \tag{74}$$

The function $U_{\mathbf{k}_{\|},\omega}(z)$ can be calculated solving Maxwell' s equations with the appropriate boundary conditions. We choose two different boundary conditions of a traveling wave coming from $z = \pm\infty$ respectively, as shown in Fig. 4. The corresponding modes are indicated with $U_{r,\mathbf{k}_{\|},\omega}(z)$

and $U_{l,\mathbf{k}_\parallel,\omega}(z)$ respectively. It is straightforward to show that

$$U_{r,\mathbf{k}_\parallel,\omega}(z) = \begin{cases} e^{ik_z z} + R_{\mathbf{k}_\parallel,\omega}e^{-ik_z z} & -\infty < z < -L/2 \\ I_{\mathbf{k}_\parallel,\omega}e^{ik'_z z} + J_{\mathbf{k}_\parallel,\omega}e^{-ik'_z z} & -L/2 < z < L/2 \\ T_{\mathbf{k}_\parallel,\omega}e^{ik_z z} & L/2 < z < +\infty \end{cases} \qquad (75)$$

where $k_z = \sqrt{\omega^2/c^2 - k_\parallel^2}$, $k'_z = \sqrt{\omega^2 n^2/c^2 - k_\parallel^2}$ and

$$I_{\mathbf{k}_\parallel,\omega} = \frac{t}{D_{\mathbf{k}_\parallel,\omega}}e^{i(k'_z - k_z)\frac{L}{2}}$$

$$J_{\mathbf{k}_\parallel,\omega} = \frac{tr}{D_{\mathbf{k}_\parallel,\omega}}e^{ik'_z L}e^{i(k'_z - k_z)\frac{L}{2}}$$

$$T_{\mathbf{k}_\parallel,\omega} = \frac{t^2}{D_{\mathbf{k}_\parallel,\omega}}e^{i(k'_z - k_z)L}$$

$$R_{\mathbf{k}_\parallel,\omega} = \frac{re^{-ik'_z L} - (1/r)e^{ik'_z L}e^{i(k'_z - k_z)L}}{D_{\mathbf{k}_\parallel,\omega}}$$

$$D_{\mathbf{k}_\parallel,\omega} = 1 - r^2 e^{2ik'_z L} \ .$$

Here we used $t^2 = 1 - r^2$. The expression for the mode $U_{l,\mathbf{k}_\parallel,\omega}$ is exactly the same as (75), with $z \to -z$ everywhere. The origin of the z axis lies at the center of the FP. It can be shown that the modes so obtained obey the appropriate orthonormality relations. Thus we can write the vector potential in second quantization form as

$$\mathbf{A}(\mathbf{x}) = \sum_{j=r,l} \sum_{\mathbf{k}_\parallel} \int_0^{+\infty} dk_z \sqrt{\frac{\hbar c}{Sn(k_\parallel^2 + k_z^2)^{\frac{1}{2}}}} \boldsymbol{\epsilon}_{\mathbf{k}_\parallel} (a_{j,\mathbf{k}_\parallel,\omega}U_{j,\mathbf{k}_\parallel,\omega}(z)e^{i\mathbf{k}_\parallel\cdot\boldsymbol{\rho}} + h.c.) \ .$$

$$(76)$$

Consequently the electric field reads

$$\mathbf{E}(\mathbf{x}) = \sum_{j=r,l} \sum_{\mathbf{k}_\parallel} \int_0^{+\infty} dk_z i\sqrt{\frac{\hbar\omega}{Sn^2}} \boldsymbol{\epsilon}_{\mathbf{k}_\parallel} (a_{j,\mathbf{k}_\parallel,\omega}U_{j,\mathbf{k}_\parallel,\omega}(z)e^{i\mathbf{k}_\parallel\cdot\boldsymbol{\rho}} - h.c.) \ . \quad (77)$$

Here S is a normalization surface which appears because of our choice of discrete quantization of the modes along the in-plane direction ($1/\sqrt{S}$ is the normalization factor of the functions $e^{\pm i\mathbf{k}_\parallel\cdot\boldsymbol{\rho}}$ along the plane). It is important to remark that the corresponding expression for the free electric field (without the FP) is the same with $e^{\pm ik_z z}$ replacing $U_{r,l}(z)$.

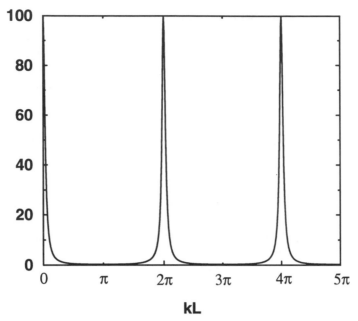

Figure 5. The square modulus of the electric field amplitude at the center of a FP as a function of frequency.

In order to visualize the density of states peaked around the FP modes, let's consider the electric field at $\mathbf{k}_{\parallel} = 0$ and $\mathbf{x} = 0$.

$$
\begin{aligned}
|\mathbf{E}(0)|^2 &= |I_\omega + J_\omega|^2 \\
&= \frac{1 - r^2}{|1 - re^{ik_z L}|^2} \\
&\simeq \sum_N \frac{1 - r^2}{r} \frac{c^2}{n^2 L^2} \frac{1}{\gamma_c^2 + (\omega - \omega_N)^2}
\end{aligned}
\tag{78}
$$

where

$$
\gamma_c = \frac{1 - r}{\sqrt{r}} \frac{nL}{c}
\tag{79}
$$

and $\omega_N = \frac{2\pi Nc}{nL}$. The last approximation leads to as a sum of lorentian peaks and is valid when $r \simeq 1$. As we can see from Fig. 5, the electric field amplitude at the center of the cavity strongly depends on the frequency. Intuitively, placing inside the cavity an exciton with energy $\hbar\omega_0$, if $\omega_0 = \omega_N$, the electric field amplitude of the vacuum field fluctuations at the exciton site will be a factor Q/π higher than $E_0 = \sqrt{(\hbar\omega_0)/(n^2 S)}$, which is defined as the amplitude of the fluctuations of the free electromagnetic field. Here Q is the cavity quality factor, defined as $Q = \omega_1/\gamma_c$ Consequently the spontaneous emission will be enhanced. If ω_0 lies in a region far from the cavity resonances, the electric field will be inhibited of the same amount with respect to E_0 and the spontaneous emission probability will be lower. The enhancement-inhibition of the spontaneous emission is proved more rigourously using Fermi golden rule. The emission rate is given by (37) where in this case H_{int} can be taken directly as the interaction hamiltonian (36) where we replace the free vector potential by (76). It can be verified that the exciton matrix element is given, in analogy to (48), as

$$\langle \Phi_{exc}(\mathbf{k}_{||})|\mathcal{H}_{int}|0\rangle = \tag{80}$$

$$E_0\sqrt{S}F_{\mathbf{k}_{||}}(0)\boldsymbol{\mu}_{cv}\cdot\epsilon\delta_{\mathbf{k}_{||},\mathbf{k}_{||ph}}\int dz c^*_{n_c}(z)v_{n_v}(z)e^{ik_z z}|I(\omega,\mathbf{k}_{||})+J(\omega,\mathbf{k}_{||})|^2$$

It is now clear how the emission probability is given by that of the free QW times the FP factor at the exciton energy. If the exciton is tuned to one of the FP modes the emission will be enhanced and it will be inhibited in the other case.

Actually, it is not possible to create an experimental situation where only excitons at a given in-plane wavevector are coherently excited. The reasons are many. First, every coherent excitation mechanism will always have a finite spread in the wavevector. Second, the interface roughness, the impurities and the coupling with phonons reduce the coherence radius of the exciton in a QW, thus increasing the spread in k-vector of the exciton state. For these reasons the most common situation is the one of a distribution of excitons with all the $\mathbf{k}_{||}$ values. The $\mathbf{k}_{||}$ dependence of the coefficients $I(\omega,\mathbf{k}_{||})$ and $J(\omega,\mathbf{k}_{||})$ is much more pronounced than that of the exciton energy. Thus if we have an exciton population at all different k-vectors, and if the FP has been designed in order to be resonant with the exciton at $\mathbf{k}_{||}=0$ for example, only a small fraction of excitons with $\mathbf{k}_{||}$ values close to zero will stay resonant with the cavity peak, all the other excitons being out of resonance. The overall spontaneous emission will not be enhanced by a factor of the order of Q but only by a "geometrical" factor which can be calculated integrating the spontaneous emission rate over the solid angle [10, 11] and has a theoretical upper limit equal to three for planar cavities. The final result is that the overall emission is enhanced by only a small factor even in very high Q planar cavities but the emission is strongly concentrated around the $\mathbf{k}_{||}$ value for which the exciton is resonant with a FP mode. In other words the spontaneous emission of these systems can be strongly collimated along any chosen direction (typically the growth direction). This fact led naturally to the idea of thresholdless lasers, namely devices in which the spontaneous emission too is directed into the laser mode. Devices of this kind have been realized [12] and are very promising for applications.

Up to now we have used the Fermi golden rule in order to obtain the spontaneous emission probability of an exciton in a model semiconductor microcavity. Actually Fermi golden rule has a well estabilished limit of validity. This limit can be explained in the following way. The coupling defines a characteristic time τ for the interaction process, which corresponds to the inverse of the transition probability found before. On the other hand there is a characteristic time τ_c of the continuum of photons, which is given by the inverse of the density of states linewidth. The time τ_c is the time necessary for a cavity photon (a photon in a quasimode of the cavity) to escape from the cavity. In order to have the decaying behaviour predicted by the Fermi golden rule, τ has to be much larger than τ_c. If it is not the case, the cavity photon will have time to be reabsorbed by the exciton before being emitted from the cavity. In this case a damped oscillation will take place instead of an exponential decay. This oscillations are the so called vacuum field Rabi oscillations. It has been showed [13, 14] that the strong coupling regime, the one in which Rabi oscillations take place, can be easily set up in semiconductor microcavities. We thus need a more complete approach to describe the exciton-photon coupling.

We introduce here the polariton model of QW excitons in microcavities in complete analogy to the model used for QW excitons in empty space. We will diagonalize the polariton hamiltonian using the Fano approach [15] and will obtain the cavity polariton dispersion relation. The polariton hamiltonian is given by

$$H = \sum_{\mathbf{k}_{\parallel}} \hbar\omega_{\mathbf{k}_{\parallel}} b^{\dagger}_{\mathbf{k}_{\parallel}} b_{\mathbf{k}_{\parallel}} + \sum_{j=r,l} \sum_{\mathbf{k}_{\parallel}} \int dk_z \hbar v |\mathbf{k}| a^{\dagger}_{j,\mathbf{k}} a_{j,\mathbf{k}} \qquad (81)$$

$$+i \sum_{j=r,l} \sum_{\mathbf{k}_{\parallel}} \int dk_z C_{j,\mathbf{k}} (a^{\dagger}_{j,\mathbf{k}_{\parallel},k_z} + a_{j,-\mathbf{k}_{\parallel},k_z})(b^{\dagger}_{\mathbf{k}_{\parallel}} - b_{\mathbf{k}_{\parallel}})$$

where $\mathbf{k} = \mathbf{k}_{\parallel} + k_z \mathbf{z}$, $v = c/n$ and

$$C_{j,\mathbf{k}} = \sqrt{\frac{\hbar}{v|\mathbf{k}|}} \omega_{\mathbf{k}_{\parallel}} (\boldsymbol{\epsilon}_{\mathbf{k}_{\parallel}} \cdot \boldsymbol{\mu}_{cv}) F_{QW}(0) \int dz \, \xi(z) U_{j,\mathbf{k}_{\parallel},k_z}(z) \ . \qquad (82)$$

We neglect the $\mathbf{A}^2$ term for reasons of simplicity. It can be shown [16] that the $\mathbf{A}^2$ term is important only far away from the polariton resonance, a situation which will not be interesting in describing cavity polaritons. All the quantities in (82) have been defined before. We emphasize the full analogy of (81) and (82) with (65) and (66) for a free QW. The only difference is in the overlap integral between exciton confinement function and the electromagnetic field which now is expressed in terms of cavity modes. The difference in the normalization constant in (82) with respect to (66) is due to the choice of quantizing the electromagnetic field in the z direction into a continuum of modes instead of a discrete set as in the other two directions. The way to proceed is now the same as that explained in the last section. In this case we are more interested in the region where $\Omega > v k_\parallel$ which is the radiative region. The dispersion relation obtained is the following:

$$
\sum_{j=r,l} \left[\int dk_z \frac{2\hbar v |\mathbf{k}| |C_{j,\mathbf{k}_\parallel,k_z}|^2}{\hbar^2 v^2 k^2 - E^2} \right.
$$

$$
\left. + \quad Z(E) \int dk_z |C_{j,\mathbf{k}_\parallel,k_z}|^2 \delta(\hbar^2 v^2 k^2 - E^2) \right] = \frac{\hbar^2 \omega_{\mathbf{k}_\parallel}^2 - E^2}{2\hbar\omega_{\mathbf{k}_\parallel}} \quad , \quad (83)
$$

where $E = \hbar\Omega$. In the last expression $Z(E)$ is an arbitrary function which comes from the theory of distributions used to diagonalize the hamiltonian. For all the details we refer to the book of Mahan [6]. The dispersion (83), because of the presence of $Z(E)$, has solutions for every value of E. This, in terms of the Fano model, means that in the radiative region mixed exciton-radiation states occur at every value of E. How can we obtain from this model a transition energy for the spontaneous emission and its related emission rate? The answer is rather tricky. The polariton states obtained are stationary states of the system. Actually, when we perform an experiment we never excite a single polariton state, but rather a continuous superposition of these states. The overall time behaviour of the projection of this state over the exciton state will be approximately given by one or more "resonances", namely by a superposition of damped oscillating functions. The frequency and damping rates of these resonances will correspond to the complex dispersion of our polariton system. In other words the Fano model describes the scattering of the photons over the exciton and shows that the scattering process is resonant at one or more given energies with corresponding resonance linewidths. The physical interpretation of the Fano model requires some deeper considerations than those given in these notes. We refer to the standard literature [6] for a more detailed treatment. Here we just state that the resonances are given as the solutions in the complex plane of the equation

$$
E - \hbar\omega_{\mathbf{k}_\parallel} + i\gamma_{exc} - \Sigma_{\mathbf{k}_\parallel}^{(ret)}(E) = 0 \tag{84}
$$

where

$$
\Sigma_{\mathbf{k}_\parallel}^{(ret)}(E) = \lim_{\eta \to 0} \sum_{j=r,l} \int dk_z \frac{2\hbar v |\mathbf{k}| |C_{j,\mathbf{k}_\parallel,k_z}|^2}{\hbar^2 v^2 k^2 - (E - i\eta)^2} \quad . \tag{85}
$$

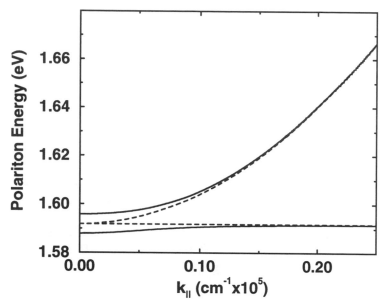

Figure 6. Polariton dispersion for a QW embedded in a planar microcavity as a function of the in-plane wavevector $k_{\parallel}$. The parameters are given in the text. The dashed lines correspond to the energies of the unperturbed exciton and cavity modes.

In the last expression we have introduced a nonradiative exciton broadening γ_{exc}. The dispersion (84) completely describes the polariton resonances in a microcavity embedded QW.

The solutions of (84) can be obtained numerically in the general case. Let's consider for example an exciton with energy $\hbar\omega_0 = 1.592eV$ and a cavity with a refraction index equal to unity and thickness λ, the wavelength corresponding to the exciton energy. The mirrors are chosen to be equal and with a reflectivity of $R = r^2 = 0.99$. The exciton oscillator strength is that typical of a GaAs/AlGaAs 75 Å QW and we choose $\gamma_{exc} = 1\ meV$.

The energy and radiative linewidth of the polariton dispersion are plotted in Fig. 6 and 7 as a function of $k_{\parallel}$. The dashed lines represent the uncoupled values relative to the exciton and cavity modes. It is very clear that the polariton coupling is responsible for a radiative shift of the levels with respect to the uncoupled ones. The radiative linewidths are also modified. When the cavity and exciton uncoupled modes are far apart in energy, the corresponding coupled levels approach the uncoupled ones. The linewidths too approach the uncoupled values as we move far from the resonant region (here resonant means that the two uncoupled levels are close in energy). If we could look at the quantum polariton states we would realize that the mixing itself is very small in the non resonant region and large in

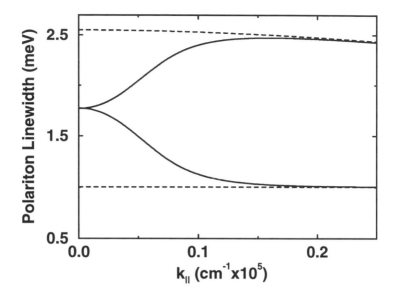

Figure 7. The linewidth corresponding to the dispersion of Fig. 6. Here too the dashed lines correspond to the unperturbed linewidths of the exciton and cavity modes.

the resonant one. We see that at $k_{||} = 0$ the radiative shift is maximum. The splitting of the two polariton modes at $k_{||} = 0$ is referred to as vacuum field Rabi splitting.

The integral in (85) can be performed in the complex plane. A very interesting result is obtained if we consider the polariton dispersion at $k_{||} = 0$ at the first order in $(\Omega - \omega_c)/\omega_c$, where ω_c is the frequency of the FP mode. In this case the formula for the dispersion becomes very simple

$$(\Omega - \omega_0)(\Omega - \omega_c + i\gamma_c) = V^2 \tag{86}$$

where γ_c is the width of the FP mode defined before in this section and

$$V = \sqrt{\frac{1+r}{r} \frac{c\Gamma_0}{n_{cav}L}} \ . \tag{87}$$

Here $\Gamma_0 = \frac{\pi}{n} \frac{\omega_0}{c} \frac{|F(0)|^2 |\mu_{cv}|^2}{\hbar}$ would be the radiative linewidth of the exciton in a free QW.

Eq. (86) can be seen as the secular equation for the coupling between two harmonic oscillators, one of which, the one corresponding to the cavity mode, is damped. Thus the physical interpretation of cavity polaritons is now clear. In the strong coupling regime we can describe the system as

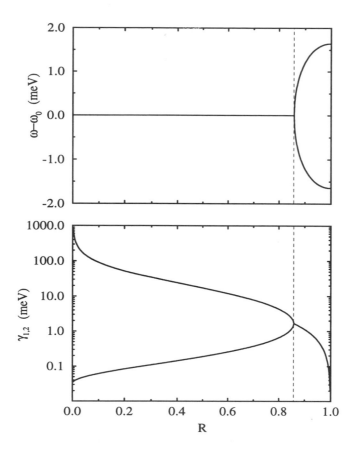

Figure 8. The solutions of eq (88) as a function of the mirrors reflectivity. The parameters are the same as in Fig. 6.

two coupled quasistationary states. We refer to some literature [17] for a detailed analysis of the implications of (86). Here we just show how, varying the reflectivity of the mirrors and consequently the width of the cavity mode, we pass from the regime of weak coupling to the one of strong coupling.

The solutions of (86) are plotted in Fig. 8 as a function of the mirrors reflectivity. The comments on these curves are the same as those for the previous dispersions. There is a critical reflectivity which marks the transition between the two regimes and is approximately given by

$$1 - R_c = 4\sqrt{\frac{2nL\Gamma_0}{c}} \quad . \tag{88}$$

Below R_c the exciton energy is not shifted and the system could be described approximately using Fermi golden rule. Above R_c The Rabi (po-

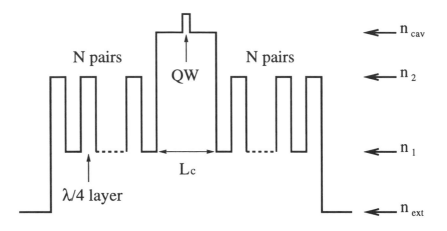

Figure 9.   A typical microcavity structure with DBRs.

lariton) splitting appears and the polariton formalism is needed.

What has been exposed up to now is an introduction to the physical properties of microcavity embedded QWs which covers the main aspects of this kind of systems. Actually there is one feature of the microcavitiy systems which still has to be addressed, namely the properties of the cavity mirrors. As already anticipated, the dielectric mirrors used in real microcavities are called DBRs and are made of several identical pairs of two $\lambda/4$ dielectric layers, where λ is the wavelength corresponding to the first FP resonance. We show the dielectric profile of a typical microcavity structure in Fig. 9. For a large number N of pairs, a DBR has a frequency region called "stop band" in which the reflectivity is close to unity, whose center lies at the frequency corresponding to the wavelength λ. FPs built with such mirrors can attain quality factors of a few thousands. We plot in Fig. 10 the modulus and phase of the reflection coefficient at normal incidence corresponding to the structure shown in Fig. 9. We see that the phase of the reflection coefficient is zero at resonance and varies strongly in the stop band region. Close to the cavity resonance, an analytical expression for this reflection coefficient holds:

$$|r|^2 = 1 - 4\frac{n_{ext}}{n}\left(\frac{n_1}{n_2}\right)^{2N} \tag{89}$$

and

$$\phi_r = \frac{nL_{DBR}}{c}(\omega - \omega_c),$$

$$L_{DBR} = \frac{\lambda_c}{2}\frac{n_1 n_2}{n(n_2 - n_1)} \tag{90}$$

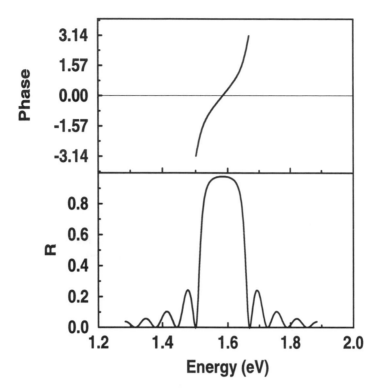

Figure 10. The phase (upper plot) and modulus (lower plot) of the reflectivity of a typical DBR as a function of frequency.

Here L_{DBR} represents a mirror penetration depth, and ϕ_r is the phase of the reflection coefficient, while n_1, n_2 and n_{ext} are the refraction indexes of the two layers and of the exterior of the FP respectively. The condition for the resonance of a DBR microcavity corresponds, as usual, in having zero phase variation of a plane wave traveling a complete round trip. The phase variation can be expressed, summing the phase variations due to the optical path and to the mirrors, as

$$
\begin{aligned}
\Delta\phi &= \frac{2\pi}{\lambda}L + \frac{nL_{DBR}}{c}(\omega - \omega_c) \\
&= (\frac{2\pi}{\lambda} - \frac{2\pi}{L})L + (\frac{2\pi}{\lambda} - \frac{2\pi}{L})L_{DBR} \\
&= (k_z - k_c)(L + L_{DBR}) \; .
\end{aligned}
\tag{91}
$$

The phase variation (91) is the same of a FP with ordinary mirrors and thickness given by $L_{eff} = L + L_{DBR}$. In ordinary structures like that in Fig. 9, L_{DBR} is much longer than L, as can be verified by using (90), and thus this effect is predominant. It can be shown [17] that all the results of this section are valid for DBR structures, provided one replaces everywhere in the formulas L by L_{eff}.

References

1. N.W. Ashcroft and N.D. Mermin, *Solid State Physics*, (Holt, Rinehart and Winston, New York 1976).
2. F.Bassani and G. Pastori Parravicini, *Electronic States and Optical Transitions in Solids*, (Pergamon Press, Oxford 1975).
3. G. Bastard, *Wave mechanics Applied to Semiconductor Heterostructures*, (Les Editions de Physique, Paris 1989).
4. R. Kubo, J. Phys. Soc. Jpn. **12** (1957), 570.
5. J. J. Hopfield, Phys. Rev. **112** (1958), 1555.
6. G. D. Mahan, *Many-Particle Physics*, Plenum Press 1981.
7. See for example S. Haroche, *Cavity Quantum Electrodynamics*, in *Fundamental Systems in Quantum Optics*, Les Houches, Session LIII, 1990, Course 13, J. Dalibard, J. M. Raimond and J. Zinn-Justin, eds., Elsevier 1992, and references therein.
8. H. Yokoyama, K. Nishi, T. Anan, H. Yamada, S. D. Bronson and E. P. Ippen, Appl. Phys. Lett. **57** (1990), 2814.
9. See for example Y. Yamamoto, S. Machida, K. Igeta and G. Bjork, in *Coherence, Amplification, and Quantum Effects in Semiconductor Lasers*, edited by Y. Yamamoto, John Wiley & Sons, 1991.
10. G. Bjork, S. Machida, Y. Yamamoto and K. Igeta, Phys. Rev. A **44** (1991), 669.
11. G. Bjork, H. Heitmann and Y. Yamamoto, Phys. Rev. A **47** (1993), 4451.
12. Y. Yamamoto and R. E. Slusher, Physics Today, June 1993, 66.
13. C. Weisbuch et al., Phys. Rev. Lett. **69**, 3314 (1992).
14. R. Houdré et al., J. de Phys. IV, Coll. C5, **3**, 51 (1993).
15. U. Fano, Phys. Rev. **124** (1961), 1866.
16. V. Savona, Z. Hradil, P. Schwendimann and A. Quattropani, Phys. Rev. B **49** (1994), 8774.
17. V. Savona, L. C. Andreani, P. Schwendimann and A Quattropani, Solid State Comm. **93** (1995), 733.

SEMICONDUCTOR QUANTUM WELLS

C. DELALANDE

Laboratoire de Physique de la Matière Condensée
de l'Ecole Normale Supérieure
24, rue Lhomond, 75005 Paris, France.

1. Introduction

In this contribution, we review some basic optical properties of semiconductor quantum wells. Emphasize is brought on the experimental point of view, since another chapter of the present volume [1] adresses the theoretical conterpart. This paper is concerned mostly by the usual absorption and photoluminescence properties of quantum wells (QW) embedded in a sample exhibiting no optical resonance. Nevertheless, an attempt will be made in each section to present some recent results concerning QW-based optical microcavities, which make up a physical system where both the electronic part and the optical part of the system are confined.

After a review of the optical consequences of the electronic confinement in various kinds of QW's and superlattices, a particular interest will be paid to the excitonic properties, which are known to dominate the absorption spectrum of QW's. In a third part, the optical properties of a QW under electric or magnetic field will be recalled, and the high excitation regime behaviour will be studied. Finally, after a brief overview of the relaxation mechanisms, a description of the photoluminescence (PL) properties concerning in particular the involved energy levels and the recombination time will be done.

A first and preliminary remark must be made: all the results summarized in the following pages would not have been possible without the impressive progresses of epitaxial techniques such as Molecular Beam Epitaxy (MBE) or Metal Organic Vapor

M. Ducloy and D. Bloch (eds.), Quantum Optics of Confined Systems, 181-199.
© 1996 *Kluwer Academic Publishers. Printed in the Netherlands.*

Phase Epitaxy (MOVPE), which allow now the realization of heterointerfaces within a regularity of one monolayer, provided that the lattice parameter difference between the succesive two semiconductor layers is not too large.

2. Confinement effects and optical absorption in quantum wells

2.1. CONDUCTION AND VALENCE BAND STATES

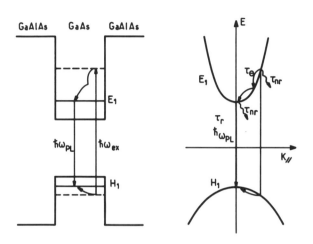

Figure 1. Left part: conduction- and valence- band edges in the growth direction. Right part: In-plane dispersion relations showing the principles of absorption, relaxation and photoluminescence in a QW.

Figure 1 shows the valence- and conduction- band edges of a semiconductor QW, i.e. a thin layer of semiconductor A (say GaAs) embedded between two thick layers of a semiconductor B (say GaAlAs). In a simple scheme where the conduction band is supposed parabolic with the same effective mass m^* in all the layers, the use of the effective mass approximation [1] leads to QW confined states E_n in the z growth direction. The in-plane electron motion remains free, characterized by an in-plane wave vector $\mathbf{k}$, with the consequence of a characteristic stair-case-like density of states per sublevel $m^* S/\pi \hbar^2$ (S is the surface of the QW). The same effect occurs in the valence band, whose top center-zone point is degenerate in the bulk material, as a consequence of symmetry considerations in zinc-blende semiconductors. But the confinement effect lifts this degeneracy, and a succession of distinct H_m heavy-hole and L_p light-hole bidimensionnal subbands appear. The problem is complicated at not-vanishing in-plane wave vector by the coupling between all these subbands [2].

2.2. OPTICAL ABSORPTION

The electromagnetic modes in free space of volume V presents a density of a quasi-continuum of states, the conduction-band and valence band as well. The Fermi Golden Rule is valid for calculating the interband absorption which promotes an electron of the valence band into the conduction band, which is almost empty in usual temperature and excitation conditions. The result near the band-edge [2] is that, apart an optical matrix element which acts on the atomic (or periodic) parts of the electronic wavefunctions and gives in particular the polarization effects, two selection rules occur. The first one is related to the overlap of the conduction and valence envelop functions in the z direction: in particular, in symmetric QW's, the transitions between the H_m and E_n sublevels are possible only if (n-m) is even. The second one results from the translational invariance in the layer plane of the QW: the difference between the conduction- and valence- in plane wave vectors must be equal to the in plane wave vector q of the absorbed photon. As the latter is very small at the scale of the first Brillioin zone, it is often claimed that the optical transitions are vertical in the reciprocal lattice. In contrast with the bulk case, only the in-plane motion is concerned in the case of QW's. Litterature abounds in experimental results showing the blue shift of the absorption edge resulting from the confinement effect [2]. Note that the order of magnitude of the absorption probability per QW for a light propagating along the growth axis is about 1%.

2.3. PHOTOLUMINESCENCE. EXCITATION SPECTROSCOPY

This low absorption probability makes necessary the use of a multi-QW sample to get reliable measurements of this quantity. Moreover, the substrate is often opaque near the QW absorption band edge. Altougth a selective etching of the substrate is often possible -but difficult-, Photoluminescence Excitation Spectroscopy (PLE), which consists in measuring the PL intensity at a given photon energy as a function of the exciting wavelength, is usually worked out (figure 2). It gives in a single QW the same

184

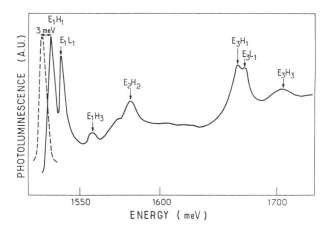

Fifgure 2: 2K photoluminescence (dashed line) and photoluminescence excitation spectra of a GaAs/GaAlAs quantum well.

informations on absorption peaks as transmission spectroscopy. This is made possible by the high efficiency of the relaxation mechanisms towards the luminescent bottom of the bands. This last point will be discussed in section 4.

2.4. EXTENSIONS

The GaAs/Ga$_{1-x}$Al$_x$As system, with x<0.4, and for instance the lattice-matched 1.55μm emiting InP/In$_{0.53}$Ga$_{0.47}$As or blue-lasing ZnSe/ZnCd$_x$Se$_{1-x}$ sytems (x about 15%) are so called type I QW's. Both constituting semiconductors provide a direct gap (maximum of the valence band and minimum of the conduction band at the center-of-zone Γ point). The minimum of the conduction band and the maximum of the valence band are spatially in the same layer, i.e. the QW (GaAs, InGaAs and ZnCdSe respectively).

It does exist type II QW's, where the minimum of the conduction band is not in the same layer than the maximum of the valence band. InP/In$_{0.52}$Al$_{0.48}$As is an example (figure 3) [3]. Obviously the absorption is lower than in a type I QW as the overlap of the envelop wavefunctions in the z direction is smaller.

The name superlattice is given to a regular series of alternate semiconductors when the thickness of the barrier layers is thin enough to allow a coupling between the adjacent wells. A miniband appears in the growth direction, with a larger energy width when the thickness or the height of the barrier is smaller. The absorption properties of such a system have been studied in particular in [4].

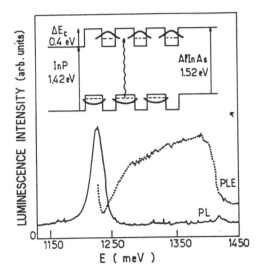

Figure 3: 2K photoluminescence (full line) and photoluminescence excitation spectra of a InP/AlInAs multiquantum well (from [3]).

3.   Excitons

3.1. EXCITONIC STATES

The absorption spectra of QW's are in fact dominated by excitonic effects even at room temperature. A well resolved and sharp (a few meV linewidth) peak appears near the absorption edge which corresponds to the 1S bound state of the exciton [1]. Knowing the effective Rydberg energy R^* of the bulk exciton (in the 5 to 10 meV range) and its effective Bohr radius a^* (in the 100Å range), the QW's exciton can be easily understood in the two QW's thickness limits.

If the thickness L of the QW is large with respect to the 3D Bohr radius a^*, the binding energy of the exciton (which is associated to the relative motion of the conduction electron and of the hole) is essentially not modified, but the z motion of the exciton center of mass becomes confined by the barriers. This gives rise to a series of absorption peaks well resolved in particular in wide CdTe QW's [5].

If on the contrary $a^*>>L$, the z motions of the conduction electron and of the hole are essentially governed by the confinement effects [2]. The solution of the in-plane coulombic motion shows an increase of the binding energy of the exciton. In ideal perfect 2D QW's, the latter is $4R^*$ and the energy levels of the nS states are $R^*/(n-1/2)^2$. In quasi 2D QW's, whith a finite L, the actual value of the binding energy R^*_{2D} is somewhat smaller.

3.2. EXCITONIC OPTICAL ABSORPTION

The new selection rules for excitonic absorption claim first that only the nS levels are optically active, with an oscillator strength proportional to $[\pi a^*(n-1/2)^3]^{-1}$ in the pure 2D case. Secondly the in-plane translational invariance of the system implies that the in-plane wave vector $\mathbf{K}$ of the exciton center of mass must be equal to the (small at the scale of the first Brillouin zone) in-plane part $\mathbf{q}$ of the absorbed photon wave vector. The lack of translational invariance in the z direction prevents any wave vector selection rule in this direction. Light interacts with a quasi-continuum of excitonic states and Fermi Golden Rule is still valid. Note that the electron-hole correlation effect also enlarges the absorption probability in the continuum of unbound excitonic states by the so-called Sommerfeld factor (a factor of 2 near the band edge in the pure 2D case).

3.4. EXCITONIC ABSORPTION IN A MICROCAVITY

A semiconductor microcavity contains one or several QW's located at the electric field antinode positions of a Fabry-Pérot-like structure. The high-reflectivity mirror between the substrate and the QW's consists usually in a so-called Bragg mirror, i.e. a stack of $\lambda/4$ layers with alternate high or low indexes, for instance GaAs (optical index n=2.96) and AlAs (n=3.54), grown monolitically with the QW. This Bragg mirror constitutes a well-known example of a one dimensionnal photonic band-gap system [6].The upper mirror is either a metallic layer or, for a very high reflectivity or a use throughout this upper mirror, a second Bragg mirror. The length of the microcavity in the z direction is a multiple of $\lambda/2$.

This microcavity selects electromagnetic modes in the z direction. If the microcavity finesse is high enough and if the excitonic transition is in resonance with a selected

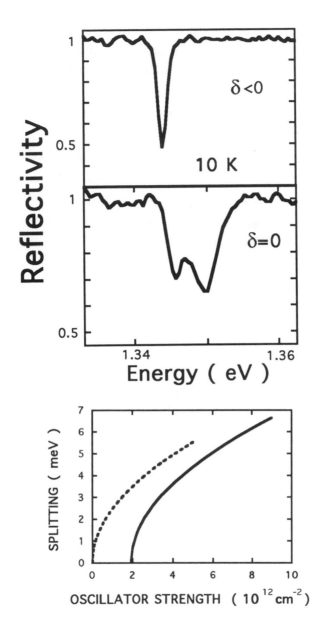

Figure 4: Upper part: reflectivity spectra of a λ GaAs microcavity containing three InGaAs QW's when the cavity mode is under the 1S excitonic mode (δ<0) and when the two modes are in resonance (δ=0). Lower part: calculated Rabi Splitting (see text). The dashed curve corresponds to a narrow excitonic line and the full one to a width of 5meV.

electromagnetic mode, the optical absorption results from the interaction of a discrete exciton state with a discrete mode of the electromagnetic field. It is known [7] that the so-called Rabi Splitting appears in the absorption spectrum. This is the strong coupling regime The first observation was made in QW's by C. Weisbuch et al. [8].

If the microcavity finesse is lower or the excitonic peak linewidth larger, this splitting does disappear. The microcavity effect is to alter the angular and frequency dependence of QW optical absorption (or emission) . This is the weak coupling regime [9]. Vertical cavity surface emitting lasers correspond to the latter case [10]. An example of a splitting measured in a reflectivity experiment is given in figure 4.

The calculation of this splitting can be performed in a pure quantum mechanical way both for the electromagnetic field and the excitonic state [1]. The same results can be found in the weak probe regime in a semiclassical scheme [11,12]: the cavity electromagnetic field is calculated by standard transfer matrix calculations and the excitonic absorption is introduced phenomenologically by a Lorentzian frequency-dependent permittivity around the E_{1S} excitonic resonance:

$$\varepsilon(\hbar\omega) = n(\hbar\omega)^2 = \varepsilon_\infty + 4\pi c \hbar^3 \, \alpha_{sf} \, f_{osc} / \, m \, L \, (E_{1S} - (\hbar\omega) - i \, \gamma \, \hbar\omega)$$

where α_{sf} if the fine structure constant, f_{osc} the 1S exciton oscillator strength per unit surface, m the electron mass, L the QW thickness. γ is the exciton linewidth. Figure 4 shows the calculated Rabi splitting Ω_{1S} as a function of the oscillator strength considered as a fit parameter. Two values of the exciton linewidth are taken: the upper one corresponds to 0.5 meV and the lower one to 5meV, the actual value found experimentally in the three $In_{0.13}Ga_{0.87}As$ QW"s embedded in a λ cavity corresponding to Figure 4. The cavity mode broadening γ_{cav} is the measured one, 1meV FWHM. It can be observed that the Rabi splitting depends on the linewidths and that a threshold exists: Ω_{1S}(without broadening) must be larger than $(\gamma_{cav} + \gamma) / 2$ [12]. Note that the Rabi splitting has been observed at room temperature in high finesse samples [12].

4.   Extensions

4.4. QUANTUM WELLS UNDER ELECTRIC FIELD

Electric field parallel to the growth axis in excess of 10^5 kV/cm can be applied in QW structures either via a Schottky barrier or by inserting the structure into the intrinsic part of a p.i.n. junction. In an isolated QW, the electric field induces a decrease of the band gap: the conduction and valence carriers are pushed towards the opposite sides of the QW. This is the so-called Quantum Confined Stark Effect, whose strength is larger when the QW thickness increases [13,14]. In contrast with the bulk case (the Franz-Keldysh effect), the carriers remain localized due to the height of the barriers, provided that the field is not too large. Many opto-electronic devices are based on this principle.

An electric field can also provide an opposite blue shift of the band gap feature in the case of a superlatice with thin enough barriers [15]. The miniband width resulting from the coupling between the wells is decreasing with the electric field as the resonance condition between the electronic levels of adjacent wells is less and less fullfilled. This is called the Wannier-Stark effect.

These two effects can be used in a microcavity in order to tune the resonance condition between the optical mode of the microcavity and the excitonic energy (cf. [16] in the weak coupling regime case).

4.2. QUANTUM WELLS UNDER MAGNETIC FIELD

A magnetic field can also be applied. We will be concerned with a magnetic field parallel to the growth axis. Neglecting in a first approach the excitonic and spin effects, the quasi-continuum of conduction band states is squeezed into discrete Landau levels whose energy with respect to the conduction band edge is $(n+1/2)\ \hbar\omega_C$, where ω_C is the cyclotron frequency eB/m^* associated to the conduction effective mass m^*. The degeneracy of each level is SeB/hc [2]. Due to heavy- and light- hole mixing, the result is not simple for the valence band [17]. Nevertheless, if the heavy- and light- hole sublevels are far one from the other, the mixing becomes less efficient and simple regularly-spaced Landau levels appear in the valence-band. This is the case if a strain effect is added to the confinement effect, like in InGaAs/GaAs quantum wells. Then the optical selection rules prescribe a conservation of the Landau index in the absorption process (figure 5).

The excitonic electron-hole correlation effect must be added, and gives indeed the predominant effect at least for the fundamental 1S line as far as the excitonic binding energy R^*_{2D} is large with respect to the Landau energy $\hbar\omega_C$. In this last case, a quadratic diamagnetic shift appears which reads: $3\ e^2\ a_{2D}^{*2}\ B^2\ /\ 16\ \mu$ where μ is the

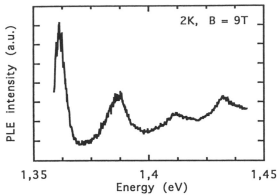

Figure 5: Photoluminescence excitation spectrum of a 75Å-thick strained $In_{0.13}Ga_{0.87}As/GaAs$ QW in a 9T magnetic field.

reduced mass and a_{2D}^* the Bohr radius of the exciton. At the second order, the Bohr radius decreases, giving rise to a corresponding increase of the oscillator strength of the magnetoexciton [18].

The semiconductor microcavity behaviour has been recently studied in the strong coupling regime [19]. Figure 6 shows some reflectivity data obtained in a λ GaAs microcavity containing three $In_{0.13}Ga_{0.87}As$ QW's. In the left column (a), the cavity mode is at lower energy than the excitonic 1S resonance ($\delta = E_{cav} - E_{exc} < 0$) and the reflectivity spectra gives the optical finesse of the Fabry-Pérot without any sample absorption (i.e. 1meV). In the middle column (b), the cavity mode and 1S excitonic mode are in resonance. A magnetic field-induced increase of the Rabi splitting is observed, in accordance with the increase of the oscillator strength due to the diamagnetic effect. Following the previous model, an oscillator strength increase by a factor of 2 is calculated at 7 Teslas. In the right column (c), the cavity mode energy is located in the quasi-continuum of unbound states of the exciton at vanishing field. The increase of the reflectivity peak width (3meV) with respect to the column (a) is related to the absorption by this quasi-continuum of states which reduces the quality factor of the cavity. By simulation of the reflectivity spectrum, this provides a simple measurement of the absorption probability per QW, 0.7%. With increasing magnetic field, some oscillations of the reflectivity peak are noticed before the observation of other reflectivity splittings.

The reflectivity peak position as a function of the detuning are shown in figure 7 for different values of the magnetic field. The new splittings correspond to the resonances of

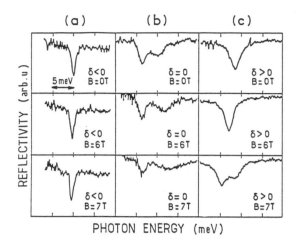

Figure 6: Reflectivity spectra of a microcavity under magnetic field (see text).

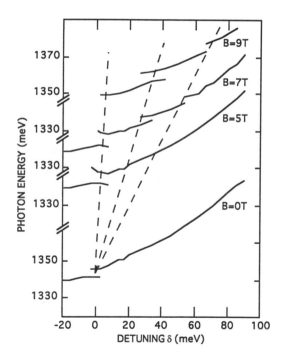

Figure 7: Measured reflectivity peak energies at various magnetic fields as a function of detuning energy.

the cavity mode with the magnetoexcitons associated with the first two excited Landau transitions. They appear at 6T. The classical model that we have presented in section 3.4 provides also a criterion for this observation. As shown in figure 4, assuming a constant linewidth, the oscillator strength of the excited magneto-excitons must be larger than $2\ 10^{12}$ cm^{-2} in order to observe the splitting. If this corresponds to the squeezing of the band to band zero field absorption p= 0.7% over an energy interval $\hbar\omega_C$, one finds a calculated critical field of 3 T for this observation, in reasonable agreement with experiment. This an evidence of a continuous tuning of an interaction between a quasi-continuum density of electronic states with a discrete electromagnetic field (weak coupling regime) to an interaction between a discrete electronic Landau level with the same electromagnetic state (strong coupling regime).

4.3. HIGH EXCITATION EFFECTS

The description of the excited levels of a QW in terms of excitons fails in the high excitation regime. The complicated theoretical considerations [20] shows that, above some critical density, no discrete bound state exist for the exciton, due to screening and band filling effects. Renormalized band to band absorption occurs with possibly a Fermi-edge singularity in the spectra. This subject has been extensively studied in QW's both in the high excitation regime [21] or in modulation doped QW's [22] where a high density of a unique kind of carriers (electrons or holes) can be obtained and continuously tuned. An approximate criterion for the disappearance of excitonic bound states is that the excitonic density n is so high that the excitons occupy all the real space, i.e. n $\approx$ $1/\pi a_{2D}^{*2}$.

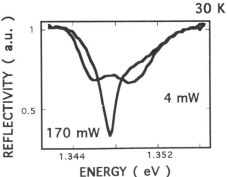

Figure 8: Relectivity spectra of a semiconductor microcavity at low and high intensity.

An increasing excitation experiment provides also a case where a continuous tuning occurs from the discrete state of the exciton (at low excitation) and the quasi-continuum of states of band-to-band like regime (at high excitation). In a microcavity, this gives rise at resonance to a tuning between the strong coupling regime to the weak coupling regime, as shown in the reflectivity spectra of figure 8. Although it is difficult to precisely estimate the carrier density in steady state experiments, one obtains a good order of magnitude for the critical excitation power, in the $1kW/cm^2$ range at resonance.

5. Relaxation

When carriers are promoted from the valence band to the conduction band with excess energy; the thermal equilibrium is restored in two steps (cf. figure 1). In a first step, the carriers relax towards the bottom of the bands, forming at low temperature an exciton. The excess electronic energy is given to acoustical or optical phonons. In a second step, photoluminescence occurs. All these processes are in competition with non-radiative processes which appear not to be efficient in QW's at low temperature.

These inelastic electron-phonon scattering processes and other elastic ones (coulombic impurities, alloy fluctuations) participate to the homogeneous linewidth of optical lines in QW's. The homogeneous linewidth of excitonic lines in GaAs is found [23] to be

$$\Gamma = \Gamma_0 + \gamma T + \Gamma_{LO} \frac{1}{e^{\hbar\omega_{LO}/k_B T} - 1}$$

with $\Gamma_0 = 0.5\text{-}1$ meV, $\gamma = 5\text{-}10\mu eV/K$ which is related to acoustical phonons, $\Gamma_{LO} = 5$meV which is related to optical phonons ($\hbar\omega_{LO} = 36$meV).

Nevertheless, it is worth noticing that, at least at low temperature, the excitonic linewidth is inhomogeneous, principally due to interface roughness.

6. Photoluminescence

6.1. ENERGY LEVELS

The metastable equilibrium of carriers at the bottom of the bands promotes the photoluminescence from electronic levels which may have a low density of states, provided that they are populated. Thus photoluminescence from shallow donors and acceptors can be observed. The luminescence energy is then lower than the excitonic one. Because of their low density of states, these extrinsic lines can be easily saturated in high excitation experiments. Note that these extrinsic lines are broad in QW's as the binding energy of donors and acceptors depends on the position of the impurity in the well [2].

In most good quality QW's, low temperature photoluminescence is excitonic. But the existence of interface roughness binds the excitons on these interface defects. Thus excitonic photoluminescence involves bound excitons whereas excitonic absorption involves free excitons. This is the explanation [24] of the few meV Stokes shift which frequently appears between absorption (or excitation) and photoluminescence (see figure 2). When k_BT is large enough with respect to the few meV binding energy of the excitons on the defects, the excitons are no more bound and free excitons are involved in photoluminescence. At higher temperature, when k_BT becomes larger than the exciton binding energy R^*_{2D}, the free exciton itself is ionized and band to band photoluminescence is recovered. This is the case in III-V systems at room temperature ($R^*_{2D} \approx 10$ meV < 25 meV).

6.2. EXCITONIC LIFETIME

As recalled in section 3.2., QW excitons interact with a quasi-continuum of photon states and Fermi Golden Rule is valid, giving rise to a finite lifetime . Assuming a thermal equilibrium of excitons in the bottom of the $\hbar^2 K^2/2M$ dispersion band related to the in-plane center of mass motion, and because only excitons with in-plane wave vector K smaller than the photon wave vector $\hbar \omega n/c$ can radiate, it can be shown [25] that the radiative lifetime of free excitons is proportional to the temperature T. On the other hand, the radiative lifetime decreases with decreasing well thickness due to a better overlap of envelop electron- and hole- wavefunctions [26]. This scheme of free excitons fails when bound localized excitons are involved and also when the characteristic coherence length of the exciton L is so small that $\hbar^2/2ML^2 > k_BT$. A temperature -independant lifetime appears, generally in the 100ps range, except in very good quality samples. In moderate quality samples, the recombination time can be governed by non radiative processes, giving rise to a recombination time decreasing with temperature, in contrast with the radiative excitonic one. Finally, the room temperature radiative process is essentially governed by a band to band bimolecular process.

6.3. MICROCAVITY EFFECTS

There is a fundamental difference between absorption and photoluminescence in a microcavity. In absorption, the sample is illuminated via one discrete cavity mode which interacts with the excitonic 1S state with the same in-plane center of mass wave vector. In photoluminescence, except in the case of resonant excitation, the excitonic band is thermally populated. Each populated excitonic state with a given in-plane wavevector $\mathbf{K}$ interacts with the cavity mode with which it is coupled. Photoluminescence occurs in all the directions, with an efficiency which is maximum in the direction where the cavity mode $E = \dfrac{\hbar c}{n}\sqrt{k_z^2 + \mathbf{k}^2}$ is in resonance with the exciton mode (figure 9).The angular dependence of the photoluminescnce spectrum reflects the cavity-polariton dispersion curve [27]. A series of photoluminescence spectra is shown in figure as well as a comparison between the experimental peak energies and the dispersion relations calculated within the same theoretical framework.

7.   Conclusion

The first scope of this contribution is to show the large versatility in the design of optical properties of QW's: versatility of the energy levels through the choice of the materials and of the QW thickness; versatility of the oscillator strength by electric field or magnetic field application. The second scope is to recall that, although the exciton is the main optical feature at low temperature and in absorption experiments, excitons disappear in the high excitation regime or in room-temperature photoluminescence measurements, at least in III-V compound systems. Third, we give some results on the optical behaviour of semiconductor microcavities in the strong coupling regime (see also [28] in the present volume. Like in "naked" QW's, these experiments are performed at low temperature or/and in absorption experiments. Unfortunately, many of the useful applications which can be conceived with semiconductor microcavities involve photoluminescence at room temperature. Work must be done either to discretize the electronic levels at room temperature (the use of large gap II-VI compounds where the exciton binding energy is larger or of quantum dots could be efficient), or to make the luminescent levels interact with a single mode of the electromagnetic field (suppression of the in-plane guided mode [29], design of curved mirrors or inhibition of in-plane light propagation through the use of suitable band-gap photonic systems).

196

The Laboratoire de Physique de la Matière Condensée de l'Ecole Normale Supérieure is Laboratoire associé au CNRS et aux Universités Paris 6 et Paris 7. We would like to thank G. Bastard, R. Houdré, R. Ferreira, Ph. Roussignol, J. Tignon, M. Voos and J. Wainstain for their contribution to the work presented here as well as for fruitful discussions.

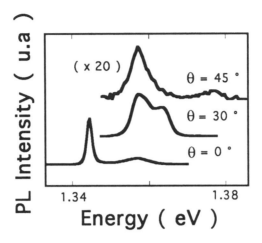

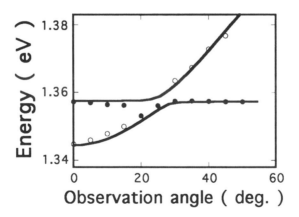

Figure 9: Upper part: photoluminescnce spectra of a microcavity as a function of angle observation. Lower part: measured (points) and calculated (lines) peak energies as a function of angle.

8. References

1. Quattropani, A., *Present volume*.

2. Bastard, G. (1988) *Wave mechanics applied to semiconductor heterostructures*, Les Editions de Physique, Les Ulis, France.

3. Lugagne-Delpon, E., Voisin, P., Vieren, J.P., Voos, M., André, J.P. and Patillon, J.N. (1992) Investigations of MOCVD-grown AlInAs-InP type II heterostructures, *Semicond. Sci. Technol.* **7**, 524.

4. Bastard, G., Delalande, C., Guldner, Y. and Voisin, P. (1988) Optical characterization of III-V and II-VI semiconductor heterolayers in *Advances in Electronics and Electron Physics*, Academic Press, New York.

5. Tuffigo, H., Cox, R.T., Lentz, G., Magnea, N. and Mariette, H. (1990) Optical properties of excitons in II-VI quantum wells: importance of center-of-mass quantization, *Journal of Crystal Growth* **101**, 778.

6. Haus, J.W., *Present volume*.

7. Raimond, J.M., *Present volume*.

8. Weisbuch, C., Nishioka, M., Ishikawa, A. and Arakawa, Y.(1992) Observation of the coupled Exciton-Photon Mode Splitting in a semiconductor Quantum Microcavity, *Phys. Rev. Lett.* **69**, 3314.

9. Brorson, S.D., Yokoyama, H. and Ippen, E.P. (1990) Spontaneous Emission Rate Alteration in Optical Waveguide Structures, *IEEE Journal of Quantum Electronics* **26**, 1492.

10. Ebeling, K.J., *Present volume*.

11. Zhu, Y., Gauthier, D.J., Morin, S.E., Wu, Q., Carmichael, H.J. and Mossberg, T.W. (1990) Vacuum Rabi Splitting as a Feature of Linear-Dispersion Theory: Analysis and Experimental Observations, *Phys. Rev. Lett.* **64**, 2499.

12. Houdré, R., Stanley, R.P., Oesterle, U., Ilegems, M. and Weisbuch, C. (1993) Room Temperature exciton-photon Rabi in a semiconductor microcavity, *Journal de Physique II* **3**, 51.

13. Vina, L., Mendez, E.E., Wang, W., Chang, L.L. and Esaki, L. (1987) Stark shifts in GaAs/GaAlAs quantum wells studied by photoluminescence spectroscopy, *J. Phys. C* **20**, 2803.

14. Miller, D.A.B., Chemla, D.S. and Schmitt-Rink, S. (1986) Relation between electroabsorption in bulk semiconductors and in quantum wells: The quantum confined Franz Keldish effect, *Phys. Rev.* **B33**, 6976.

15. Bleuse, J., Bastard, G. and Voisin P. (1988) Electric-Field-Induced Localization and Oscillatory Electro-optical Properties of Semiconductor Superlattices, *Phys. Rev. Lett.* **60**, 220.

16. Ochi, N., Shiotani, T., Yamanishi, M., Honda, Y. and Suemune I. (1991), Controllable enhancement of excitonic spontaneous emission by quantum confined Stark effect in GaAs quantum wells embedded in quantum microcavities, *Appl. Phys. Lett.* **58**, 2735.

17. Delalande, C., Brum, J.A., Orgonasi, J., Meynadier, M.H., Bastard, G., Maan, J.C, Weimann, G. and Schlapp, W. (1987) Landau levels and magnetoluminescence of n-type GaAs-Ga(Al)As modulation doped quantum wells, *Superlattices and Microstructures* **3**, 29.

18. Ferreira, R., Soucail, B., Voisin, P. and Bastard, G. (1990) Dimensionality effects on the interband magnetoelectroabsorption of semiconductor superlattices, *Phys. Rev.* **B42**, 11404.

19. Tignon, J., Voisin, P., Delalande, C., Voos, C. Houdré, R., Oesterle, U. and Stanley, R.P. (1995) From Fermi's Golden Rule to the Vacuum Rabi Splitting: Magnetopolaritons in a Semiconductor Optical Microcavity, *Phys. Rev. Lett.* **74**, 3967.

20. Bauer, G.E.W. and Ando, T. (1986) Theory of band gap renormalization in modulation-doped quantum wells, *J. Phys.* **C19**, 1537.

22. Tränkle, G., Leier, H., Forchel, A., Haug, H., Ell, C. and Weimann, G. (1987) Dimensionality Dependence of the Band-Gap Renormalization in Two- and Three-Dimensional Electron-Hole Plasmas in GaAs, *Phys. Rev. Lett.* **58**, 419.

22. Delalande, C., Bastard, G., Orgonasi, J., Brum, J.A., Liu, H.W., Voos, M., Weimann, G. and Schlapp, W. (1987) Many-Body Effects in a Modulation-Doped Semiconductor Quantum Well, *Phys. Rev. Lett.* **59**, 2690.

23. Schultheiss, L., Honold, A., Kuhl, J., Köhler, K. and Tu, C.W. (1986) Optical dephasing of homogeneously broadened two-dimensional exciton transitions in GaAs quantum wells, *Phys. Rev.* **B34**, 9027.

24. Bastard, G., Delalande, C., Meynadier, M.H., Frijlink, P.M. and Voos, M. (1984) Low-temperature exciton trapping on interface defects in semiconductor quantum wells, *Phys. Rev.* **B29**, 7042.

25. Andreani, L.C., Tassone, F. and Bassani, F. (1991) Radiative lifetime of free excitons in quantum wells, *Solid State Commun.* **77**, 642.

26. Feldman, J., Peter, G., Göbel, E.O., Dawson, P., Moore, K., Foxon, C., and Elliot, R.J. (1987) Linewidth Dependence of Radiative Excitons in Quantum Wells, *Phys. Rev. Lett.* **59**, 2337

27. Houdré, R., Weisbuch, C., Stanley, R.P., Oesterle, U., Pellandi, P. and Ilegems, M. (1994) Measurement of Cavity-Polariton Curve from Angle-resolved Photoluminescence Experiments, *Phys. Rev. Lett.* **73**, 2043.

28. Yamamoto, Y., *Present volume*.

29. Abram, I., Iung, S., Kuszelewicz, R., Le Roux, G., Licoppe, C., Oudar, J.L., Bloch, J.I., Planel, R. and Thierry-Mieg, V. (1994) Nonguiding half-wave semiconductor microcavities displaying the excton-photon mode splitting, *Appl. Phys. Lett.* **65**, 2516.

SQUEEZING AND CAVITY QED IN SEMICONDUCTORS

YOSHIHISA YAMAMOTO
ERATO Quantum Fluctuation Project
E. L. Ginzton Laboratory, Stanford University
Stanford, CA 94305, U.S.A.

M. Ducloy and D. Bloch (eds.), Quantum Optics of Confined Systems, 201-281.
© 1996 *Kluwer Academic Publishers. Printed in the Netherlands.*

1. Introduction

The purpose of this lecture is to give an overview of the control of quantum noise (squeezing) and the control of spontaneous emission (cavity quantum electrody-namics) in semiconductors. In the second section, the generation and application of squeezed states of light with a semiconductor laser are discussed. The third section treats the suppression of transport and tunneling noise of electrons in mesoscopic systems, and introduces a single photon state generating turnstile de-vice. In the fourth section, the principle of cavity QED in low-Q regime and its application to microcavity semiconductor laser are described. The last section is devoted to the discussions of the principle of cavity QED in high-Q regime and its application to quantum well exciton systems.

2. Squeezed State Generation by Semiconductor Lasers—"Teaching Noisy Photons to Follow Quiet Electrons"

2.1. QUADRATURE AMPLITUDE SQUEEZED STATES VS. NUMBER-PHASE SQUEEZED STATES

A minimum uncertainty state which satisfies the Heisenberg's uncertainty princi-ple,

$$\left[\hat{O}_1, \hat{O}_2\right] = i\hat{O}_3 \longrightarrow \langle\Delta\hat{O}_1^2\rangle\langle\Delta\hat{O}_2^2\rangle \geq \frac{1}{4}\left|\langle\hat{O}_3\rangle\right|^2 \quad , \tag{1}$$

with an equality sign is mathematically defined as an eigenstate of a non-Hermitian operator, that is,

$$\left(e^r\hat{O}_1 + ie^{-r}\hat{O}_2\right)|\chi\rangle_{\mathrm{MUS}} = \left(e^r O_1 + e^{-r} O_2\right)|\chi\rangle_{\mathrm{MUS}} \quad . \tag{2}$$

Here $\hat{O}_1$ and $\hat{O}_2$ are conjugate observables, like position and momentum, and O_1 and O_2 are the expectation values of the corresponding observables for the state $|\chi\rangle_{\mathrm{MUS}}$. r is called a squeezing parameter which determines the distribution of the minimum uncertainty product over the two conjugate observables:

$$\langle\Delta\hat{O}_1^2\rangle = \frac{1}{2}\left|\langle\hat{O}_3\rangle\right| e^{-2r}$$

$$\langle\Delta\hat{O}_2^2\rangle = \frac{1}{2}\left|\langle\hat{O}_3\rangle\right| e^{2r} \quad . \tag{3}$$

If we consider the two conjugate observables in the eigenvalue problem (2) as a position $\hat{q}$ and momentum $\hat{p}$ of a free particle, $\hat{O}_3$ is a c-number ($= \hbar$) and a minimum uncertainty wavepacket is constructed as a linear superposition state of momentum eigenstates (de Broglie waves) $|p\rangle$:

$$|\chi\rangle_{\mathrm{MUS}} = \int (2\pi\Delta p^2)^{-\frac{1}{4}} \exp\left[-\frac{i}{\hbar}\bar{q}(p - \bar{p}) - \frac{(p - \bar{p})^2}{4\Delta p^2}\right] |p\rangle dp \quad . \tag{4}$$

As shown in Fig. 1 (a), the constructive and destructive interferences between the different de Broglie waves localize the position of the free particle within the constraint of the minimum uncertainty product. However, if the particle evolves in time according to the Hamiltonian, $\hat{\mathcal{H}} = \frac{\hat{p}^2}{2m}$, each de Broglie wave acquires a different phase shift $\exp\left(i\frac{p^2}{2\hbar m}t\right)$ and the constructive and destructive interferences between the different de Broglie waves result in a squeezed noise distribution as shown in Fig. 1(b). The result can be understood intuitively as follows: A larger momentum component of the particle wavepacket moves faster and a smaller momentum component moves slower. A free particle prepared in a minimum uncertainty state at $t = 0$ ceases to be a minimum uncertainty state at a later time.

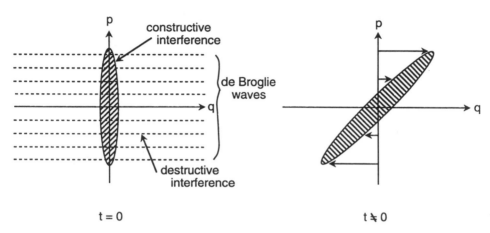

t = 0   t ≠ 0

Fig. 1. Time evolution of a free particle initially prepared in a minimum uncertainty wavepacket at $t = 0$.

However, the situation would be completely different, if the particle is subject to a harmonic potential. A mechanical harmonic oscillator with the Hamiltonian, $\hat{\mathcal{H}} = \frac{\hat{p}^2}{2m} + \frac{1}{2}k\hat{q}^2$, can preserve the minimum uncertainty product due to the contractive function of the restoring harmonic potential. In general, the mechanical harmonic oscillator prepared in a minimum uncertainty state preserves the minimum uncertainty product but pulsate periodically as shown in Fig. 2. Such a solution was first found by Schrödinger.[1]

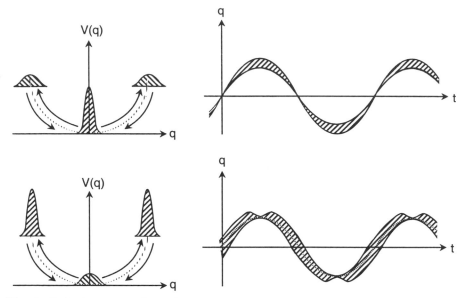

Fig. 2. Pulsating wavepackets of a minimum uncertainty state of a mechanical harmonic oscillator.

An electromagnetic field is mathematically equivalent to a mechanical harmonic oscillator, so similar minimum uncertain states should exist. If we consider the two conjugate observables in the eigenvalue problem (2) as two quadrature amplitudes $\hat{a}_1$ and $\hat{a}_2$ for an electric field operator, i.e. $\hat{E} = \mathcal{E}(\hat{a}_1 \cos \omega t + \hat{a}_2 \sin \omega t)$, $\hat{O}_3$ is a c-number $(= \frac{1}{2})$ and a corresponding minimum uncertainty state is a quadrature amplitude squeezed state.[2]–[4] Here $\mathcal{E} = \sqrt{\frac{\hbar \omega}{2\varepsilon_0 V}}$ is the electric field amplitude of a vacuum state $|0\rangle$, V is a cavity volume and ω is an angular frequency. When the squeezing parameter r is equal to zero, a quadrature amplitude squeezed state is reduced to a coherent state.[5] If we consider the two conjugate observables in the eigenvalue problem (2) as a photon number operator $\hat{n}$ and a sine operator $\hat{S} = \frac{1}{2i} \left[(\hat{n} + 1)^{-\frac{1}{2}} \hat{a} - \hat{a}^\dagger \left(\hat{n} + \frac{1}{2} \right)^{-\frac{1}{2}} \right]$, $\hat{O}_3$ is a cosine operator $\hat{C} = \frac{1}{2} \left[(\hat{n} + 1)^{-\frac{1}{2}} \hat{a} + \hat{a}^\dagger (\hat{n} + 1)^{-\frac{1}{2}} \right]$[6] and a corresponding minimum uncertainty state is a number-phase squeezed state.[7]

Table I summarizes the generation schemes based on unitary time evolution, non-unitary state reduction via quantum measurement and dissipative process in open systems for the two kinds of squeezed states. The quadrature amplitude squeezed state is generated by a quadratic Hamiltonian, which indicates simultaneous emission and absorption of two photons.[2][4] Physical realization of this Hamiltonian calls for a degenerate parametric amplifier, second harmonic generation and four wave mixing.[8] A degenerate parametric amplifier is the most successful device for generating the quadrature amplitude squeezed state so far. The number-phase squeezed state is generated by a quartic Hamiltonian, which indicates self-phase modulation, followed by a displacement operation.[9] A nonlinear fiber loop mirror driven by optical solitons is a modification of this scheme and generates a squeezed vacuum state.[10]

If a coherent state in a signal channel is amplified by a non-degenerate parametric amplifier with a vacuum state in an idler channel, quantum correlated photon twins are generated in the signal and idler channels. When the quadrature amplitude of the idler output is measured by an optical homodyne detector, the signal output is non-unitarily reduced to the quadrature amplitude squeezed state.[11] On the other hand, when the photon number of the idler output is measured by a photon counter, the signal output is reduced to the number-phase squeezed state.[11] In order to generate the two kinds of squeezed states continuously by such a measurement process, a feedforward (or feedback) control of the eigenvalue of the reduced signal wavepackets based on the measurement result for the idler channel is indispensible.[12]

Direct generation of the two kinds of squeezed states calls for a degenerate parametric oscillator[13] and a pump-noise-suppressed laser.[14] This latter scheme is physically realized by a constant current driven semiconductor laser and is the most successful method for generating the number-phase squeezed states so far.

Table I. Various generation schemes of squeezed states of light

Two Kinds of Squeezed States	Unitary Evolution	Nonunitary State Reduction	Dissipation Process in Open Systems
Quadrature Squeezed State $[\hat{a}_1, \hat{a}_2] = \frac{i}{2} \Rightarrow \Delta a_1 \Delta a_2 = \frac{1}{4}$	$\hat{\mathcal{H}}_I = \hbar\kappa(\hat{a}^{\dagger 2} + \hat{a}^2)$ · degenerate parametric amplification · second harmonic generation · four wave mixing	$\hat{\mathcal{H}}_I = \hbar\kappa(\hat{a}_s^{\dagger}\hat{a}_i^{\dagger} + \hat{a}_s\hat{a}_i)$ · nondegenerate parametric amplification $|\alpha_1\rangle_{i\;i}\langle\alpha_1|$ · optical homodyne detection of idler output	degenerate parametric oscillator
Number-Phase Squeezed State $[\hat{n}, \sin\hat{\phi}] = i\cos\hat{\phi}$ $\Rightarrow \Delta n \Delta(\sin\phi) = \frac{1}{2}|\langle\cos\phi\rangle|$	$\hat{\mathcal{H}}_I = \hbar\kappa\hat{a}^{\dagger 2}\hat{a}^2$ · self-phase modulation $\hat{\mathcal{H}}_I = \hbar g(\hat{a} + \hat{a}^{\dagger})$ · displacement (interference)	$\hat{\mathcal{H}}_I = \hbar\kappa(\hat{a}_s^{\dagger}\hat{a}_i^{\dagger} + \hat{a}_s\hat{a}_i)$ · nondegenerate parametric amplification $|n\rangle_{i\;i}\langle n|$ · photon counting of idler output	pump-noise-suppressed laser

2.2. QUANTUM NOISE OF A PUMP-NOISE-SUPPRESSED LASER

A laser is a non-equilibrium open dissipative system and its quantum statistical properties can be elucidated by using a so-called "reservoir theory." Only two sub-systems of a laser, such as a cavity internal field and an assembly of inverted

206

atoms are treated as "systems" of primary interest and all the other sub-systems of a laser including an external field and pump source are considered as "reservoirs" of secondary interest, as shown in Fig. 3. The systems (internal field and inverted atoms) dissipate to the respective reservoirs and the reservoirs inject noise back into the systems. These two processes are uniquely related by the so-called fluctuation-dissipation theorem. The cavity internal field decays with a finite photon lifetime $\left(\frac{\omega}{Q}\right)^{-1}$ due to output coupling to an external field via an end mirror, while a vacuum fluctuation of the external field is coupled into the cavity internal field through the same mirror. The inverted atoms decay with a finite atom lifetime τ_{sp} due to spontaneous emission into nonlasing radiation modes with continuous spectra. The pump process is considered as a reverse process of such a spontaneous decay process and carries the independent noise. A laser oscillator is driven by such three independent noise sources as a vacuum field fluctuation coupled to a lasing mode, spontaneous emission noise and pump noise.

A macroscopic quantum coherence established in a physical system exhibiting the phase transition is determined by the balance between the system's ordering force and reservoirs fluctuating forces. In the case of a laser oscillator, gain saturation induced by a phase coherent stimulated emission counteracts the noise driving sources and establishes a stabilized amplitude and localized phase.

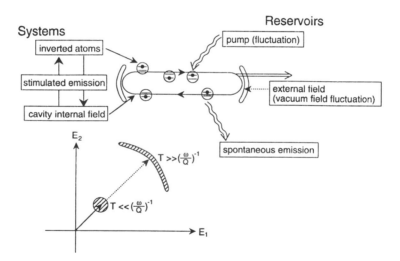

Fig. 3. A reservoir model of a laser oscillator with a system's ordering force (stimulated emission) and reservoirs' fluctuating forces. The quantum states of the output field for $T \ll \left(\frac{\omega}{Q}\right)^{-1}$ and $T \gg \left(\frac{\omega}{Q}\right)^{-1}$.

The amplitude noise spectra of the output field from an ordinary laser with shot-noise-limited pump fluctuation and a pump-noise-suppressed laser are compared in Fig. 4(a) and (b).[15] The amplitude noise of a laser operating at just above the threshold ($P/P_{th} \geq 1$) is far above the shot noise value due to "amplified spontaneous emission noise." The system's ordering force is weak and the noise driving forces dominate in this pump region. The resonant peak is due to a so-called relaxation oscillation between the field and the inverted atoms. The amplitude noise of an ordinary laser operating at far above the threshold ($P/P_{th} \gg 1$) approaches the standard shot noise value as shown in Fig. 4(a). The origin for this shot noise at below the cavity cut-off frequency $\left(\Omega < \frac{\omega}{Q}\right)$ is identified as the (shot-noise-limited) pump noise and the origin for this shot noise at above the cavity cut-off frequency $\left(\Omega > \frac{\omega}{Q}\right)$ is attributed to the vacuum field fluctuation. The third noise force of spontaneous emission noise is completely quenched at such a far above the threshold pump region due to strong stimulated emission.[15] On the other hand, the amplitude noise of a pump-noise-suppressed laser operating at far above the threshold is decreased to below the shot noise value at $\Omega < \frac{\omega}{Q}$ but is equal to the shot noise value at $\Omega > \frac{\omega}{Q}$ as shown in Fig. 4(b). This is an expected result from the above identification of the origins for the shot noise of an ordinary laser.

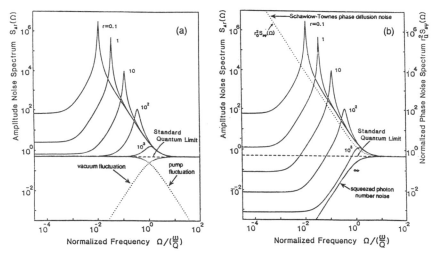

Fig. 4. (a) Amplitude noise spectra for various pump levels, $r \equiv P/P_{th} - 1$, in a laser with shot-noise-limited pump-amplitude fluctuation and the origins for the standard quantum limit (dotted lines). (b) Amplitude noise spectra in a laser with suppressed pump-amplitude fluctuation and normalized phase noise spectrum (dotted line). (Ref. [25]).

The normalized phase noise spectrum of a pump-noise-suppressed laser is also plotted in Fig. 4(b). The normalized phase noise increases in proportion to $\left(\frac{\omega}{Q}\right)^2/\Omega^2$ due to a so-called Schawlow-Townes phase diffusion process. The decreased amplitude noise to below the shot noise value at $\Omega < \frac{\omega}{Q}$ is compensated for by the increased phase (diffusion) noise, and so the Heisenberg uncertainty principle (complimentarity relation) is secured.

The quantum state of a pump-noise-suppressed laser at far above the threshold is, therefore, a number-phase squeezed state when a measurement time T is much longer than a cavity photon lifetime $\left(\frac{\omega}{Q}\right)^{-1}$ (in a low-frequency region) but is a coherent state when T is much shorter than $\left(\frac{\omega}{Q}\right)^{-1}$ (in a high-frequency region).[15] This is schematically illustrated in Fig. 3. The effect of a cavity (mirror) is twofold in the complimetarity relation in such a process of generation these two kinds of quantum states: the cavity mirror provides an optical feedback to induce phase coherent stimulated emission and also introduces a jitter in output coupling of photons. With increasing the cavity photon lifetime $\left(\frac{\omega}{Q}\right)^{-1}$ and thus the coherent photon strage time, the phase noise is decreased due to increased phase coherent stimulated emission relative to random phase spontaneous emission, but the amplitude noise is increased due to increased randomness in output coupling.

2.3. COLLECTIVE COULOMB BLOCKADE IN A CONSTANT-CURRENT-DRIVEN PN JUNCTION

The pump process of an ordinary laser is a photon absorption process by a gain medium, which is a random-point-process and carries a full-shot-noise-limited fluctuation. On the other hand, the pump process in a semiconductor laser is a thermionic emission of electrons and holes across the depletion layer potential of a pn heterojunction. The thermionic emission rate under a constant-current-driving condition is given by[16]

$$
\begin{aligned}
\kappa(t) &= \kappa(0) \exp\left\{\frac{1}{V_T}\left[V(t) - V_{\mathrm{bi}} - \frac{e}{2C}\right]\right\} \\
&= \kappa(0) \exp\left[\frac{t}{\tau_{\mathrm{te}}} - rn(t)\right] .
\end{aligned}
\tag{5}
$$

Here $V_T = \frac{k_B T}{e}$ is a thermal voltage, $V(t)$ is a junction voltage, V_{bi} is a built-in potential, C is a junction capacitance, $\tau_{\mathrm{te}} = k_B T C/eI$ is a thermionic emission time, $r = \left(\frac{e^2}{C}\right)/k_B T$ is the ratio of single electron charging energy to thermal energy which is much smaller than one for a normal semiconductor laser, and $n(t)$ is the number of electrons emitted across the depletion layer between $t = 0$ and $t = t$. A single electron charging time $\tau_e = e/I$ and the thermionic emission time τ_{te} are related by $\tau_e/\tau_{\mathrm{te}} = r$, which indicates the thermionic emission rate is modulated appreciably only after $1/r(\gg 1)$ electrons are accumulated in the junction. Equation (5) suggests that the thermionic emission rate $\kappa(t)$ is modulated by the two competing processes: continuous charging of the junction by a constant current I (the first term in (5)) and discrete thermionic emission of electrons across the depletion layer (the second term in (5)). The junction voltage drop e/C due to single electron thermionic emission event is much smaller than the thermal voltage V_T in a macroscopic pn junction with $r \ll 1$. This situation is schematically shown in Fig. 5. Individual electron thermionic emission event is a totally random Poisson-point-process even though an external circuit (relaxation) current is completely constant. This constant external circuit current is a direct consequence of large source resistance R_s (or long CR_s time constant) and has nothing to do with a random electron thermionic emission event inside a pn junction.[17] However, the number of electrons thermionically emitted during a time interval τ much longer than the thermionic emission time τ_{te} obeys sub-Poissonian statistics. This is because the junction voltage is modulated to the order of V_T by the collective behavior of many electrons ($\simeq 1/r$) which provides a sufficient negative feedback to stabilize the thermionic emission events. In this way, the pump noise of a constant-current-driven pn junction is reduced to below the shot noise value in a low-frequency region below $\omega_c = \frac{1}{\tau_{te}} = \frac{eI}{k_B TC}$ but is shot noise limited in a high frequency region above ω_c. This is called a "collective Coulomb blockade" effect.[16]

If the same pn junction is driven by a (low-impedance) constant voltage source, an external circuit (relaxation) current pulse flows quickly and recovers a constant junction voltage immediately after each electron thermionic emission event, so there is no feedback mechanism in this case and the pump process (electron emission across the depletion layer) is a totally random Poisson-point-process for any measurement time interval, as shown in Fig. 6.[17]

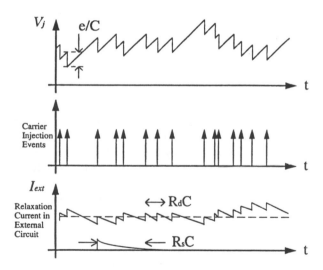

Fig. 5. The junction voltage $V_j(t)$ and circuit current $I_{ext}(t)$ driven by (sub-Poissonian) electron thermionic emission events in a high-impedance constant-current-driven pn junction.

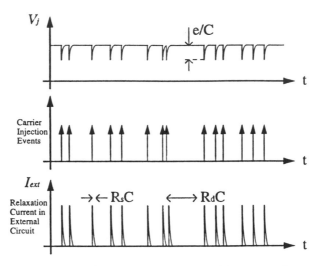

Fig. 6. The junction voltage $V_j(t)$ and circuit current $I_{ext}(t)$ driven by (Poissonian) electron thermionic emission events in a low-impedance constant-voltage-driven pn junction.

The squeezing bandwidth dependence on a current I, temperature T and junction capacitance C due to the collective Coulomb blockade effect was confirmed experimentally using GaAs light emitting diodes (LEDs).[18] The light intensity fluctuation from the LED is decreased to below the shot noise value within the pump squeezing bandwidth $\omega_c = \frac{eI}{k_B T C}$ if the radiative recombination lifetime is much shorter than the inverse of ω_c. Figure 7 shows the measured squeezing bandwidths of four LEDs with different junction capacitances C vs. a junction current at room temperature. The dotted lines are the theoretical squeezing bandwidths ω_c due to the collective Coulomb blockade effect and the dashed lines are the squeezing bandwidths determined by a finite radiative recombination lifetime. The experimental results agree well with the theoretical predictions (solid lines) including the two effects of the collective Coulomb blockade and the finite radiative recombination lifetime. The measured squeezing bandwidths at various temperatures are plotted in Fig. 8 as a function of a junction current. It was confirmed that the squeezing bandwidths at small junction currents are determined by the collective Coulomb blockade effect, i.e. they are proportional to the junction current I and inversely proportional to the junction capacitance C and the temperature T.

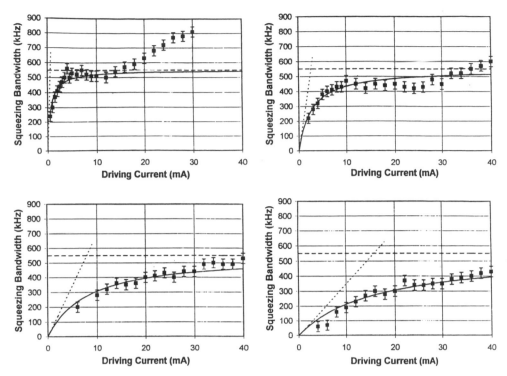

Fig. 7. The measured squeezing bandwidth as a function of a driving current for various light emitting diodes (LEDs) at room temperature. Broken lines show the radiative recombination lifetime limitation of about 560 kHz (carrier lifetime of 290 nsec). Dotted lines are the squeezing bandwidths due to collective Coulomb blockade effect. The solid lines are the expected overall squeezing bandwidth. Areas of the LED's (capacitance values to fit the data) are (a) 0.073 mm^2 (6.5 nF), (b) 0.423 mm^2 (30 nF), (c) 1.00 mm^2 (90 nF) and (d) 2.10 mm^2 (180 nF). (Ref. [18]).

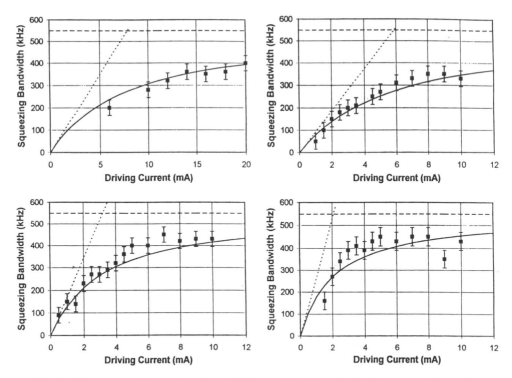

Fig. 8. The measured squeezing bandwidth as a function of a driving current for various temperatures. The area of the LED was 1.00 mm^2. The temperatures are (a) 295 K (identical to Fig. 7(c)), (b) 220 K, (c) 120 K and (d) 78 K. Broken lines show the radiative recombination lifetime limitation of about 560 kHz.. Dotted lines are squeezing bandwidths due to collective Coulomb blockade effect and the solid lines are the overall bandwidths. The junction capacitance of 90 nF and carrier lifetime of 290 nsec, estimated from room temperature measurement (Fig. 7), were used. (Ref. [18]).

2.4.   SQUEEZING AND QUANTUM CORRELATION

The intensity noise spectrum of a semiconductor laser and the corresponding shot noise value can be measured simultaneously using a delayed balanced detector shown in Fig. 9.[19] The semiconductor laser output is equally divided by a 50-50%

beam splitter and the two beams are independently detected by two photodiodes. One photodiode output current I_1 is delayed by a time τ and recombined with the other photodiode output current I_2 via a differential amplifier or a 180° hybrid. The noise value displayed on RF spectrum analyzer at a frequency $\Omega = 2N\pi/\tau$ (N: integer) corresponds to the fluctuation of the difference of the two currents $I_1 - I_2$, which "measures" the vacuum fluctuation incident on the 50-50% beam splitter from an open port. This noise value is precisely equal to the shot noise value irrespective of the laser noise value. The noise value at a frequency $\Omega = (2N+1)\pi/\tau$, on the other hand, corresponds to the fluctuation of the sum of the two currents $I_1 + I_2$, which "measures" the laser intensity noise itself. Therefore, if the laser intensity fluctuation has excess noise (anti-squeezing), the measured noise at $\Omega = (2N+1)\pi/\tau$ is higher than the shot noise value at $\Omega = 2N\pi/\tau$. On the other hand, if the laser intensity fluctuation is reduced to below the shot noise value (squeezing), the measured noise at $\Omega = (2N+1)\pi/\tau$ is lower than the shot noise value at $\Omega = 2N\pi/\tau$. This "inverted modulation" in the noise spectrum is an unmistakable mark for intensity squeezing. As shown in Fig. 9, the measured intensity noise spectrum for a GaAs semiconductor laser operating at just above the threshold features normal modulation (anti-squeezing), while the measured intensity noise spectrum at far above the threshold features inverted modulation (squeezing).

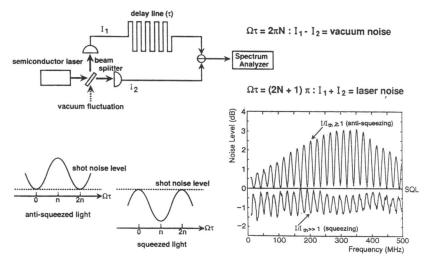

Fig. 9. A delayed balanced detector for simultaneously measuring the intensity fluctuation of a semiconductor laser and the shot noise value. Normal modulation (anti-squeezing) and inverted modulation (squeezing) observed for a GaAs transverse junction stripe (TJS) semiconductor laser at just above the threshold and far above the threshold. (Ref. [19]).

The measured low-frequency intensity noise spectral density of a constant-current driven GaAs semiconductor laser is plotted in Fig. 10 as a function of a

214

pump rate.[20] The excess intensity noise above the shot noise value was observed at a pump rate just above the threshold. This excess intensity noise is attributed to amplified spontaneous emission. On the other hand, the intensity noise at a pump rate far above the threshold is decreased to below the shot noise value as predicted by the theory.[15] The maximum observed squeezing was -8.6 dB below the shot noise value, which corresponds to -14 dB squeezing at a semiconductor laser output facet.[20] The observed maximum squeezing bandwidth was 1.1 GHz, which is not determined by the collective Coulomb blockade pump squeezing bandwidth but is determined by the relaxation oscillation frequency (combined stimulated emission lifetime and photon lifetime).[21]

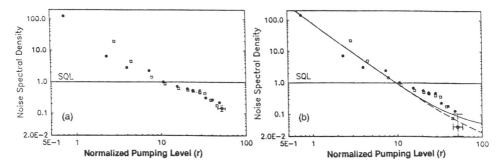

Fig. 10. Normalized intensity fluctuation spectral density at 100 MHz vs normalized pump rate $r = P/P_{th} - 1$. Solid circles are the experimental results for the LED and laser coupled to the different detectors, while the open squares are those for the laser and LED coupled to the same detector. (a) The measured intensity noise normalized by the shot noise value. (b) The calibrated intensity noise at the laser front facet (after correction for a detector quantum efficiency). Solid and dashed lines are theoretical estimates with different internal losses. If $R_1 = 0.01$ and $R_2 = 0.99$, then the minimum noise level is 2×10^{-3}. (Ref. [20]).

As shown in Fig. 5, or indicated by eq. (5), the electron thermionic emission events and the junction voltage under a constant-current-operation are negatively correlated, i.e. if there are more than average electron thermionic emission events, the junction voltage drops to less than average and vice versa. The electron thermionic emission events (pump process) and the photon emission events are positively correlated at below the relaxation oscillation frequency, so the output intensity fluctuation and the junction voltage fluctuation are negatively correlated in such a low-frequency region. This negative correlation was observed in a GaAs semiconductor laser as shown in Fig. 11.[22] The semiconductor laser intensity noise is lower than the shot noise value calibrated by a GaAs light emitting diode.

If this squeezed intensity noise is combined with the delayed junction voltage noise, the combined noise features an oscillatory behavior, which is the manifestation of the quantum correlation between the laser intensity noise and the junction voltage noise. The minimum combined noise is reduced to below the (original) squeezed intensity noise. This means that the negative correlation indeed penetrates into a quantum domain.

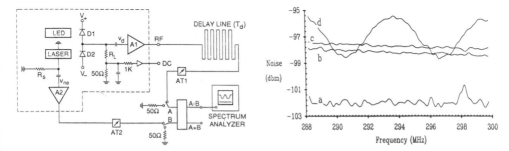

Fig. 11. (a) Basic experimental arrangement. The apparatus inside the dotted lines was enclosed in a closed-cycle refrigerator and the temperature was maintained at 66 K. Photocurrent noise v_d and the junction-voltage noise v_{ne} are combined using a wide band 180° hybrid. Attenuators AT1 and AT2 were used to equalize the individual noise spectra observed on the spectrum analyzer. At the output labeled DC, the detector current generated by either the LED or the laser was obtained. (b) Curve c, the SQL produced by the LED; curve b, laser intensity noise; curve a, junction-voltage fluctuation; and curve d, combined laser intensity noise and junction voltage noise. In all the traces the respective dark noise levels were subtracted and the spectrum repeatedly filtered with a Gaussian of full width 81 kHz. Resolution bandwidth of the spectrum analyzer was 100 kHz and the video bandwidth was 30 Hz. The pump level was $r = 9.6$. (Ref. [22]).

2.5. SUPPRESSION OF MODE COMPETITION NOISE BY AN INJECTION-LOCKING TECHNIQUE

The gain bandwidth of a semiconductor laser is usually broader than a longitudinal mode separation, so a few longitudinal modes compete to oscillate with each other and sometimes can oscillate simultaneously even under a stable *cw* pump operation. Spontaneously emitted photons primarily go to a particular mode closest to the gain peak, but at some other time spontaneous photons primarily go to the other longitudinal mode. Such a longitudinal mode competition due to stochastic spontaneous emission coupling was observed in many semiconductor lasers and is known to introduce excess intensity noise.[23] Figure 12(a) shows the theoretical

216

lasing characteristics of three longitudinal modes in a homogeneous gain medium with slightly different stimulated emission gain coefficients which are proportional to the spontaneous emission coupling coefficients $\beta_i(i = 1, 2, 3)$ due to the Einstein relation for A and B coefficients. The multimode Langevin equations in a photon number representation was used for this calculation.[24] The solid lines correspond to a relatively broad gain bandwidth, in which the two side modes have the spontaneous emission coefficients $\beta_1 = \beta_3$ of $(1 - 5 \times 10^{-5})$ times smaller than the spontaneous emission coefficient β_2 of the central mode. The three modes simultaneously oscillate at above the threshold but the central mode starts to dominate with increasing a pump rate due to gain saturation. The dashed lines correspond to a relatively narrow gain bandwidth, in which the two side modes have the spontaneous emission coefficients $\beta_1 = \beta_3$ of $(1 - 5 \times 10^{-3})$ times smaller than the spontaneous emission coefficient β_2 of the central mode. A single mode oscillation with large side mode suppression is realized even at a pump rate just above the threshold in this case. As shown in Fig. 12(b), the total intensity noise of the three modes is reduced to below the shot noise value at a pump rate $P/P_{th} \geq 3$, where P_{th} is the threshold pump rate, but the intensity noise of individual mode is much higher than the shot noise value. This excess noise is due to the mode competition noise. There exists a strong negative correlation between the central mode intensity and the two side mode intensities, so the total intensity noise can be squeezed even though individual mode intensities are noisy.

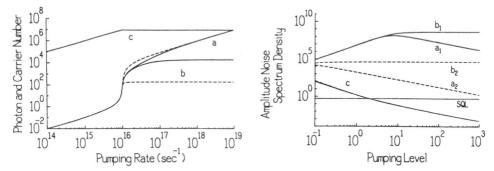

Fig. 12. (a) The average photon numbers of main (a) and side (b) longitudinal modes and the carrier number (c) vs. pump rate (electron/sec) for a semiconductor laser with a homogeneous gain. (b) The total intensity noise and the intensity noise of individual mode vs. pump rate. Solid lines correspond to $\beta_1 = \beta_3 = (1 - 5 \times 10^{-5})\beta_2$ and dashed lines correspond to $\beta_1 = \beta_2 = (1 - 5 \times 10^{-3})\beta_2$. (Ref. [24]).

This negative intensity correlation was directly observed in a GaAs semiconductor laser.[24] The longitudinal mode spectrum is plotted in Fig. 13(a), in which the three longitudinal modes simultaneously oscillate. The total intensity noise is reduced to below the shot noise value at a pump rate $P/P_{th} \geq 3$, as shown in Fig. 13(b) even though individually measured each mode intensity is extremely noisy. Figure 13(c) and (d) demonstrate the negative intensity correlation between the central mode and the two side modes. If a frequency dependent loss exists either inside or outside a semiconductor laser, this cancellation of mode competition noise via negative intensity correlation easily fails and the intensity squeezing is destroyed.

To suppress the longitudinal mode competition noise, a mode selective element can be incorporated into a semiconductor laser, such as a monolithic distributed feedback laser[25] and external grating feedback.[26][27] An external injection-locking technique using an independent single-frequency master oscillator is particularly effective to suppress the mode competition noise and also useful for various applications to spectroscopy and interferometry. The quantum noise of an injection-locked laser can be calculated using the Fokker-Planck equation and Langevin equation.[28] The uncertainty product of the suppressed phase diffusion noise and the residual squeezed intensity noise of an injection-locked semiconductor laser was calculated by the Langevin equation.[29] The experimental result shown in Fig. 14 demonstrates that the excess intensity noise of a free-running semiconductor laser due to the mode competition noise and frequency dependent loss is suppressed by an external injection-locking technique.[30] In this experiment, the locking bandwidth of about 10 GHz was achieved. The maximum observed squeezing in an injection-locked semiconductor laser starting from an excess intensity noise in a free-running laser was -3 dB below the shot noise value.[31]

A semiconductor lasers exhibits the polarization mode competition as well as the longitudinal mode competition. An edge-emitting semiconductor laser usually oscillates in a TE mode due to a higher reflectivity of the TE mode, but a (nonlasing) TM mode still fluctuates the TE mode intensity. A vertical cavity surface emitting semiconductor laser features much stronger mode partition noise between the two orthogonal polarization modes and also between different transverse modes. An external injection-locking is also effective to suppress such polarization mode and transverse mode partition noise.

218

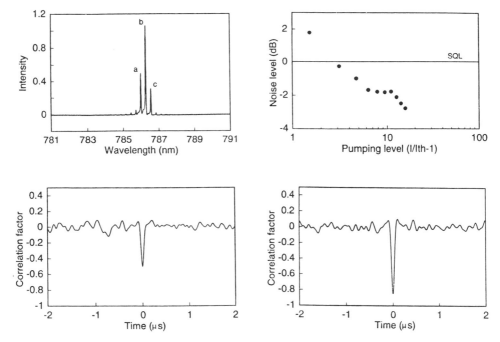

Fig. 13. (a) The longitudinal mode spectrum of a multimode semiconductor laser. (b) The total intensity noise vs. normalized pump rate. (c) The normalized intensity correlation between the central mode b and the side mode c vs. delay time. (d) The normalized intensity correlation between the central mode b and the side mode a vs. delay time. (Ref. [24]).

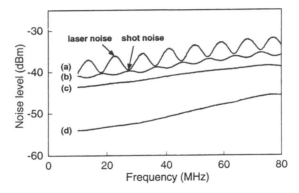

Fig. 14. Intensity-noise spectra for (a) the free-running slave laser, (b) the injection-locked slave laser, (c) the noise output of single detector, and (d) the thermal noise. (Ref. [30]).

2.6. APPLICATIONS OF SQUEEZED STATES TO SPECTROSCOPY AND INTERFEROMETRY

A squeezed vacuum state generated by a degenerate parametric amplifier was applied to improve the sensitivity in spectroscopy[32] and interferometry.[33] A number-phase squeezed state generated by a semiconductor laser can also be used to increase the sensitivity in FM modulation spectroscopy and dark fringe interferometry.

Figure 15(a) shows such an experimental setup of an FM modulation spectroscopy of Rb atoms in a cell using a squeezed semiconductor laser.[34] A single-frequency semiconductor master laser with an external grating feedback has a narrow spectral linewidth (< 1 MHz) and is directly frequency modulated at 80 MHz via its drive current.[35] This FM modulated master laser output is injected into a *cw* pumped slave semiconductor laser. The FM modulation of the master laser output is replicated onto the intensity squeezed output from the slave laser without degradation because the modulation frequency of 80 MHz is much smaller than the locking bandwidth (~ 10 GHz) of the slave laser. The spurious intensity modulation of the master laser output associated with a direct current modulation is suppressed by more than 50 dB due to the gain saturation of the slave laser. The strong carrier component of the master laser output saturates the transition line from $2S_{1/2}(F = 2)$ to $2P_{1/2}(F = 3)$ for the Rb atoms with a longitudinal velocity $v_z \simeq 0$ as shown in Fig. 15(b). The lower FM side band of the counter-propagating slave laser output is attenuated by the (saturated) absorption of this transition line, which results in the FM-to-AM conversion signal. The Doppler-free information about the resistive and reactive responses of the Rb atom transition line can be extracted by detecting the in-phase and quadrature-phase components of this FM-to-AM conversion signal.[36]

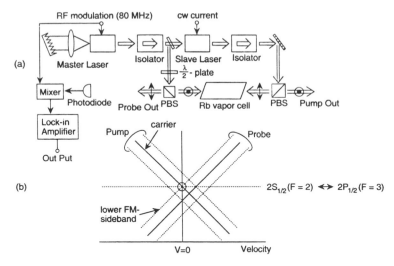

Fig. 15. (a) An experimental setup for FM modulation spectroscopy of Rb atoms using an injection-locked semiconductor laser. (b) Interaction of Rb atoms with counter-propagating pump (master) and probe (slave) signals.

Figure 16 shows the observed FM-to-AM conversion signal intensity vs. the center frequency of the master/slave lasers.[34] The Doppler-free FM-to-AM conversion signal with a natural linewidth ($\simeq$ 5 MHz) was observed within the non-linear saturation window and below the shot noise value. Since the slave semiconductor laser intensity is squeezed, such a weak FM modulation spectroscopy signal can be retrieved. The FM noise spectroscopy using a broadband FM noise of a semiconductor laser instead of an artificial FM modulation was also demonstrated by a squeezed semiconductor laser.[34]

Figure 17 shows an experimental setup for a dark fringe Michelson interferometer using a squeezed semiconductor laser.[37] A high-power single-frequency semiconductor master laser with an external grating feedback drives a Michelson interferometer. The two arms of the Michelson interferometer have exactly equal lengths, so all the input laser beam is reflected back to the input port. If there is a small arm length difference, a small leakage field emerges at the output port connecting into a photodetector. The noise superimposed on this leakage signal field is a part of vacuum fluctuation incident on the open port of the 50-50% beam splitter. This vacuum fluctuation results in the standard quantum limit for the minimum detectable phase ($\Delta\phi_{\min} = 1/\sqrt{n}$, n: photon number) in this dark fringe Michelson interferometer.

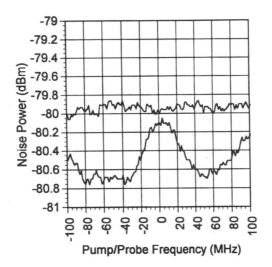

Fig. 16. Sub-shot-noise spectroscopy signal of rubidium atoms. The traces were taken by sweeping the master laser center frequency and keeping the modulation frequency fixed at 80 MHz. The grey trace is the shot-noise level, taken by substracting the photocurrents in the double-balanced setup. The black trace is the saturated transition surrounded by another transition (left peak) and the Doppler profile (right peak). The resolution bandwidth was 300 KHz and the video bandwidth was 300 Hz with 10 averages. (Ref. [34]).

In the experimental set up shown in Fig. 17, the open port of the beam splitter is driven by the number-phase squeezed state from an injection-locked slave semiconductor laser. In such a case, the incident squeezed state emerges at the output port connecting into the photodiode. The leakage field of the master laser beam due to a small arm length difference is coherently superposed on the amplitude squeezed slave laser field. The signal-to-noise ratio is improved by squeezing the amplitude of the slave laser field as indicated in Fig. 18.[37] An amplitude squeezed state with enhanced phase noise can improve the sensitivity of a phase measuring interferometer, because there is a $\frac{\pi}{2}$ phase difference between the master and slave laser fields at the input into the beam splitter and thus the amplitude noise of the slave laser field acts as the phase noise for the master laser field after recombined by the beam splitter.

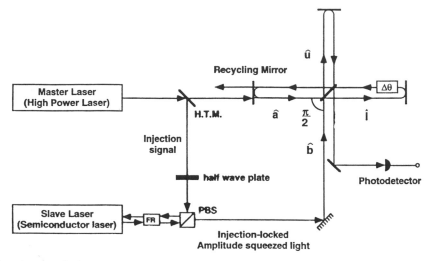

Fig. 17. A sub-shot noise dark fringe Michelson interferometer using an injection-locked semiconductor laser.

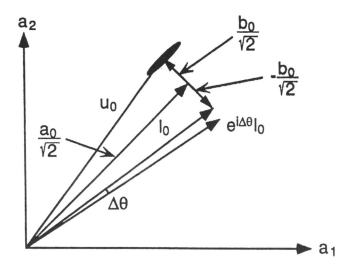

Fig. 18 The signal $(a_0 \pm b_0)$ and noise powers of a sub-shot noise dark fringe interferometer. An amplitude squeezed signal $\hat{b}$ from the injection-locked laser is converted to the phase squeezed signal by the addition of the master laser signal $\hat{a}$.

3. Shot Noise Suppression in Mesoscopic Transport and Tunneling—"Dissipation Enhances Squeezing for Fermi Particles (Electrons)"

3.1. CLASSICAL PARTITION VS. QUANTUM PARTITION

When a classical particle is incident on a 50-50% beam splitter as shown in Fig. 19(a), the particle is detected in either one of two detectors. The average counts of the two detectors for N incident particles are equal to one-half of the total particle number, i.e. $\langle n \rangle = \frac{N}{2}$. The variance of each detector count is half the Poisson noise, i.e. $\langle \Delta n^2 \rangle = \frac{1}{2} \langle n \rangle$. That is, division of incident particles by a 50-50% beam splitter is a stochastic process and introduces partition noise. In a general case of the beam splitter with a transmission coefficient T, the average count and the variance of the transmitted particles are $\langle n \rangle = TN$ and $\langle \Delta n^2 \rangle = (1 - T)TN = (1 - T)\langle n \rangle$, respectively. If we prepare the ensemble of exactly N particles and divide each ensemble member by a beam splitter with a very small transmission coefficient $T \ll 1$, the ensemble of transmitted particles carry a full shot noise or a Poisson distribution even though the incident ensemble do not have any noise. If an incident particle is either a boson (like a photon) or a fermion (like an electron), the quantum mechanically calculated partition noise is identical to the above classical partition noise.[38][39] A constant photon or electron stream acquires a full shot noise (Poisson noise) if it encounters high attenuation.

When two particles are incident on a 50-50% beam splitter simultaneously from two input ports as shown in Fig. 19(b), classical, boson and fermion particles feature different partition properties.[40] For the classical particles, the probabilities of detecting $(2, 0)$, $(1, 1)$ and $(0, 2)$ particles by the two detectors are 1/4, 1/2 and 1/4, respectively. This result reflects the fact that two particles are divided by the beam splitter "independently". For the boson particles, the probabilities of detecting $(2, 0)$, $(1, 1)$ and $(0, 2)$ particles by the two detectors are 1/2, 0 and 1/2, respectively. The joint probability $(1, 1)$ of each detector counting one particle is identically equal to zero. This is because the probability amplitudes of the two particles simultaneously transmitting through the beam splitter and simultaneously reflecting from the beam splitter, both of which result in the same result $(1, 1)$ count, destructively interfere to nullify the total probability of $(1, 1)$ count result. This theoretically prediction was confirmed experimentally.[41] Such a destructive interference mathematically originates from the "symmetric" nature of the wavefunction for two boson particles.[40] Physically, this is a clear demonstration of the bunching property of boson particles. For the fermion particles, the probabilities of detecting $(2, 0)$, $(1, 1)$ and $(0, 2)$ particles in the two detectors are 0, 1 and 0, respectively. The joint probability $(1, 1)$ of each detector counting one particle is now equal to one. This is because the probability amplitudes of the two particles simultaneously transmitting through the beam splitter and simultaneously reflecting from the beam splitter now constructively interfere. A completely opposite interference from a bosonic case is due to the "anti-symmetric" nature of the wavefunction for two fermion particles.[40] Physically, this anti-bunching property is a clear demonstration of Pauli's exclusion principle.

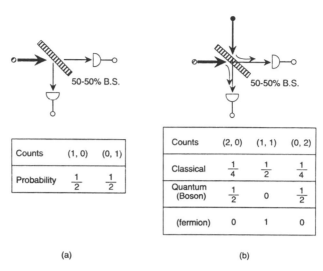

(a)   (b)

Fig. 19. (a) A single particle incident on a 50-50% beam splitter and the probabilities of counting (1, 0) and (0, 1) particles by the two detectors for classical, boson and fermion particles. (b) Two particles simultaneously incident on a 50-50% beam splitter and the probabilities of counting (2, 0), (1, 1) and (0, 2) particles by the two detectors for classical, boson and fermion particles.

It is well known that shot noise is generated by ballistic division or partial tunneling of electrons in mesoscopic systems. This shot noise originates from a stochastic single particle partition process shown in Fig. 19(a). "Electron optics" are essentially identical to "photon optics" in such a coherent and ballistic regime. It is also known that such shot noise is partially suppressed by distributed elastic scattering and fully suppressed by distributed inelastic scattering of electrons in mesoscopic systems. This suppression of shot noise has the physical origin as a deterministic two fermion particles partition process shown in Fig. 19(b). "Electron optics" now departs from "photon optics" when two electrons are involved in the scattering process and Pauli's exclusion principle plays a role. Squeezing in photon optics is always destroyed by dissipation, but squeezing in electron optics can be enhanced by dissipation.

3.2. QUANTUM PARTITION NOISE IN PHOTON OPTICS AND ELECTRON OPTICS

When a bose-condensed field (like a photon number state field $|n\rangle_1$) is incident on a beam splitter as shown in Fig. 20(a), not only this incident field (denoted by a Heisenberg operator $\hat{a}_1$) but also a vacuum field $|0\rangle_2$ incident from an open port (denoted by $\hat{a}_2$) must be taken into account to satisfy the boson commutator bracket properly in the two output ports:[38]

$$\begin{pmatrix} \hat{a}_3 \\ \hat{a}_4 \end{pmatrix} = \begin{pmatrix} \sqrt{T} & \sqrt{1-T} \\ -\sqrt{1-T} & \sqrt{T} \end{pmatrix} \begin{pmatrix} \hat{a}_1 \\ \hat{a}_2 \end{pmatrix}$$
$$\Longrightarrow [\hat{a}_i, \hat{a}_i^\dagger]_- = \hat{a}_i\hat{a}_i^\dagger - \hat{a}_i^\dagger\hat{a}_i = 1 \quad (i = 3 \text{ or } 4) \quad . \tag{6}$$

If the incident field $\hat{a}_1$ is in a photon number eigenstate, i.e. $|n\rangle_1|0\rangle_2$, the joint-correlated output field states can be conveniently described by a collective angular momentum operator $\hat{J} = (\hat{J}_x, \hat{J}_y, \hat{J}_z)$ just like an assembly of n identical two-level atoms.[42] Figure 20(b) shows a collective angular momentum Bloch sphere, in which the joint-correlated output field states are represented by a spin coherent state (or Bloch state) $|\theta, \gamma\rangle$.[42] When the beam splitter transmission coefficient $T = 1$, the output state is $|n\rangle_3|0\rangle_4$ which corresponds to the spin state $|J, J\rangle$ located at the north pole. Similarly, the output state $|0\rangle_3|n\rangle_4$ for the case of $T = 0$ is represented by the spin state $|-J, J\rangle$ located at the south pole. With increasing T from 0 to 1, the spin coherent state $|\theta, 0\rangle$ make a rotation from the south pole ($\theta = \pi$) to north poles ($\theta = 0$). At a special case of a 50-50% beam splitter ($T = \frac{1}{2}$), the spin coherent state is located on the equator ($\theta = \pi/2$). The relevant commutator bracket and Heisenberg uncertainty relation for this state $|\frac{\pi}{2}, 0\rangle$ are given by

$$[\hat{J}_y, \hat{J}_z]_- = i\hat{J}_x \Longrightarrow \langle \Delta \hat{J}_y^2\rangle\langle\Delta\hat{J}_z^2\rangle \geq \frac{1}{4}|\langle\hat{J}_x\rangle|^2 \quad . \tag{7}$$

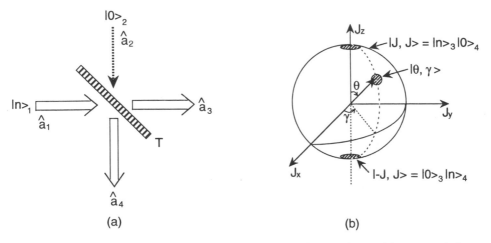

(a)

(b)

Fig. 20. (a) Two bosonic fields $\hat{a}_1$ and $\hat{a}_2$ scattered by a beam splitter with a transmission coefficient T. (b) A collective angular momentum operator sphere and spin coherent states describing joint-correlated boson output states.

The photon number difference operator $\hat{n}$ between the two outputs is given by $2\hat{J}_z = \hat{a}_3^\dagger\hat{a}_3 - \hat{a}_4^\dagger\hat{a}_4 = 2(\hat{a}_{1c}\hat{a}_{2c} + \hat{a}_{1s}\hat{a}_{2s})$, where $\hat{a}_{ic}$ and $\hat{a}_{is}$ are the (Hermitian) cosine and sine quadrature components of the non-Hermitian operator $\hat{a}_i (i = 1$ or $2)$. The photon number difference noise $\langle\Delta\hat{n}^2\rangle$ is mathematically equal to the uncertainty $4\langle\Delta\hat{J}_z^2\rangle$ of the angular momentum operator $\hat{J}_z$ and physically originates from beat noise between coherent excitation of the input number state field $\hat{a}_1$ and in-phase fluctuation of the input vacuum state field $\hat{a}_2$. The phase difference operator $\hat{\phi}$ between the two outputs is given by $\hat{J}_y/|\langle\hat{J}_x\rangle| = -i(\hat{a}_3^\dagger\hat{a}_4 - \hat{a}_4^\dagger\hat{a}_3) = 2i(\hat{a}_{1c}\hat{a}_{2s} - \hat{a}_{1s}\hat{a}_{2c})$. The phase difference noise $\langle\Delta\hat{\phi}^2\rangle$ is mathematically identical to the uncertainty $\langle\Delta\hat{J}_y^2\rangle/|\langle\hat{J}_x\rangle|^2$ of the angular momentum operator $\hat{J}_y$ and physically originates from beat noise between coherent excitation of the input number state field $\hat{a}_1$ and quadrature-phase fluctuation of the incident vacuum state field $\hat{a}_2$. The uncertainty product between the photon number difference noise $\langle\Delta\hat{n}^2\rangle$ and the phase difference noise $\langle\Delta\hat{\phi}^2\rangle$ is calculated using (7):

$$\langle\Delta\hat{n}^2\rangle\langle\Delta\hat{\phi}^2\rangle = 4\langle\Delta\hat{J}_z^2\rangle\frac{\langle\Delta\hat{J}_y^2\rangle}{|\langle\hat{J}_x\rangle|^2} \geq 1 \quad , \tag{8}$$

where $\langle\Delta\hat{n}^2\rangle = \langle n\rangle$ and $\langle\Delta\hat{\phi}^2\rangle = \frac{1}{\langle\hat{n}\rangle}$ for $T = \frac{1}{2}$. The photon number difference noise is measured by detecting the two output field $\hat{a}_3$ and $\hat{a}_4$ by photon counters, while the measurement of the phase difference noise requires the construction of an interferometer with a $90°$ phase shift.[38]

When a fermion field (like an electron field) denoted by a Heisenberg operator $\hat{b}_1$ is incident on a beam splitter as shown in Fig. 21(a), a fermion vacuum field denoted by $\hat{b}_2$ must also be taken into account to satisfy the proper fermion anticommutator bracket in two output ports:[39]

$$\begin{pmatrix} \hat{b}_3 \\ \hat{b}_4 \end{pmatrix} = \begin{pmatrix} \sqrt{T} & \sqrt{1-T} \\ -\sqrt{1-T} & \sqrt{T} \end{pmatrix} \begin{pmatrix} \hat{b}_1 \\ \hat{b}_2 \end{pmatrix}$$
$$\Longrightarrow [\hat{b}_i, \hat{b}_i^\dagger]_+ = \hat{b}_i\hat{b}_i^\dagger + \hat{b}_i^\dagger\hat{b}_i = 1 \quad (i = 3 \text{ or } 4) \quad . \tag{9}$$

The joint-correlated output field states can be described by a Pauli spin operator $\hat{\sigma} = (\hat{\sigma}_x, \hat{\sigma}_y, \hat{\sigma}_z)$ just like a single two-level atom. Figure 21(b) shows a Bloch vector sphere, in which the joint-correlated output field states are represented by a spin coherent state. When $T = 1$, the output state is $|1\rangle_3|0\rangle_4$ which corresponds to the spin state $|\frac{1}{2}, \frac{1}{2}\rangle$. When $T = 0$, the output state is $|0\rangle_3|1\rangle_4$ and the corresponding spin state is $|-\frac{1}{2}, \frac{1}{2}\rangle$. In a special case of $T = \frac{1}{2}$, the output state is $\frac{1}{\sqrt{2}}(|1\rangle_3|0\rangle_4 + |0\rangle_3|1\rangle_4)$ which corresponds to the spin state $|\frac{\pi}{2}, 0\rangle$. The relevant commutator bracket and Heisenberg uncertainty relation for this state $|\frac{\pi}{2}, 0\rangle$ are given by

$$[\hat{\sigma}_y, \hat{\sigma}_z]_+ = i\hat{\sigma}_x \Longrightarrow \langle\Delta\hat{\sigma}_y^2\rangle\langle\Delta\hat{\sigma}_z^2\rangle \geq \frac{1}{4}|\langle\hat{\sigma}_x\rangle|^2 \quad . \tag{10}$$

So far there is no difference between the partitions of a single photon eigenstate field and a single electron field.

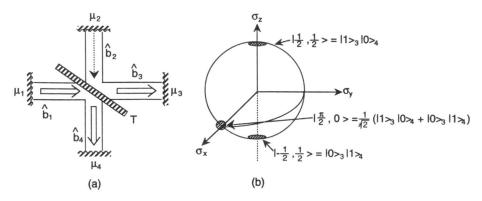

Fig. 21. (a) Two fermionic fields scattered by a beam splitter with a transmission coefficient T. The chemical potential μ_1 of a reservoir electrode 1 is higher than the chemical potentials of the other electrodes ($\mu_2 = \mu_3 = \mu_4$). (b) A Bloch vector sphere and spin coherent states describing joint-correlated fermion output fields.

The collective electron number difference operator $\hat{N}$ for N independently incident electrons is given by $2\sqrt{N}\hat{\sigma}_z = \sqrt{N}(\hat{b}_3^\dagger\hat{b}_3 - \hat{b}_4^\dagger\hat{b}_4) = 2\sqrt{N}(\hat{b}_{1c}\hat{b}_{2s} + \hat{b}_{1s}\hat{b}_{2c})$, where $\hat{b}_{ic}$ and $\hat{b}_{is}$ are the (Hermitian) cosine and sine components of $\hat{b}_i (i = 1$ or $2)$. The electron number difference operator $\langle\Delta\hat{N}^2\rangle$ is mathematically given by $4N\langle\Delta\hat{\sigma}_z^2\rangle$ and physically originates from beat noise between coherent excitation of the single electron field $\hat{b}_1$ and quadrature-phase fluctuation of the vacuum field $\hat{b}_2$. The different physical origin for an electron partition noise (quadrature-phase fluctuation instead of in-phase fluctuation) is due to the anti-commutation relation (instead of the commutation relation).[43] The collective phase difference operator for N independently incident electrons is given by $\hat{\Phi} = \frac{1}{\sqrt{N}}\frac{\hat{\sigma}_y}{|\langle\hat{\sigma}_x\rangle|} = \frac{-i}{\sqrt{N}}(\hat{b}_3^\dagger\hat{b}_4 - \hat{b}_4^\dagger\hat{b}_3) = \frac{2i}{\sqrt{N}}(\hat{b}_{1c}\hat{b}_{2c} - \hat{b}_{1s}\hat{b}_{2s})$. The phase difference noise $\langle\Delta\hat{\Phi}^2\rangle$ is mathematically given by $\frac{\langle\Delta\hat{\sigma}_y^2\rangle}{N|\langle\hat{\sigma}_x\rangle|^2}$ and physically originates from beat noise between coherent excitation of the single electron field $\hat{b}_1$ and in-phase fluctuation of the vacuum field $\hat{b}_2$. The uncertainty product between the collective electron number difference noise $\langle\Delta\hat{N}^2\rangle$ and the collective phase difference noise $\langle\Delta\hat{\Phi}^2\rangle$ is calculated using (10):

$$\langle\Delta\hat{N}^2\rangle\langle\Delta\hat{\Phi}^2\rangle = 4N\langle\Delta\hat{\sigma}_z^2\rangle\frac{\langle\Delta\hat{\sigma}_y^2\rangle}{N|\langle\hat{\sigma}_x^2\rangle|^2} \geq 1 \quad , \tag{11}$$

where $\langle\Delta\hat{N}^2\rangle = N$ and $\langle\Delta\hat{\Phi}^2\rangle = \frac{1}{N}$ for $T = 1/2$. The electron number difference noise and phase difference noise can be measured in a similar way as the photon case.[43]

3.3. EQUILIBRIUM AND NON-EQUILIBRIUM NOISE IN MESOSCOPIC ELECTRON TRANSPORT AND TUNNELING

Let us consider a single-mode electron waveguide shown in Fig. 22, in which an electron wavepacket is emitted by one reservoir electrode and propagates toward another reservoir electrode without scattering. A conductance of such a ballistic single electron channel is given by $G_Q = e^2/h$, if an electron spin degeneracy is not taken into account.[44] The reason why the ballistic channel without scattering has a finite conductance is that only one electron can be accommodated for each electron wavepacket (degree-of-freedom) which satisfies the minimum time-energy uncertainty product, $\Delta E \Delta t = \frac{\hbar}{2}$, and the number of degrees-of-freedom for a given time internal Δt is finite for a given chemical potential difference, $\mu_2 - \mu_1 = \frac{\Delta E}{e}$, where μ_1 and μ_2 are the chemical potentials of the two electrodes.

If there is no chemical potential difference (zero external bias voltage) between two reservoir electrodes as shown in Fig. 22, an average current is zero but there is a finite noise current. The electron energy distributions in both electrodes obey a Fermi-Dirac distribution f and electrons in both electrodes near a Fermi energy are stochastically emitted according to the partition theorem with an emission rate of $0 < f < 1$. The variance of emitted electron number for N electron wavepackets is $\langle \Delta N^2 \rangle = Nf(1-f)$. This stochastic emission of electrons due to the Fermi-Dirac distribution at a finite temperature θ results in thermal noise with a current spectral density $S_i^{th}(\omega) = 4k_B\theta G_Q$.[45] There is another noise mechanism which exists even at zero temperature ($\theta = 0$); When the electron wavepackets above and below the Fermi energy (one empty and one occupied) are absorbed by the receiving reservoir, beating between these two wavepackets results in a noise current at a frequency determined by the energy difference between the two wavepackets. This beating between an occupied low-energy electron wavepacket and an empty high-energy electron wavepacket results in quantum noise with a current spectral density $S_i^q(\omega) = 2\hbar\omega G_Q$. The total noise current spectral density is identical to "generalized Nyquist noise" spectral density:[46]

$$S_i(\omega) = 4\hbar\omega \left[\frac{1}{\exp(\hbar\omega/k_B\theta) - 1} + \frac{1}{2} \right] G_Q \quad . \tag{12}$$

If there is a finite chemical potential difference between two reservoir electrodes, an average current increases with a bias voltage $\langle i \rangle = G_Q(\mu_1 - \mu_2)$ but a noise current remains constant. This is because the stochastic emission of electrons due to Fermi-Dirac distribution (thermal noise) and the beating between high-energy empty electron wavepacket and low-energy occupied electron wavepacket (quantum noise) occur in the two reservoir electrodes independently, so the chemical potential difference does not affect.

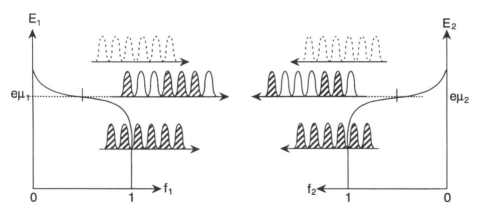

Fig. 22. Equilibrium noise in a ballistic electron channel. The stochastic emission of electrons near the Fermi energy due to Fermi-Dirac distribution results in thermal equilibrium noise and the beating betwen high-energy empty electron wavepacket and
low-energy occupied electron wavepacket results in quantum equilibrium noise.

Next let us consider an electron channel with a single elastic scattering center
between two reservoir electrodes as shown in Fig. 23. An elastic scattering center with a finite transmission T and reflection $(1 - T)$ coefficients, is physically
implemented into an electron channel by unintentional ionized impurity located
either inside or outside of the channel, artificially introduced channel constriction
and tunnel barrier. When there is no chemical potential difference between the
two electrodes ($\mu_1 = \mu_2$), the generalized Nyquist formula (12) still holds if the
quantum unit of conductance G_Q is replaced by TG_Q.[47] As far as a system is
in thermal equilibrium, the generalized Nyquist formula (12) holds irrespective
of the details of a dissipative element, i.e. independent of its material, shape
and size. The scattering-free ballistic conduction, diffusive conduction with many
elastic scatterings and dissipative conduction with many inelastic scatterings in
mesoscopic systems are not an exception of this general principle.[46]

However, when there is a chemical potential difference between the two electrodes ($\mu_1 > \mu_2$) as shown in Fig. 23, non-equilibrium excess noise is generated.
This is because a transmitted occupied electron wavepacket (from the left electrode) and reflected empty electron wavepacket (from the right electrode) beat
with each other, just like the quantum partition noise in the electron beam splitter situation discussed in the previous section. Similarly, a reflected occupied electron wavepacket and transmitted empty electron wavepacket produce independent
excess beat noise. This non-equilibrium excess noise has the similar quantum mechanical origin as the high-frequency quantum noise $2\hbar\omega G_Q$ discussed above but
is frequency independent white noise.

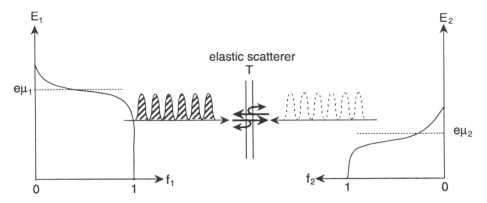

Fig. 23. Non-equilibrium noise in a ballistic electron channel with a single elastic scatterer. A transmitted (reflected) occupied electron wavepacket from the left electrode and reflected (transmitted) empty electron wavepacket from the right electrode beat with each other to produce the quantum non-equilibrium noise.

Figure 24 shows the noise current spectral density at 1 KHz normalized by the equilibrium thermal noise $4\,k_B\theta \cdot TG_Q$ vs. normalized bias voltage $eV/k_B\theta$.[48] The transmission coefficient T of an elastic scatterer is assumed to be 0.5. When a chemical potential difference $(V = \mu_1 - \mu_2)$ is smaller than the thermal voltage $k_B\theta/e$ that determines the region of stochastic emission of electrons due to Fermi-Dirac distribution, the non-equilibrium excess noise of quantum mechanical origin is smaller than thermal noise and the total noise current spectral density is dominated by thermal noise. However, at a chemical potential difference larger than the thermal voltage, $V > k_BT/e$, the non-equilibrium excess noise exceeds thermal noise and the total noise current spectral density is just one-half of full shot noise $S_i(\omega) = e\langle i \rangle = \frac{1}{2}eG_QV$. As the transmission coefficient T decreases, the total noise current spectral density approaches to a full shot noise, i.e. $S_i(\omega) = 2e(1 - T)TG_QV \longrightarrow 2e\langle i \rangle$ $(T \ll 1)$. Figure 25 shows the noise current spectral density at a bias voltage $eV/k_B\theta = 602$ vs. normalized frequency $\hbar\omega/k_B\theta$.[48] In a low-frequency region $(\hbar\omega \leq eV)$, the noise current spectral density is dominated by the non-equilibrium quantum partition noise with a spectral density of $S_i(\omega) = 2e(1 - T)TG_QV = e\langle i \rangle$ (half shot noise). On the other hand, in a high-frequency region $(\hbar\omega \geq eV)$, the noise current spectral density is dominated by the equilibrium quantum beat noise with a spectral density of $S_i(\omega) = 2\hbar\omega TG_Q = \hbar\omega G_Q$.

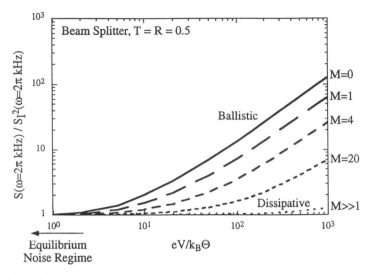

Fig. 24. For an electron 50-50% beam splitter, the current noise at 1 kHz (normalized by the equilibrium noise, eq. (12) with $G = G_Q/(M + 1)$) is plotted as a function of normalized bias voltage. As inelastic scattering increases (increasing the number of voltage probe reservoirs M), the excess noise is suppressed to the generalized Nyquist noise limit. (Ref. [48]).

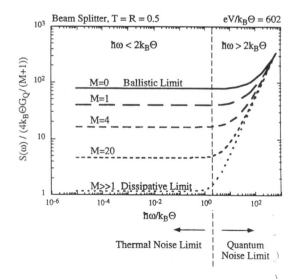

Fig. 25. The normalized current noise for an electron 50-50% beam splitter is plotted as a function of normalized frequency. For no inelastic scattering ($M = 0$), quantum non-equilibrium noise induces a large deviation from the generalized Nyquist noise. This is suppressed with increasing M to the Nyquist noise limit. (Ref. [48]).

3.4. SUPPRESSION OF ELECTRON NUMBER PARTI- TION NOISE BY INELASTIC SCATTERING

If an electron is scattered by a lattice vibration (phonon), the electron loses part of its kinetic energy by phonon emission and is scattered into either forward or backward propagating state. This inelastic scattering process in a mesoscopic electron channel is macroscopically modeled by a "high-impedance voltage probe reservoir."[49] In this voltage probe model, all incident electrons are absorbed by the reservoir and form a thermal equilibrium Fermi-Dirac distribution char- acterized by a local chemical potential as shown in Fig. 26. The local chemical potential is determined in such a way that a total outgoing current is equal to a total incoming current. The two key features in a dissipative electron transport are implemented in this voltage probe model and uniquely determine the chemical potential: Pauli exclusion principle leads to a (local) thermal equilibrium electron energy distribution and Coulomb-Coulomb interaction results in a (local) current conservation.[49]

When an external voltage V is applied to two terminal electrodes $(\mu_1 - \mu_2 = V)$ and there is no scatterer between the two, the channel conductance is $G_Q = \frac{e^2}{h}$ and the average current is $\langle i \rangle = G_Q V$ with thermal equilibrium noise $S_i(\omega)$ given by (12). If there is an electron beam splitter with $T = \frac{1}{2}$ and $V \gg \frac{k_B \theta}{e}$, the channel conductance is $G = \frac{1}{2} G_Q$ and the average current is $\langle i \rangle = \frac{1}{2} G_Q V$ with non-equilibrium noise $S_i(\omega) = e \langle i \rangle$ (one-half of full shot noise). If we add one voltage probe reservoir after the beam splitter as shown in Fig. 26, the average chemical potential of the voltage probe reservoir is just one-half between the two terminal chemical potentials, $\mu_{VP} = \frac{1}{4} \mu_1 + \frac{3}{4} \mu_2$. This average chemical potential results in average forward and backward currents with equal magnitudes. The channel conductance is $G = \frac{1}{4} G_Q$ because just one-half of the incident current into the voltage probe reservoir reaches the receiving terminal reservoir and the remaining half is reflected back. The noise current incident into the voltage probe model is also halved into the forward and backward directions, so the new noise current spectral density is $S_i(\omega) = \frac{1}{2} e \langle i \rangle$ where $\langle i \rangle = \frac{1}{4} G_Q V$. In general, with M cascade voltage probe reservoirs, the noise current spectral density is reduced to $S_i(\omega) = \frac{e \langle i \rangle}{M+1}$ where $\langle i \rangle = \frac{G_Q}{2(M+1)} V$. With increasing the number of voltage probe reservoirs, the channel conductance and non-equilibrium noise are suppressed by a same factor of $(M+1)^{-1}$. A physical origin of non-equilibrium noise suppression is a self-feedback mechanism produced by an inelastic scattering. When a current incident on the inelastic scattering reservoir increases, the chemical potential μ_{VP} increases and the backward current increases. The increased backward current counteracts the increased incident current and this compensation effect results in the suppression of forward current noise.

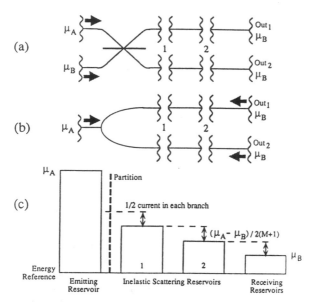

Fig. 26. An electron 50-50% beam splitter geometry (a) and an impedance matched symmetric 50-50% Y-branch (b) followed by two inelastic scattering reservoirs in each output branch. Solid arrows indicate the incident occupied electron wavepackets, and the hatched arrows indicate the incident empty electron wavepacket. (c) A schematic of the zero temperature energy distributions for electrons emitted from each of the reservoirs for the geometries above.

Figures 24 and 25 show how the non-equilibrium partition noise is suppressed and the equilibrium noise (generalized Nyquist noise) is recovered with increasing the number M of voltage probe reservoirs.[48] This (artificial) mathematical model of voltage probe reservoirs has a connection to a real physical situation if the number of voltage probe reservoirs is replaced by the channel conductance $G = \frac{G_Q}{2(M+1)}$.[50] In the case of a beam splitter configuration, the suppression of non-equilibrium noise has a linear dependence on the channel conductance as shown in Fig. 27(a). On the other hand, when an electron is split into two output ports by a Y-branch, the suppression of non-equilibrium noise has a (stronger) nonlinear dependence on the channel conductance as shown in Fig. 27(b).[50] This nonlinear suppression in the Y-branch is due to a push-pull compensation process introduced by inelastic back-scattering. The chemical potential fluctuations in the voltage probe reservoirs of the two arms are anti-correlated and a direct return current flows between the two voltage probe reservoirs suppress the noise current more efficiently than the beam splitter case.

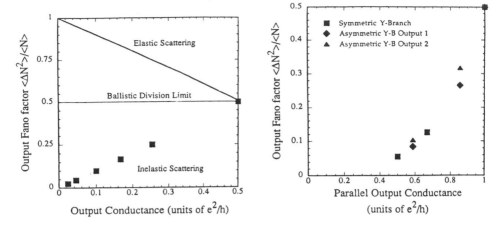

3.5. MICROSCOPIC MODEL FOR SUPPRESSION OF ELECTRON NUMBER PARTITION NOISE AND LOSS OF ELECTRON WAVE PHASE COHERENCE

The inelastic scattering process in a mesoscopic single electron channel is microscopically modeled by a Monte-Carlo numerical simulation.[51] An external bias voltage V uniquely determines the chemical potential difference between two terminal electrodes. The corresponding electron wavenumber, which contributes a current transport at $T = 0$, is divided into small segments with a wavenumber interval of Δk. A minimum uncertainty electron wavepacket with a wavenumber spread Δk and position spread $\Delta x \left(\Delta k \Delta x = \frac{1}{2}\right)$ is assumed to propagate in the channel. The system is divided into many degrees of freedom, as shown in Fig. 28, which accommodate a single electron per DOF. At $T = 0$, every electron wavepacket emitted from the left electrode is occupied. We put an electron beam splitter with a transmission coefficient T to introduce a non-equilibrium partition noise. When an electron moves from one DOF to next DOF with the same wavenumber, a computer-generated random number q between 0 and 1 is compared with a prescribed phonon scattering probability p. If the random number is greater than the scattering probability and the electron final state is empty, the electron emits an acoustic phonon with an energy of about 0.5 meV and is back-scattered (the process A in Fig. 28). Otherwise the electron moves to a next DOF with keeping its wavenumber, which includes the two cases: the random number q is smaller than the scattering probability (the process B in Fig. 28) or q is greater than p but the final state is occupied (the process C in Fig. 28). A noise current is measured at an input plane of either one terminal electrodes.

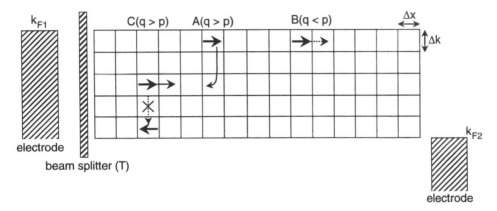

Fig. 28. A Monte-Carlo simulation model for an electron transport with inelastic scattering.

Figure 29 shows the non-equilibrium noise current normalized by full shot noise value vs. channel conductance with distributed inelastic scattering, obtained by such numerical simulation.[51] In the case of zero inelastic scattering probability, with decreasing the beam splitter transmission coefficient, the channel conductance decreases according to TG_Q and the normalized noise current increases according to $S_i(\omega)/2e\langle i \rangle = (1 - T)$. In the case of high inelastic scattering probability, the non-equilibrium noise is completely suppressed. This noise suppression originates from the final-state occupancy dependent scattering rate. In a very high inelastic scattering probability, both forward and backward propagating electrons obey thermal equilibrium Fermi-Dirac distribution rather than non-equilibrium distribution originally produced by the beam splitter. In this high inelastic scattering limit, the (macroscopic) voltage probe model and the (microscopic) Monte-Carlo simulation model give the same result. It is interesting to note that the noise current without elastic scattering center ($T = 1$) increases with an inelastic scattering probability initially but decreases with further increase in an inelastic scattering probability. Such a behavior is not predicted by the voltage probe model.

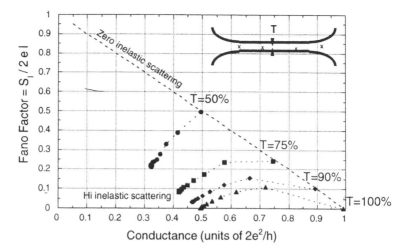

Fig. 29. Normalized noise current spectral density $S_i(\omega)/2eI$ vs. channel conductance in a mesoscopic electron channel with a beam splitter with a transmission coefficient T and distributed inelastic scattering.

Figure 30 shows the non-equilibrium noise current normalized by full shot noise value vs. channel conductance with distributed elastic scattering.[51] Irrespective of an initial noise value, the final noise current is always reduced to one-third of full shot noise in the limit of strong elastic scattering. The Pauli exclusion principle is effective to suppress partially the nonequilibrium noise even if there is no energy dissipation associated with electron back scattering.

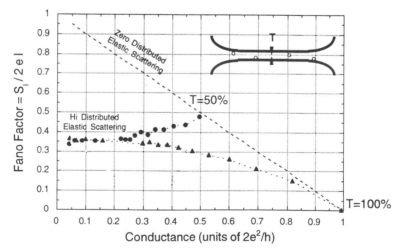

Fig. 30. Normalized noise current spectral density $S_i(\omega)/2eI$ vs. channel conductance in a mesoscopic electron channel with and without a 50-50% beam splitter and distributed elastic scatterings.

Since the non-equilibrium partition noise introduced, by a 50-50% beam splitter originates from Heisenberg uncertainty relation between the electron number difference and electron phase difference between two arms, suppression of the non-equilibrium partition noise must accompany increase in the electron phase difference noise (loss of coherence). Figure 31(a) shows an electron wave (Mach-Zehnder) interferometer model for quantum Monte-Carlo wavefunction study[52]–[54], in which an upper arm introduces only phase shift but a lower arm couples to phonon reservoirs. Ten electrons are injected one by one into the interferometer and electron numbers are counted in the two output channels as a function of acoustic phonon emission probability. Figure 31(b) shows the oscillatory behavior of the electron wave interferometer output $(N_1 - N_2)/(N_1 + N_2)$ for small and large phonon emission probabilities.[55] Figure 31(c) shows how visibility is degraded with increasing system-reservoir interaction length.[55] A discrete drop in the visibility by 1/10 is due to the emission of a single phonon in the lower arm, which localizes one electron out of total ten electrons into the lower arm. A continuous decrease in the visibility is due to no emission result of a phonon in the lower arm, which localizes an electron into the upper arm. In the example of Fig. 31(c), six electrons are found in the lower arm and four electrons must have propagated in the upper arm.

The same mechanism of phonon emission simultaneously suppresses the electron number partition noise through Pauli exclusion principle and increases the electron phase partition noise (loss of coherence). The complementarity between ΔN and $\Delta\Phi$ is preserved in this way.

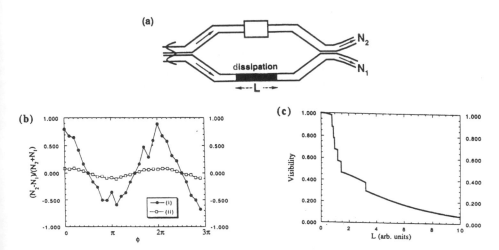

Fig. 31. Suppression of the electron wave intereference by dissipative electron-phonon coupling in an electron-interferometer: (a) Schematic structure, (b) the interferometer output vs. phase shift of one arm for two values of the length L of the dissipative region: (i) $L = 0.1\hbar k\Gamma^{-1}(k)/m^*$, and (ii) $L = 0.5\hbar k\Gamma^{-1}(k)/m^*$, (c) the visibility as a function of L. Each data point is obtained by averaging over 10 injected electrons. (Ref. [55]).

3.6. COULOMB BLOCKADE IN MESOSCOPIC ELECTRON TUNNELING AND SINGLE PHOTON TURNSTILE DEVICE

In a mesoscopic pn junction with a junction capacitance $C < e^2/2k_B\theta$, driven by a high-impedance constant current source, a single electron Coulomb blockade effect regulates individual electron tunneling or thermionic emission process. Figure 32 shows the junction voltage oscillation at a frequency $f = I/e$ due to regulated single electron thermionic emission with a time separation of $\tau = e/I$, where I is a drive current.[56] A similar behavior is expected in a pn mesoscopic tunnel junction.[57] As shown in this figure, a source resistance R_S must be large compared to a differential resistance $R = \frac{V_T}{I} = \frac{k_B T}{eI}$ of a pn junction (constant current operation) to regulate an individual electron emission or tunneling process. However, a parasitic capacitance (so-called electromagnetic environment effect) usually shunts such a small junction capacitance, which makes an ideal constant current operation of such a mesoscopic junction very difficult.

A double-barrier mesoscopic p-i-n tunnel junction shown in Fig. 33(a) has a unique feature of regulating a single electron and single hole injection process even under constant voltage operation.[58] At a junction voltage $V = V_0$, the resonant tunneling condition for an electron sub-band state is satisfied and a single electron tunnels through a potential barrier. When a single electron is captured by a quantum well, the electron resonant tunneling peak is shifted to $V_0 + e/C_1$, due to a Coulomb blockade effect where C_1 is an electron tunnel barrier capacitance, and subsequent electron tunneling is inhibited. Then we switch a junction voltage to $V_0 + \Delta V$, at which the resonant tunneling condition for a hole sub-band state is satisfied. Once a single hole tunnels through a potential barrier, the hole resonant tunneling peak is shifted to $V_0 + \Delta V + e/C_2$, where C_2 is a hole tunnel barrier capacitance, and subsequent hole tunneling is inhibited. In this way, a single electron and single hole can be injected into the quantum well per a period of junction voltage modulation. If an electron-hole pair radiative recombination lifetime is much shorter than the period of voltage modulation, a single photon is generated per a period. Figure 33(b) shows the Monte-Carlo simulation result, in which a voltage spike in each junction voltage corresponds to a single electron or single hole tunneling event and each cross indicates a time of photon emission.

Such a regulated single photon stream with a well-defined time clock may find applications in high precision current standard technologies and quantum cryptography communications.

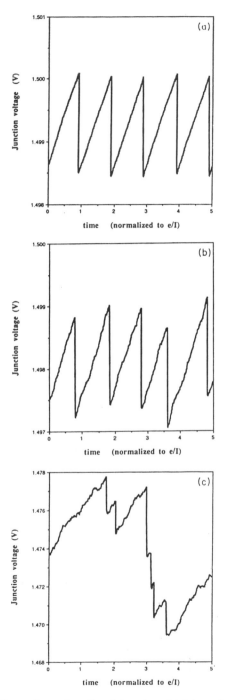

Fig. 32. The junction voltage of a pn junction driven by a high impedance constant current source as a function of time for $R_s = 300$ MΩ, $C_s = 0$, $V_{in} = 1.6$ V ($e/\langle I \rangle = 500$ ps), and (a) $T = 0.3$ K, (b) $T = 3$ K, and (c) $T = 30$ K. (Ref. [56]).

240

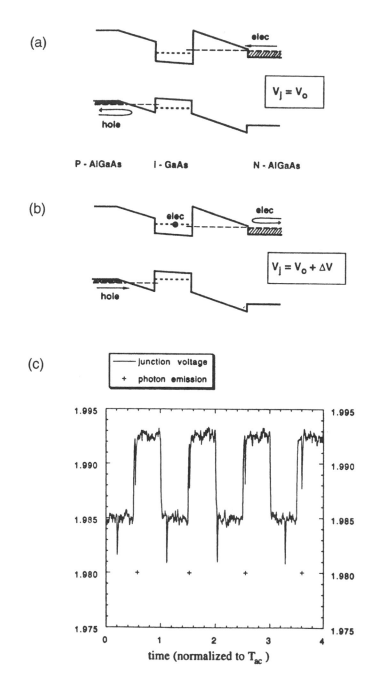

Fig. 33. The energy-band diagram of the $p - i_p - i - i_n - n$ AlGaAs-GaAs heterojunction, with applied voltage (a) $V_j = V_0$ and (b) $V_j = V_0 + \Delta V$. For $V_j = V_0 (V_j = V_0 + \Delta V)$, the Fermi energy of the $N-$ $(P-)$ type AlGaAs layer is at least $e^2/2C_{ni}(e^2/2C_{pi})$ higher (lower) than the energy of the quantum-well electron (hole) subband. (c) The junction voltage as a function of time shown together with the accompanying photon emission events. The *voltage spikes* correspond to electron and hole tunneling events. (Ref. [58]).

4. Semiconductor Cavity QED in Low-Q Regime— "Squeezing Vacuum Fluctuations in Frequency and Space"

4.1. CAVITY QED IN LOW-Q AND HIGH-Q REGIMES

Spontaneous emission of an atom is not an immutable property of an atom but is a consequence of atom-vacuum field coupling. For instance, an isotropic radiation pattern of a randomly oriented atomic dipole originates from isotropic density of states for the radiation fields in free space. An irreversible exponential decay of atom's excitation is due to continuous spectra of the radiation field modes in free space. However, such uniformity of the free-space radiation fields in space and frequency can be modified by use of a cavity wall, which results in modification of atom's spontaneous emission. This is the basic principle of cavity quantum electrodynamics (cavity QED).[59]

There are two distinctly different regimes for modification of spontaneous emission: cavity QED in low-Q (weak coupling) regime and in high-Q (strong coupling) regime. A decisive parameter to distinguish the two regimes is a normal mode splitting (vacuum Rabi frequency) given by

$$\Omega_R = \frac{d}{\hbar}\mathcal{E}_{\text{vac}} = \frac{d}{\hbar}\sqrt{\frac{\hbar\omega}{2\varepsilon_0 V}} \quad . \tag{13}$$

Here d is an atomic dipole moment, ω is an oscillation frequency and V is an optical cavity volume. When Ω_R is smaller than the decay rate of a cavity internal field γ_{photon} and that of an atomic dipole γ_{atom}, the spontaneous emission is still an irreversible process of exponential decay. However, a spontaneous emission radiation pattern and decay rate can be modified.[60]–[64] A cavity QED laser with such modified spontaneous emission has an enhanced spontaneous emission coupling efficiency and thus a reduced threshold pump rate.[64]–[66] When N atoms are confined in a volume smaller than optical wavelength and are excited coherently, a collective behavior of N atoms enhances a spontaneous emission decay rate; the phenomenon is known as a superradiance.[67] A cavity can further enhance such a superradiance decay, specifically when N atoms form a coherent dipole sheet perpendicular to the cavity axis.[68] This is indeed the situation for a resonantly excited quantum well microcavity.

When Ω_R is larger than γ_{photon} and γ_{atom}, the spontaneous emission becomes a (partially) reversible process. The atom and the cavity field coherently exchange an energy back and forth with a period of π/Ω_R. As a result of phase-reversed oscillating spontaneous emission, the spectrum features a doublet structure ("suppressed-carrier AM sidebands") with a frequency separation of $2\Omega_R$.[69]–[72] This is a general property (normal mode splitting) of two coupled harmonic oscillators. When N atoms collectively couple with a single-mode cavity field, the normal mode splitting is enhanced in proportion to $\sqrt{N}$. The single two-level atom, described by

Pauli spin operator, coupled with a single-mode cavity field is a highly nonlinear system due to the fermionic feature of Pauli spin operator. Theoretical and experimental studies on such atomic and semiconductor cavity QED systems are summarized in the recent monograph.[73] On the other hand, an assembly of N two-level atoms, coupled with a single-mode cavity field described by a collective angular momentum operator,[74] and is a mostly linear system due to the bosonic feature of collective angular momentum operator. A quantum well Wannier exciton in a semiconductor microcavity is such an example. Semiconductor cavity QED in high-Q regime is the main subject of the next chapter.

4.2. CAVITY-QED LASER

A spontaneous emission coupling efficiency β of a laser is defined as a fractional spontaneous emission rate into a single lasing mode divided by a total spontaneous emission rate. If a laser cavity volume is much larger than the cube of optical wavelength ($V \gg \lambda^3$), there are many cavity modes in the gain bandwidth of an active medium and a spontaneous emission coupling efficiency into each cavity mode is identical. The spontaneous emission coupling efficiency β for such a case is simply an inverse of number of cavity modes within the gain bandwidth and is given by[75]

$$\beta = \frac{\lambda^4}{4\pi^2 V n^3 \Delta\lambda} \quad , \tag{14}$$

where $\Delta\lambda$ is the gain bandwidth and n is the refractive-index of the cavity medium. An ordinary laser has an extremely small β value ($\simeq 10^{-5} \sim 10^{-10}$) due to either a large cavity volume or broad gain bandwidth or both. In a laser with $\beta = 10^{-10}$, for instance, only if 10^{10} photons are emitted spontaneously per cavity photon lifetime, one photon is captured by the cavity and this single photon serves as a seed for stimulated emission. This is the threshold condition of a laser, i.e. $P_{\text{th}} = \gamma/\beta$, where $\gamma = 1/\tau_{\text{ph}}$ is the cavity photon decay rate. If an optical cavity volume is decreased (to an optical wavelength size), the number of cavity modes is decreased and thus β can be increased. Moreover, by modifying a spontaneous radiation pattern to couple into a lasing mode preferentially, β can be further enhanced. Such a laser with an enhanced β is called a cavity QED laser.

The rate equations for excited atom population N and photon number n are[76]

$$\frac{d}{dt}N = P - \frac{N}{\tau_{\text{sp}}} - g(N - N_0)n \quad , \tag{15}$$

$$\frac{d}{dt}n = \beta\frac{N}{\tau_{\text{sp}}} - \frac{n}{\tau_{\text{ph}}} + g(N - N_0)n \quad , \tag{16}$$

where P is a pump rate, τ_{sp} is a spontaneous lifetime of the atoms, g is a gain coefficient and N_0 is unexcited atom population. The spontaneous emission rate into a lasing mode should be equal to the stimulated emission rate into the same mode induced by one photon, i.e. $\beta\frac{N}{\tau_{\text{sp}}} = gN$. This is Einstein's relationship between A and B coefficients.[77] A conventional definition of the laser threshold is such that a (unsaturated) net stimulated emission gain is equal to a cavity loss: $g(N - N_0) = \gamma$. This definition leads to a following expression of the threshold pump rate,

$$P_{\text{th},1} = \frac{\gamma}{\beta}(1 + \xi)(1 + \beta) \quad . \tag{17}$$

Here $\xi = \frac{\beta N_0 \tau_{\text{ph}}}{\tau_{\text{sp}}}$ is a photon number of the lasing mode at a transparency point $(N = N_0)$.[76] A somewhat new definition of the laser threshold is such that a photon number of the lasing mode becomes one: $n = 1$, where phase coherent stimulated emission exceeds incoherent spontaneous emission. This second definition leads to a different expression of the threshold pump rate,

$$P_{\text{th},2} = \frac{\gamma}{2\beta}[1 + \beta + \xi(1 - \beta)] \quad . \tag{18}$$

When $\xi \ll 1$ (due to small β or small N_0 or small τ_{ph} or large τ_{sp}), the two definitions of the laser threshold (17) and (18) give identical results within a factor of two.

Figure 34(a) shows the average photon number of a lasing mode vs. the pump current for a semiconductor as a function of β,[76] where $N_0 = 10^3$, $\tau_{\text{sp}} = 10^{-9}$ s, $\tau_{\text{ph}} = 10^{-12}$ s and thus $\xi = \beta \leq 1$. The laser phase transition behavior is obvious from a jump in a differential quantum efficiency due to onset of stimulated emission. When $\beta \ll 1$, the expression (17) is reduced to $P_{\text{th},1} = \frac{\gamma}{\beta}$ which gives an upper bound on the laser threshold (the approaching point of a differential quantum efficiency to an above-threshold value of one), while the expression (18) is reduced to $P_{\text{th},2} = \frac{\gamma}{2\beta}$ which gives a lower bound on the threshold (the departure point of a differential quantum efficiency from a sub-threshold value β). The laser threshold is decreased inversely proportional to β, and when $\beta = 1$, the threshold behavior disappears. This behavior is referred to as a thresholdless (or zero-threshold) laser.[65] The low-frequency intensity noise spectral density normalized by a standard shot noise value vs. the pump current is shown in Fig. 34(b) as a function of β.[76] It is assumed that the pump current carries a full shot noise, and thus the noise spectral densities well below and well above the threshold are shot-noise-limited. The excess intensity noise observed at a threshold is due to amplified spontaneous emission. The threshold behavior also disappears in the intensity noise characteristic when $\beta = 1$. It is this disappearance of the threshold behaviors in the average and fluctuation of the laser intensity that a cavity QED laser with $\beta = 1$ is called a thresholdless device.[78]

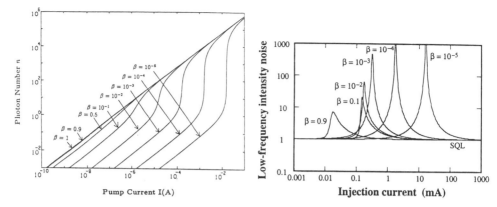

Fig. 34. (a) Internal photon number n versus pump current I as a function of spontaneous emission coefficient β. $n_{\mathrm{sp}} = 1$ and $\gamma = 10^{12}$ s^{-1}. (b) Output intensity noise spectral density at low-frequency vs. pump current I as a function of β.

However, if the stimulated emission gain (proportional to excited atom population) and the spectral linewidth of the field are plotted as a function of pump current (Fig. 35 (a) and (b)), we can immediately know that a definite laser threshold exists even for a cavity QED laser with $\beta = 1$.[76] A dominant decay process for excited atom population is a spontaneous emission with a constant lifetime τ_{sp} at below the threshold but is a stimulated emission with a decreasing lifetime $\tau_{\mathrm{st}} = 1/gn = \frac{\tau_{\mathrm{sp}}}{\beta n}$ at above the threshold. Thus, the excited atom population increases linearly with pump rate at below the threshold but saturates to the value, $N_0 + \gamma \tau_{\mathrm{sp}}/\beta$, at above the threshold due to a decreasing stimulated emission lifetime, as shown in Fig. 35 (a). The cavity QED laser with $\beta = 1$ is a linear amplifier at below the threshold but a nonlinear (saturated) oscillator at above the threshold.

The photon is emitted mainly by incoherent spontaneous emission at below the threshold but by phase coherent stimulated emission at above the threshold. Formation of phase coherence, which is manifested by a decreased linewidth, is another signature of the laser phase transition. As shown in Fig. 35 (b), the linewidth of a lasing mode is equal to a cold cavity linewidth at below the threshold, which indicates the generated light is simply a cavity-filtered incoherent spontaneous emission. On the other hand, the linewidth at above the threshold decreases inversely proportional to the pump current (characteristic of Schawlow-Townes linewidth), which indicates the coherent excitation is produced in the lasing mode. This distinct threshold behavior is observed even in a cavity QED laser with $\beta = 1$.[76]

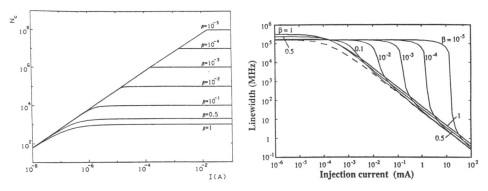

Fig. 35 (a) Electron number N_c versus pump current I as a function of spontaneous emission coefficient β. $n_{sp} = 1$. $\gamma = 10^{12}$ s^{-1} and $\tau_{sp} = 10^{-9}$ s. (b) Spectral linewidth vs. pump current I as a function of β.

The numerical examples in Fig. 34 and Fig. 35 correspond to a microcavity semiconductor laser with small active volume ($V \simeq 1\mu m \times 1\mu m \times 0.01\mu m = 10^{-14}$cm^3) and a small transparency carrier density ($N_0/V = 10^{17}$cm^{-3}). In such a case, a population inversion ($N > N_0$) is formed in an active region in an entire pump current region of interest and thus $\xi \leq 1$. In a conventional edge-emitting or large area vertical cavity surface-emitting semiconductor laser with a large active region ($V \simeq 10^{-11}$cm^3) and a large transparency carrier density ($N_0/V = 10^{18}$cm^{-3}), the threshold pump rate is not decreased by increasing β but is determined by an inversion condition. Figure 36(a) and (b) show such a numerical example,[79] in which the threshold pump rate for $10^{-1} > \beta > 10^{-4}$ is independent of β and is given by the condition of creating a population inversion $P_{th} = N_0/\tau_{sp}$. The two expressions (17) and (18) for a laser threshold are reduced to this population inversion condition when $\xi \gg 1$ and $\beta < 1$. It is not very useful to increase β by a cavity QED technique for such a semiconductor laser with a large active volume, unless β is made very close to one. When β is made very close to one, a new regime of so-called "laser without inversion" emerges, which will be discussed later.

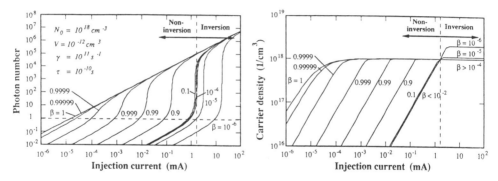

Fig. 36. (a) The mean photon number in a lasing mode (left axis) vs. the pump current. It is assumed that $\xi = 10^5\beta$, so ξ is larger than unity for $\beta > 10^{-3}$. (Ref. [79]). (b) The mean carrier density vs. the pmup current. It is assumed that $N_0 = 10^{18}$ cm^{-3}. (Ref. [79]).

4.3. ENHANCED SPONTANEOUS EMISSION COUPLING EFFICIENCY IN MICROCAVITIES

Let us consider a confocal optical cavity with a fixed mode acceptance angle Ω in which excited atoms (or electron-hole pairs) are injected into the center of the cavity. If the atomic transition frequency is on-resonance with the cavity mode and the atomic linewidth is narrower than the cavity (mode) linewidth as shown in Fig. 37, the spontaneous emission coupling efficiency β into a single cavity mode is given by[80]

$$\beta = \frac{1}{2}\frac{\Gamma_{\text{cavity}}}{\Gamma_{\text{cavity}} + \Gamma_{\text{free space}}} = \frac{1}{2}\frac{\frac{3}{8\pi}\Omega\frac{1}{1-R}}{\frac{3}{8\pi}\Omega\frac{1}{1-R} + \left(1 - \frac{3}{8\pi}\Omega\right)} \quad . \tag{19}$$

The spontaneous emission rate into a single cavity mode is enhanced by a factor of $(1 - R)^{-1}$, while the spontaneous emission rate into (open) free space is unaltered. The effective solid angle of spontaneous emission in free space available for a randomly oriented atomic dipole is not 4π but $\frac{8\pi}{3}$ steradians, and thus the fractional spontaneous emission rate into the cavity mode with a solid acceptance angle Ω without taking into account the cavity enhancement effect is $\frac{3}{8\pi}\Omega$. Equation (19) approaches one-half when the mirror reflectivity R goes to one. A factor $1/2$ in (19) originates from identical spontaneous emission coupling into two degenerate polarization modes. To achieve this ultimate β value, however, the cavity linewidth which is decreased by a factor of $(1 - R)^{-1}$ must be broader than the atomic linewidth. This requires a very small cavity length (of an optical wavelength size specifically for a broad semiconductor gain medium).

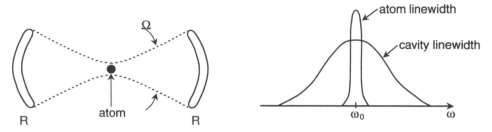

Fig. 37. A confocal optical cavity with enhanced spontaneous emission rate into a lasing mode, in which the atomic linewidth is narrower than the cavity linewidth.

It is easily understood that a spontaneous decay rate $1/\tau_{\rm sp}$ is also enhanced by a factor of $1+\frac{3}{8\pi}\Omega\left(\frac{R}{1-R}\right)$. However, once the cavity linewidth becomes comparable to the atomic linewidth, the enhancement of the spontaneous emission coupling efficiency β and spontaneous decay rate $1/\tau_{\rm sp}$ stops.

A simplest semiconductor microcavity is a planar microcavity with a single quantum well located at the center of a half-wavelength ($\lambda/2$) or one-wavelength (λ) optical cavity layer which is sandwiched by top and bottom distributed Bragg reflectors (DBRs), as shown in Fig. 38. The spontaneous emission coupling efficiency β for a horizontal dipole (dipole moment parallel to a plane of a quantum well) in such a GaAs/AlGaAs planar microcavity is plotted in Fig. 39 as a function of the normalized atomic linewidth $\Delta\lambda_e/\Delta\lambda_c$, where the cavity linewidth $\Delta\lambda_c$ is kept constant ($= 0.4$ nm).[81] A maximum spontaneous emission coupling efficiency $\beta_{\rm max}$ attainable in this GaAs/AlGaAs planar microcavity is ~ 0.05. This value is independent of the mirror reflectivity R in contrast to the confocal cavity, because the peak spontaneous emission rate into a cavity normal direction is enhanced by increasing R but the cavity mode acceptance angle Ω decreases, and thus the total integrated spontaneous emission intensity over the mode acceptance angle is constant. Most of spontaneous emission is radiated into a plane of a quantum well in such a planar microcavity structure. However, $\beta_{\rm max}$ can be increased by increasing a refractive-index difference Δn of the DBR materials, as shown in Fig. 40, because the mode acceptance angle Ω is a monotonically increasing function of Δn. The polarization factor $1/2$ is not included in Fig. 40, so the quantum efficiency counts for the two orthogonal polarization modes.[82] For instance, a GaAs/air ($n = 1$) planar microcavity has a β value close to 0.5 (theoretical limit) for a horizontal dipole and 0.4 for a random dipole.

248

When the atomic linewidth becomes broader than the cavity linewidth, β decreases in proportion to $(\Delta\lambda_e/\Delta\lambda_c)^{-2}$ as shown in Fig. 39. To obtain a high β value in a planar microcavity, the mirror reflectivity cannot be increased arbitrarily.

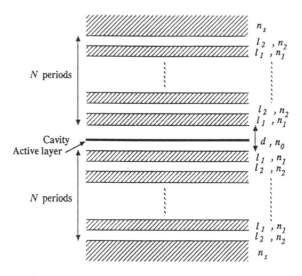

Fig. 38. A microcavity structure with a single quantum well and one-dimensional distributed Bragg reflectors.

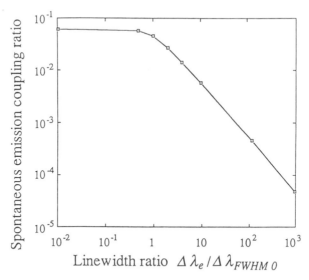

Fig. 39. The spontaneous emission coupling factor β of a GaAs/AlGaAs planar microcavity as a function of the emission linewidth normalized by the cavity line width $\Delta\lambda_{em}/\Delta\lambda_{FWHM0}$. (Ref. [81]).

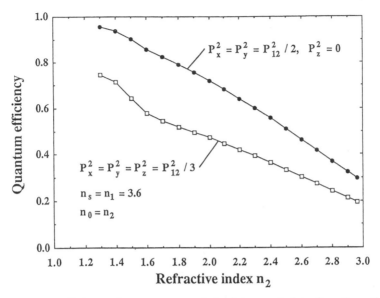

Fig. 40. Coupling efficiency of spontaneous emission into a main lobe (two times the spontaneous emission coefficient β) versus refractive index n_2 of DBR material, in which the other DBR material is GaAs ($n_1 = 3.6$). It is assumed that $\Delta\lambda_{em} \ll \Delta\lambda_{FWHM0}$. (Ref. [82]).

As shown in Fig. 34 (a) and Fig. 35 (b), the sudden increase in the differential quantum efficiency and sudden decrease in the spectral linewidth at a laser threshold are equal to the spontaneous emission coupling efficiency β. We can measure a β value of a microcavity using these two relations. Figure 41 shows the observed output power vs. the pump power of a single quantum well one-wavelength GaAs/AlGaAs microcavity at 4 K.[83][84] The β value was estimated ~ 0.01 from this result. Figure 42 shows the spectral linewidth vs. the pump power of the same sample.[84] The same β value (~ 0.01) was obtained from the independent experiment. The emission linewidth (~ 2 nm) at 4 K normalized by the cold cavity linewidth (~ 0.4 nm) is about 5 and this measurement result ($\beta \sim 0.01, \Delta\lambda_e/\Delta\lambda_c \sim 5$) is compatible with the theoretical curve shown in Fig. 39.[85]

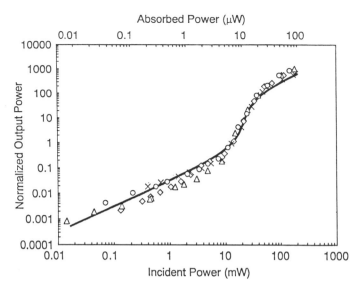

Fig. 41. Some overlaid 3.4 μm diameter microlaser input/output data and the theoretical curve for $\beta = 9 \times 10^{-3}$ fitted to the threshold power. The ordinate is normalized to unity at threshold, and thus gives the theoretical number of photons in the lasing mode. The threshold values of different lasers scatter by ±40%. (Ref. [84]).

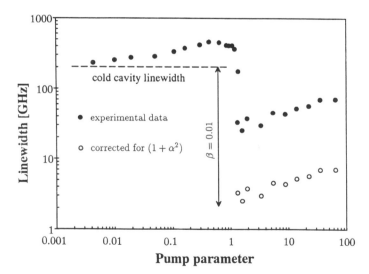

Fig. 42. Laser linewidth of the planar microcavity vs. pump power (normalized to unity at threshold). (Ref. [84]).

The (absorbed) threshold pump power for an excitation spot size of 3.4 μm in diameter was 6.7 μW (Fig. 41),[84] which can be compared to the theoretical value $P_{\text{th}} = \frac{\gamma}{2\beta}\hbar\omega \simeq 6\mu\text{W}$ for $\gamma = 4 \times 10^{11} 1/s (\tau_{\text{ph}} = 2.5 \text{ psec})$ and $\beta = 10^{-2}$. This threshold optical pump power corresponds to an electrical current of $4\mu\text{A}$.

To obtain a high β value, a full three-dimensional confinement of the electro-magnetic field is preferred. A micropost Fabry-Perot laser is fabricated by an electron beam lithography and (ECR or RIE) dry etching technique and the post size is as small as 0.5 μm in diameter. It is theoretically predicted that β can be increased to the theoretical limit (~ 0.5) with decreasing the post diameter.[86] A hemispherical Fabry-Perot laser is fabricated by a photolithography and (RIE) dry etching technique and the top microlens structure is as small as a few μm in diameter.[87] It is theoretically predicted that β is as high as ~ 0.5 and no degradation of β due to a finite atomic linewidth is realized up until $\Delta\lambda_e/\Delta\lambda_c \simeq 10$.[81] A micro-disk laser with a whispering-gallery mode resonator is fabricated by a selective wet etching technique and is expected to have a very high β value (~ 0.7).[88]

4.4. LASERS WITHOUT INVERSION

The two definitions of the laser threshold, eqs. (17) and (18), predict very different results when $\beta \simeq 1$ and $\xi \gg 1$. According to the conventional definition (gain = loss) (17), the laser threshold in this limit is given by $P_{th,1} \simeq N_0/\tau_{sp}$. This is the condition for creating population inversion in a gain medium. On the other hand, the new definition (stimulated emission = spontaneous emission) (18) results in $P_{th,2} \simeq \gamma$. This is the condition for populating one photon in a lasing mode. When $\beta \simeq 1$ and $\xi \gg 1$, the second threshold pump rate is much smaller than the first threshold pump rate, $P_{th,2} \ll P_{th,1}$, i.e. the lasing mode acquires more than one photon before the medium is inverted.

As shown in Fig. 36 (a) and (b), the two distinct features of the laser phase transition, sudden increase in the differential quantum efficiency and clamping of the carrier population, are observed in the pump region where the medium is "not" inverted, if β is greater than 0.1. These numerical results suggest the new definition (18) is more appropriate to describe a laser threshold in the regime of $\beta \simeq 1$ and $\xi \gg 1$. A cavity QED laser with $\beta \simeq 1$ can lase without population inversion.[79][89] A physical mechanism for the laser without inversion is a "photon recycling" which is a repetition of (real) absorption by a non-inverted medium and single-mode stimulated emission induced by a cavity mode field. As far as β is close to one, this photon recycling is a dissipation-free process and effectively increases a carrier lifetime $\tau_{sp,eff}$. The carrier population $P\tau_{sp,eff}$ is increased to the threshold value N_0 even for a small pump rate P due to this increase in an effective carrier lifetime $\tau_{sp,eff}$, as shown in Fig. 36 (b).

Study on intensity squeezing and quantum correlation between carrier population and photon number also suggests the laser phase transition should be associated with the new definition (18) and supports "laser without inversion" in microcavities.[89]

4.5. CAVITY ENHANCED EXCITONIC SUPERRADI-ANCE

A quantum well Wannier exciton is a collective excitation of many unit cells (dipole moments). If an exciton is resonantly excited by a coherent optical field and an excitation spot size exceeds an exciton Bohr area, even a single exciton acquires a spatially extended coherent dipole moment and thus a radiative lifetime is decreased with increasing an excitation spot size.[90]–[92] An uncorrelated electron-hole pair has an effective dipole moment corresponding to a single unit cell (~ 1 Å for a GaAs) which results in a radiative lifetime of electron-hole pairs (~ 3 nsec for a GaAs). On the other hand, a quantum well Wannier exciton has a much shorter radiative lifetime (~ 20 psec for a GaAs) due to the above mentioned excitonic superradiance effect.[90]–[92] Even if a very large area of the quantum well is coherently excited, the excitonic superradiance lifetime is not shortened to below the certain limit. An exciton superradiance lifetime is independent of the excitation spot size once it exceeds an optical wavelength in a medium. This is because the increase in a dipole moment is compensated for by the decrease in the field density of states available for photon emission. The isotropic field density of states is fully available for a point dipole because the radiation pattern is also isotropic. However, a large area exciton dipole has a narrow radiation pattern along the normal direction and thus the relevant final field density of states is relatively small.

If a quantum well is embedded inside a planar microcavity in which the field density of states is highly concentrated in the normal direction, it is expected that an excitonic superradiance lifetime is decreased with increasing an excitation spot size. Figure 43 (solid lines) shows the decay rate (radiative lifetime) of a coherently excited exciton vs. a half-wavelength microcavity mirror transmittivity as a function of an excitation spot size. When the mirror reflectivity is $R = 0$ (bare quantum well), the radiative lifetime is independent of an excitation spot size as mentioned above. With increasing R, the radiative lifetime is decreased to less than 1 psec (depending on the excitation spot size).[68] Dashed lines in Fig. 43 correspond to the case that there is inhomogeneous broadening for a quantum well exciton frequency. A cavity photon lifetime τ_p is also plotted in Fig. 43. If the excitonic radiative lifetime becomes shorter than the photon lifetime (corresponding to a high reflectivity region, i.e. left side of the crossover point), the system belongs to cavity-QED in high-Q regime and the exciton and the field exchange energy back and forth coherently. Thus, the effective excitation decay rate becomes slower.

This crossover characteristic from a cavity enhanced excitonic superradiance in low-Q regime to a microcavity exciton polariton splitting in high-Q regime is theoretically studied.[93] We will discuss the coupled exciton-field system in high-Q regime in detail in the next section.

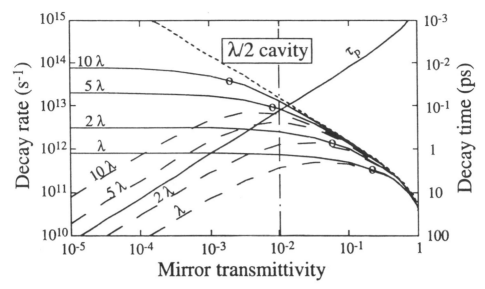

Fig. 43. The decay time as a function of the mirror reflectivity for a half-wavelength-long cavity. The cavity photon lifetime τ_p is also indicated. The circles indicate the highest reflectivity where it is still possible to form an excitation spot with the given radius by optical means. To the left of the circles the lateral exciton confinement must be achieved by other means. The dashed lines correspond to the decay time of an inhomogeneously broadened exciton transition line with a 2.5 nm linewidth. In the region to the left of the vertical dash-dotted line it is not possible to excite the entire spectral range of the exciton line due to the chromatic filtering of the cavity. (Ref. [68]).

4.6. APPLICATIONS OF CAVITY QED LASERS

Short distance optical communications, such as a local area network, board-to-board connection, chip-to-chip connection and gate-to-gate connection, do not require a high laser power, because such a short distance communication link does not have high transmission loss. An on-off modulated signal with a Poisson photon distribution must have only 21 photons to achieve a bit error rate of less than 10^{-9}, if there is no additional noise in a receiver. If we impose 10 dB power penalty due to an optical loss between a laser and a receiver and thermal noise of the receiver circuit, we still need only 210 photons for an on-pulse. This receiver level corresponds to an optical power of 10 nW if a bit rate is ~ 1 Gbit/s. It is easily understood that a conventional edge-emitting semiconductor laser, which dissipates an input energy of ~ 10 mW into spontaneous emission before reaching a laser threshold, is not suitable for those applications. On the other hand, a cavity QED laser with $\beta \simeq 1$ and $\gamma \simeq 10^{11}$ s^{-1} should have a laser threshold current of ~ 16 nA and dissipates an energy of ~ 20 nW before reaching the threshold. Low power dissipation and low heat generation are specifically important for large scale integration of two-dimensional array lasers. For most of the above applications, a laser is not necessary. A cavity QED light emitting diode with $\beta \simeq 1$ (operating below a laser threshold) is free from various nonlinear effects such as mode competition noise and relaxation oscillation and thus preferable.

If a microcavity laser is integrated on 5 inch wafer with a $10\mu m$ spacing, a total number of lasers on the chip is 10^8. Even though a yield (probability of making a good laser) is 1% and a wafer process time is 10 hours, an effective fabrication time per laser is only ~ 0.04 sec.

5. Semiconductor Cavity QED in High-Q Regime— "Dressing Excitons with Electromagnetic Vacuum"

5.1. DRESSED FERMION VS. DRESSED BOSON

When a single two-level atom is resonantly coupled with a single-mode field, such a coupled atom-field system forms a new normal mode ("dressed fermion") which is an eigenstate of the (total) system Hamiltonian:

$$\mathcal{H} = \frac{1}{2}\hbar\omega\sigma_z + \hbar\omega\left(a^\dagger a + \frac{1}{2}\right) + \hbar\Omega(a\sigma_+ + a^\dagger\sigma_-) \quad . \tag{20}$$

Here the field operator $a, a^\dagger$ and the atom operator σ_+, σ_- satisfy the boson commutation relation, $[a, a^\dagger] = 1$ and fermion anti-commutation relation, $\{\sigma_+, \sigma_-\}_+ = 1$, respectively.

Figure 44 shows the eigen-frequencies of such dressed fermions of $N = 0, N = 1$ and $N = 2$ excitation manifolds. The $N = 0$ excitation manifold has a single state $|0\rangle_f|g\rangle_a$ of vacuum field and ground state atom. The $N = 1$ excitation manifold has two degenerate bare states $|1\rangle_f|g\rangle_a$ and $|0\rangle_f|e\rangle_a$, which are split into the two non-degenerate dressed states $|1, -\rangle = \frac{1}{\sqrt{2}}(|1\rangle_f|g\rangle_a - |0\rangle_f|e\rangle_a)$ and $|1, +\rangle = \frac{1}{\sqrt{2}}(|1\rangle_f|g\rangle_a + |0\rangle_f|e\rangle_a)$ with frequency separation of 2Ω. The emission spectrum from the $N = 1$ to the $N = 0$ excitation manifold has double peaks, which is called a vacuum Rabi splitting or normal mode splitting. The $N = 2$ excitation manifold has two degenerate bare states $|2\rangle_f|g\rangle_a$ and $|1\rangle_f|e\rangle_a$, which are split into the two non-degenerate dressed states $|2, -\rangle = \frac{1}{\sqrt{2}}(|2\rangle_f|g\rangle_a - |1\rangle_f|e\rangle_a)$ and $|2, +\rangle = \frac{1}{\sqrt{2}}(|2\rangle_f|g\rangle_a + |1\rangle_f|e\rangle_a)$ with frequency separation of $2\sqrt{2}\Omega$. The emission spectrum from the $N = 2$ to $N = 1$ excitation manifold has four peaks which correspond to the four possible transition lines as shown in Fig. 45. In general, the dressed states of the N excitation manifold are split by $2\Omega\sqrt{N}$ and the emission spectrum from the N to the $N - 1$ excitation manifold makes a continuous transition from the four peaks to the three peaks (Mollow's triplet) with increasing N, because the two central transition lines become degenerate. The dressed fermion is a highly nonlinear system and the nonlinearity stems from the fermion feature of the atomic spin operator.

When two harmonic oscillators are resonantly coupled, such a coupled harmonic oscillator system also forms a new normal mode ("dressed boson"), which is an eigenstate of the (total) system Hamiltonian:

$$\mathcal{H} = \hbar\omega \left(a^\dagger a + \frac{1}{2} \right) + \hbar\omega \left(b^\dagger b + \frac{1}{2} \right) + \hbar\Omega(ab^\dagger + a^\dagger b) \quad . \tag{21}$$

Here both field operators satisfy the boson commutation relation $[a, a^\dagger] = [b, b^\dagger] = 1$. If $M(\gg 1)$ identical atoms are resonantly coupled with a single-mode field and the excitation is much smaller than the number of atoms ($N \ll M$), the above dressed boson is a valid concept. A coupling constant Ω in (21) is not that of a single atom-field coupling constant Ω_s but is enhanced to $\Omega_s\sqrt{M}$ by a collective behavior of M atoms. A Wannier exciton in a semiconductor quantum well is similar to such an assembly of M identical two-level atoms. Each lattice site having a unit dipole moment corresponds to an individual two-level atom. The dressed bosons up to the $N = 1$ excitation manifold are identical to the corresponding dressed fermions as shown in Fig. 44. The $N = 2$ excitation manifold has three degenerate bare states $|2\rangle_a|0\rangle_b$, $|1\rangle_a|1\rangle_b$, and $|0\rangle_a|2\rangle_b$, which are split into the three non-degenerate dressed states $|2, -\rangle = \frac{1}{2}|2\rangle_a|0\rangle_b - \frac{1}{\sqrt{2}}|1\rangle_a|1\rangle_b + \frac{1}{2}|0\rangle_a|2\rangle_b$, $|2, 0\rangle = -\frac{1}{\sqrt{2}}|2\rangle_a|0\rangle_b + \frac{1}{\sqrt{2}}|0\rangle_a|2\rangle_b$ and $|2, +\rangle = \frac{1}{2}|2\rangle_a|0\rangle_b + \frac{1}{\sqrt{2}}|1\rangle_a|1\rangle_b + \frac{1}{2}|0\rangle_a|2\rangle_b$, with frequency separation of 2Ω. The emission spectrum from the $N = 2$ to $N = 1$ excitation manifold is, however, identical to the emission spectrum for the $N = 1$ to $N = 0$ manifold (double peak), even though there are six possible transition lines. This is because the two pairs of the central transition lines are degenerate in frequency and the two out-most transition lines have a zero transition matrix element. The emission spectrum always has double peaks irrespective of the excitation manifold because all the transition lines except the degenerate two central lines have a zero transition matrix element. The dressed boson is a perfectly linear system.

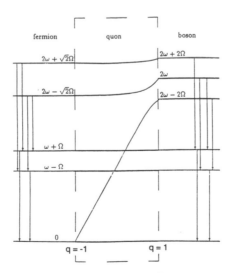

Fig. 44. The eigen-frequencies of a dressed fermion, dressed quon and dressed boson in the $N = 1$ and $N = 2$ excitation manifolds.

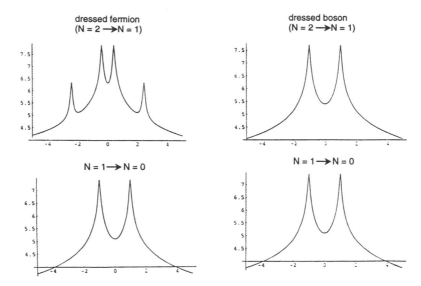

Fig. 45. The emission spectra of the dressed fermion and dressed boson from the $N = 2$ to $N = 1$ manifold and the $N = 1$ to $N = 0$ manifold.

5.2. MICROCAVITY EXCITON POLARITON —DISPERSION, NORMAL MODE SPLITTING AND OSCILLATING SPONTANEOUS EMISSION—

A microcavity exciton polariton is a normal mode of a strongly coupled QW Wannier exciton and microcavity field (dressed boson). A particular microcavity structure used in the experiment has a 200 Å GaAs single quantum well (SQW) embedded at the center of one optical wavelength cavity layer which is sandwiched by the top (bottom) distributed Bragg reflectors consisting of 19 pairs (30 pairs) of AlAs/AlGaAs quarter-wavelength stacks. The samples were grown either by MBE or MOCVD technique. In general, MBE samples have smaller inhomogeneous broadening of QW Wannier excitons than MOCVD samples. The sample was installed in a 4 K cryostat and optically pumped either resonantly (direct excitation of $k_{//} \simeq 0$ excitons) or non-resonantly (excitation of electron-hole pairs at high above the band edge). The three lasers, such as a femtosecond-pulse Ti:Al$_2$O$_3$ laser, single-frequency tunable ring-cavity Ti:Al$_2$O$_3$ laser and single-mode external grating feedback semiconductor laser, were used as a pump source. The pump wave was incident on the sample with a slight off-set angle ($3 \sim 5°$) and the emission from the microcavity was measured along the cavity normal direction, unless stated otherwise. The emission spectrum from the microcavity was measured by a spectrometer with a spectral resolution of 0.02 nm and the temporal behavior was measured by a streak camera with a time resolution of 1.5 psec.

The dispersion characteristics (energy $\hbar\omega$ vs. in-plane wavenumber $k_{//}$) of upper and lower microcavity exciton polaritons as well as a bare microcavity photon and QW Wannier exciton are illustrated in Fig. 46 as a function of a normalized exciton-photon coupling constant $g = (\Omega/\omega)^2$. The microcavity photon acquires a "mass" of $m_{ph} \simeq 3.5 \times 10^{-35}$ Kg due to the cavity dispersion, i.e. it features a parabolic dispersion curve near $k_{//} = 0$. The QW Wannier exciton has a much heavier mass of $m_{ex} = 5 \times 10^{-31}$ Kg and so the parabolic dispersion character is hardly seen in Fig. 46. These two bare states form two dressed states, upper and lower exciton polaritons. At $k_{//} = 0$ (resonant point), the upper and lower exciton polaritons are "half-photon" and "half-exciton". With increasing $k_{//}$, the upper exciton polariton approaches the cavity photon dispersion (photon-like polariton) and the lower exciton polariton approaches the QW Wannier exciton dispersion (exciton-like polariton). The normal mode splitting 2Ω is calculated by[94]–[96]

$$2\Omega = \omega \sqrt{\frac{4\pi e^2}{\varepsilon_r m_0 \omega^2 L_c} \frac{f}{A}} \quad , \tag{22}$$

where ω is the center angular frequency, L_c is the effective cavity length and $\frac{f}{A}$ is the oscillator strength per area of the quantum well. The collective enhancement of the atom-field coupling constant by N atoms (N unit cell dipoles in the present case) is included in the oscillator strength per area $\frac{f}{A}$ in (22). The number of unit cell dipoles in a typical planar microcavity mode area of 6 μm in diameter is $N \simeq 10^8$. A GaAs SQW one wavelength microcavity has a normal mode splitting on the order of $2\hbar\Omega \simeq 3$ meV.[94]–[96] If N_{QW} quantum wells are embedded at the anti-mode position of the cavity field, the normal mode splitting is enhanced by a factor of $\sqrt{N_{QW}}$. The normal mode splitting of a half-wavelength cavity is slightly larger than that of a one-wavelength cavity because of smaller effective cavity length.

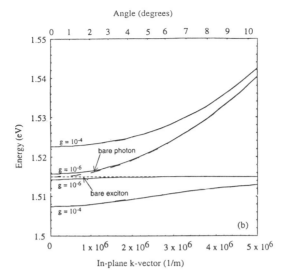

Angle (degrees)

In-plane k-vector (1/m)

Fig. 46. The dispersion for an ideal planar microcavity exciton-polariton for two different coupling constants $g = (\Omega/\omega_0)^2$. The solid lines denote the coupled mode dispersion of upper and lower exciton-polariton branches. The dotted lines denote bare exciton dispersion and the dashed line the cavity-mode dispertion. The longitudinal (optical inactive) exciton branch coincides with the bare exciton (dotted line) to within the resolution of the figure. At the top of the graph the propagation angle inside the GaAs microcavity is shown.

If the microcavity photon field frequency ω_p is not resonant with the $k_{//} = 0$ QW Wannier exciton frequency ω_{ex} and both bare systems dissipate to their reservoirs by respective decay rates γ_p and γ_{ex}, the normal mode splitting is given by[96]

$$2\Omega' \simeq \sqrt{4\Omega^2 + (\omega_p - \omega_{ex})^2 - \frac{1}{2}(\gamma_p - \gamma_{ex})^2} \quad . \tag{23}$$

When the QW Wannier exciton frequency ω_{ex} is inhomogeneously broadened due to QW thickness fluctuation and the inhomogeneous width exceeds the (potential) normal mode splitting (22), the actual normal mode splitting is decreased because the oscillator strength per area $\frac{f}{A}$ is decreased due to decreased number of unit dipole cells which effectively couples with the microcavity field. Moreover, the emission spectra of different QW exciton sub-groups further smear out the normal mode splitting with each other due to distributed center frequencies. In this way, the normal mode splitting is decreased drastically once the inhomogeneous broadening width exceeds the (potential) normal mode splitting (22). On the other hand, if the inhomogeneous broadening width is smaller than the (potential) normal mode splitting (22), the inhomogeneous broadening hardly affects the magnitude of the normal mode splitting. This is also the case for the effect of the homogeneous broadening width on the normal mode splitting (see eq. (23)).

Figure 47 show the experimental and theoretical emission spectrum from the MBE grown GaAs SQW one wavelength microcavity under non-resonant (200 fsec) pulse excitation. The observed normal mode splitting of ~ 1.3 meV is consistent with a (potential) normal mode splitting of ~ 3 meV and an independently measured exciton inhomogeneous linewidth of ~ 5 meV. When the excitation spot is shifted on the sample, the microcavity photon frequency is shifted with keeping the exciton center frequency fixed and the emission spectra feature asymmetric characteristics (stronger intensity in a photon-like polariton). In this case the upper and lower exciton polaritons are populated by thermalization of hot excitons with large in-plane wavenumber via optical and acoustic photon emission process. The theoretical curves (solid lines in Fig. 47) taking into account decreased oscillator strength per area $\frac{f}{A}$ due to inhomogeneous broadening explain the experimental data well. The dispersion characteristics of the photon-like and exciton-like polaritons can be clearly seen in such a non-resonant excitation case.

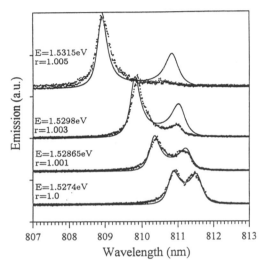

Fig. 47. Off resonance excitation spectrum of the MBE grown sample at different positions. Experimental points and theoretical fits are shown. $r = L_c/\lambda_c$, where $\lambda_c = 212.16$ nm and L_c is the physical size of the quantum well and its adjacent buffer layers. $E = \hbar\omega_0$ is the center frequency of the exciton. Parameters are $\hbar\gamma_{ex} = 0.6$ meV and $4\pi\beta = 0.3 \times 10^{-3}$. (Ref. [96]).

When the emission spectrum is observed at a different slanting angle, the in-plane dispersion characteristics of the upper and lower polaritons can be directly measured.[97] This is because the emitted photon from the microcavity preserves the in-plane momentum of the exciton-polariton inside the cavity.

Figure 48 show the emission spectra from the MOCVD grown GaAs SQW one wavelength microcavity under resonant (200 fsec) pulse excitation.[98][99] The observed normal mode splitting (~ 0.32 meV) is much smaller than that for the MBE sample. This is due to a much larger inhomogeneous width of the MOCVD sample.

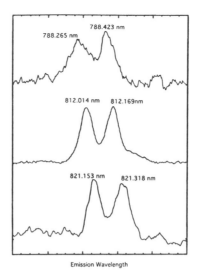

Fig. 48. Observed normal mode splitting of the MOCVD grown sample at different exciton center wavelengths $\lambda = 788$ nm, 812 nm and 821 nm. (Ref. [99]).

Figure 49 shows the time evolution of emission intensity measured by a streak camera from the MOCVD sample under resonant (200 fsec) pulse excitation.[98][99] The peak at $t = 0$ is an intentionally introduced (reflected) pump pulse which serves as a time marker. The first, second and third emission peaks appear at 7 psec, 21 psec and 35 psec, which correspond to the peak of $\sin^2 \Omega t$ function at $t = \frac{\pi}{2\Omega}, \frac{3\pi}{2\Omega}$ and $\frac{5\pi}{2\Omega}$. In this experiment, the QW Wannier exciton with $k_{//} \geq 0$ is excited by a slight slanting angle (off set angle of 3° in an air) pulse excitation at $t = 0$. That excitation quickly relax to $k_{//} = 0$ Wannier exciton state by an elastic scattering process. Therefore, the Rabi oscillation starts with the initially excited Wannier exciton and microcavity vacuum field. This is the reason why the sine-type oscillation of the photon field was observed instead of the cosine-type oscillation. If a microcavity is excited by a normally incident pulse, the Rabi oscillation starts with the initially excited microcavity field and vacuum Wannier exciton and the cosine-type oscillation of the photon field is observed.[100] The normal mode splitting (~ 0.32 meV) and the oscillating period (~ 14 psec) satisfy the Fourier transform relation, if we recall that the photon field is phase-reversed for each cycle (the first peak and the second peak are 180° out of phase in Fig. 49). The vacuum Rabi splitting is the so-called suppressed-carrier AM sidebands for a phase-reversed amplitude modulated signal.

The normal mode splitting and the oscillation period are independent of the excitation (pump power) over two orders of magnitude, as shown in Fig. 50 (a) and (b). The experimental results support a completely linear dressed boson picture for a microcavity exciton polariton. No nonlinearity was observed up to the polariton excitation level of $\sim 10^9$ (cm^{-2}). A GaAs QW bare exciton starts to feature the exciton-exciton scattering effect at the exciton density exceeding $\sim 10^9$

(cm^{-2}).[101] The experimental result shown in Fig. 50 suggests that the exciton in such a highly inhomogeneous (MOCVD grown) QW is weakly localized and thus more robust for exciton-exciton scattering. In fact, the linewidth of the upper and lower exciton polaritons (Fig. 48) is 0.1 $\sim$ 0.2 meV, which is much narrower than the linewidth (0.5 $\sim$ 1 meV) of the exciton polaritons of the MBE sample.

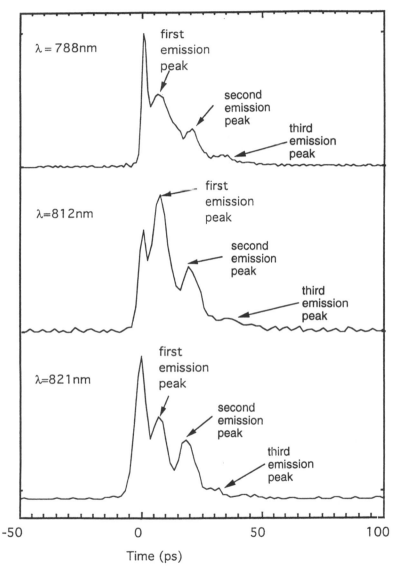

Fig. 49. Observed temporal evolution of microcavity exciton-polariton emission of the MOCVD grown sample at different exciton center wavelengths $\lambda = 788, 812$ and 821 nm. (Ref. [99]).

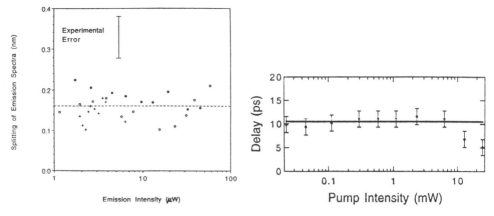

Fig. 50. (a) Spectral aplitting as a function of the emission intensity at three different exciton wavelengths. Cross: $\lambda = 788$ nm, circle: $\lambda = 812$ nm, triangle: $\lambda = 821$ nm. The number of excitons per Bohr area $(N_{ex}\pi a_B^2)$ are about 10^{-3}, 10^{-2}, and 10^{-1} for the emission intensity of 2, 20 and 200 μW, respectively. (Ref. [99]). (b) Exciton-polariton oscillation half period vs. pump intensity at $\lambda = 812$ nm. The number of excitons per Bohr area are about 10^{-3}, 10^{-2} and 10^{-1} for the pump intensity of 0.1, 1 and 10 mW. (Ref. [98]).

5.3. SUPPRESSION OF ACOUSTIC PHONON SCATTERING

A microcavity exciton polariton has a very light mass (twice of a photon mass: $m_{ep} \simeq 2m_{ph} = 7 \times 10^{-35}$ Kg) near $k_{//} = 0$, which is four orders of magnitude lighter than a bare exciton mass $(m_{ex} \simeq 5 \times 10^{-31}$ Kg). This means that the density of states for the microcavity exciton polariton is four orders of magnitude smaller than that of the bare exciton near $k_{//} = 0$. At low temperatures, a dominant inelastic scattering process of a QW Wannier exciton at $k_{//} = 0$ is absorption of acoustic phonons with small energy $\hbar\omega_{ap} \leq k_B T$. The inelastic scattering rate for a GaAs QW Wannier exciton is shown in Fig. 51 using a deformation potential and unaltered final densities of states for higher energy excitons.[102] The scattering rates are continuously increasing from $10^{10}(1/s)$ at 1K to $5 \times 10^{12}(1/s)$ at 300 K.

On the other hand, the scattering rate for a microcavity exciton polariton near $k_{//} = 0$ is partially suppressed due to small densities of states for slightly higher energy exciton polaritons. Figure 51 shows the inelastic scattering rate of a lower exciton polariton as a function of a normalized exciton-photon coupling constant $g = (\Omega/\omega)^2$. With increasing g, the exciton polariton mass is modified in a wider region of in-plane wavenumber $k_{//}$ as shown in Fig. 46 and the energy separation of the $k_{//} = 0$ lower exciton polariton and the optically inactive (bare) exciton increases, and so the acoustic phonon scattering rate is decreased. This suppression of acoustic phonon scattering is analogous to the suppression of spontaneous emission of an atom by reducing the field density of states with a cavity well, discussed in Sec. 4 (cavity QED in low-Q regime). At high temperatures, the inelastic scattering by optical phonon absorption dominates over that by acoustic phonon absorption. Since there is no modulation in the density of states for excitons with such large in-plane wavenumbers, the inelastic scattering rate is not altered at high temperatures.

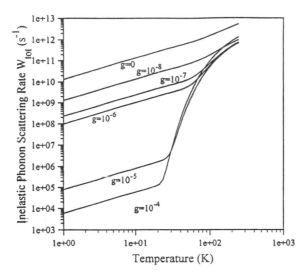

Fig. 51. The lower branch polariton scattering rate $W_{tot}(T)$ as a function of temperature for different coupling constant $g = 0$ (bare exciton), 10^{-8}, 10^{-7}, 10^{-6}, 10^{-5}, 10^{-4}. (Ref. [102]).

A similar suppression is expected for low-energy polariton-polariton scattering. These suppressed polariton-phonon scattering and polariton-polariton scattering due to modified density of states are key physical origins for observing such new coherent phenomena as the boser phase transition and non-bosonic (nonlinear) spectrum of a microcavity exciton polariton in high excitation regions, which will be discussed in the following sections.

5.4. MICROCAVITY EXCITON POLARITON BOSER

The final state stimulation in bosonic systems create many interesting physical phenomena, such as lasing, superconductivity, and Bose-Einstein condensation.[103] For a massive bosonic field, quantum statistics become important when the thermal de Broglie wavelength $\lambda_{dB} = \frac{h}{\sqrt{2k_B T m}}$ exceeds the interparticle spacing, where m is a particle mass. Optically inactive excitons in semiconductors have a long lifetime and thus offer an ideal system for the study of Bose-Einstein condensation in a 3D bulk system and Kosterliz-Thouless phase transition in a 2D quantum well system. Recent experiments on para-excitons in Cu_2O bulk crystal[104] and on magneto-excitons in GaAs quantum wells[105] have demonstrated the presence of such condensation effects.

A completely opposite approach to generate a coherent population of nonequilibrium excitons is to dress an optically active exciton with an electromagnetic vacuum field in a microcavity. In this system, an exciton is strongly coupled to a microcavity optical field but the optical field is trapped by a high-Q microcavity and the exciton can effectively have a long lifetime due to coherent exchange of excitation energy between the exciton and the field (Rabi oscillation). A relevant quasi-particle in this case is not a bare exciton but an exciton polariton. Moreover, due to their light mass, microcavity exciton polaritons with $k_{//} = 0$ have a very long thermal de Broglie wavelength ($\lambda_{\mathrm{dB}} = 7\mu m$ at 4 K) compared to a much shorter bare exciton thermal de Broglie wavelength ($\lambda_{\mathrm{dB}} = 0.07\mu m$ at 4 K). Thus it is relatively straightforward to obtain exciton polariton densities with much smaller interparticle spacings than a thermal de Broglie wavelength, but with much larger interparticle spacings than the critical values for the nonideal effects such as exciton-exciton collisions and phase-space-filling effects. The exciton-exciton collision and phase-space-filling effects become important with interparticle spacings less than 0.3 μm and 0.03 μm, respectively for bare excitons.[101] Therefore, at reasonably high exciton-polariton densities, an exponential growth of exciton-polariton occupancy due to final state stimulation without non-ideal scattering effects is expected.

Figure 52 shows the energy flow in the microcavity exciton polariton boser, which consists of the following four processes: (1) nonresonant excitation of electron-hole pairs, (2) formation of hot excitons with large in-plane wavevectors, (3) spontaneous and stimulated exciton relaxation into upper and lower exciton polariton states via acoustic phonon emission, and (4) photon leakage from the microcavity. It is this phonon emission process (3) into one of the two exciton polariton states that is enhanced by final state stimulation. On the other hand, the photon leakage from the microcavity is a completely linear process. The rate equations for the hot (reservoir) exciton population N_0, the upper polariton population N_u and the lower polariton population N_ℓ are given by [102][106]

$$\frac{d}{dt}N_0 = P_0 - \frac{N_0}{\tau_0} - C_u[(n_u + 1)(N_u + 1)N_0 - n_u N_u(N_0 + 1)]$$
$$-C_\ell[(n_\ell + 1)(N_\ell + 1)N_0 - n_\ell N_\ell(N_0 + 1)]$$
$$+D_u N_u N_0(N_0 + 1)^2 + D_\ell N_\ell N_0(N_0 + 1)^2 \quad , \tag{24}$$

$$\frac{d}{dt}N_u = P_u - \frac{N_u}{\tau_u} + C_u[(n_u+1)(N_u+1)N_0 - n_u N_u(N_0+1)] - D_u N_u N_0(N_0+1)^2 \quad , \tag{25}$$

$$\frac{d}{dt}N_\ell = P_\ell - \frac{N_\ell}{\tau_\ell} + C_\ell[(n_\ell+1)(N_\ell+1)N_0 - n_\ell N_\ell(N_\ell+1)] - D_\ell N_\ell N_0(N_0+1)^2 \quad . \tag{26}$$

Here P_i and $\tau_i(i = 0, u, \ell)$ are the (direct) external pump rate and the radiative lifetime of the hot exciton, upper polariton and lower polariton, $C_{u(\ell)}$ is the phonon assisted upper (lower) polariton emission coefficient, $n_{u(\ell)}$ is the phonon population for the energy difference between the hot exciton and upper (lower) polariton and $D_{u(\ell)}$ is the hot exciton-polariton scattering coefficient.

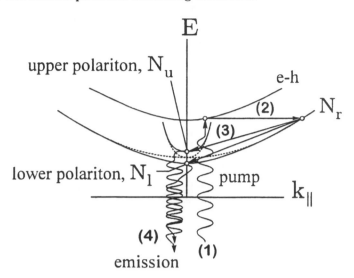

Fig. 52. Schematic energy flow in the microcavity exciton polariton boser, which consists of (1) non-resonant excitation of electron-hole pairs, (2) formation of hot excitons, (3) exciton relaxation into two polariton states via phonon emission and (4) photon leakage from the cavity.

Figure 53 (a) shows the measured gain/loss for a resonantly excited microcavity exciton polariton vs. external pump rate for a non-resonantly excited hot exciton. A single-frequency 0.78 μm Ti:Al$_2$O$_3$ laser and 0.81 μm diode laser simultaneously excite the hot exciton N_0 and the exciton polariton $N_{u(\ell)}$.[106] When the pump rate for the exciton polariton $N_{u(\ell)}$ is small (0.13 μW/μm^2), the exciton polariton population is increased due to the final state stimulation with increasing the pump rate for the hot exciton N_0 but saturates at higher pump rates. This saturation is due to the onset of exciton-polariton scattering. When the pump rate for the exciton polariton is increased (0.35 μW/μm^2), amplification of the exciton polariton population is observed only in weak pump rates for the hot exciton. If the pump rate for the exciton polariton is further increased (0.78 μW/μm^2), the exciton polariton population is monotonically decreased with increasing the pump rate for the hot exciton. The theoretical curves (Fig. 53 (b)) calculated by the rate equations (24)-(26) with variable P_0 and $P_{u(\ell)}$ qualitatively explain the observed phenomenon. In an optical laser amplifier, a stimulated emission gain is also saturated by a strong input signal but the input signal is never attenuated in an inverted medium. The transition from amplification to attenuation of the exciton polariton is due to the exciton-polariton scattering effect, which makes an exciton polariton boser amplifier distinctly different from an optical laser amplifier.

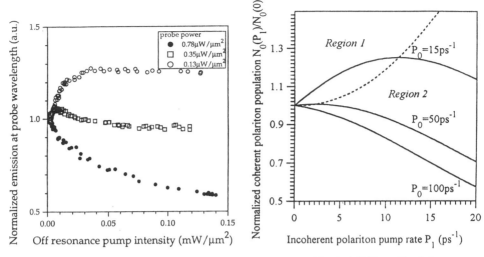

Fig. 53. (a) Normalized experimental emission intensity, $I(P_1 \neq 0)/I(P_1 = 0)$, of the upper branch polariton as a function of off resonance CW pump power, P_1 for three different on resonance pump powers. (b) Theoretical plot of normalized upper branch polariton N_0 versus off resonance CW pump rate P_1 for different on resonance pump rate P_0. The parameters are $C = 1/8$ ps^{-1}, $D = 1/3000$ ps^{-1}, $\tau_0 = \tau_1 = 1$ ps, $\bar{n} = 5$, $M_1 = 4$ and $M_2 = 15$. Dashed line separates the regions where dN_0/dP_1 changes sign. (Ref. [106]).

Figure 54 shows the emission spectra of the two exciton polaritons with $k_{//} = 0$ under non-resonant excitation of the hot exciton (no resonant pumping of the exciton polaritons). The sample was grown by a MBE technique and has a normal mode splitting of ~ 1.5 meV. The linewidth of the exciton polariton is rather broad (0.5 $\sim$ 1.0 meV). At low pump power, the emission intensities from the upper and lower polaritons separated by 1.5 meV increase linearly with the pump power, and the intensity difference is well described by thermal equilibrium Boltzmann distributions of the two exciton polariton states. However, at above a certain threshold, a nonequilibrium buildup of the upper-polariton population is observed, i.e., the population of the upper polariton starts to increase nonlinearly and the population of the lower polariton starts to saturate as shown in Fig. 55 (a). This particular data were taken at a sample position where the upper polariton is a photon-like state and the lower polariton is an exciton-like state. When an excitation spot is shifted to an oppositely detuned position, we observed the nonequilibrium build-up of the photon-like branch (lower polariton in this case) as shown in Fig. 55 (b). A photon-like state always dominates an exciton-like state because a photon-like state has a longer thermal de Broglie wavelength and smaller exciton scattering rate. Such an exciton polariton boser phase transition behaviour (stimulated emission of a lasing-photon-like polariton and gain clamping of a non-lasing exciton-like polariton) can be qualitatively explained by the rate equations (24)-(26). Figure 55 (c) shows the calculated emission intensities for the (photon-like) upper and (exciton-like) lower polaritons, which corresponds to the situation of Fig. 55 (a).

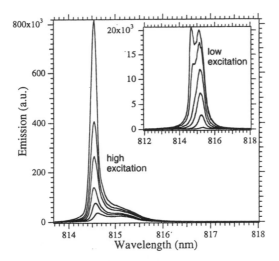

Fig. 54. Emission spectra from off resonance pulse excitation at power of 0.079, 0.1, 0.13, 0.19, 0.24, 0.32 mW/μm^2. Inset shows the emission spectra at lower excitation powers: $1/38 \times 10^{-3}$, 7.4×10^{-3}, 1.95×10^{-2}, 3.2×10^{-2}, 4.9×10^{-2}, and 6.3×10^{-2} mW/μm^2. Symmetric emission occurs near the excitation power of 6.3×10^{-2} mW/μm^2.

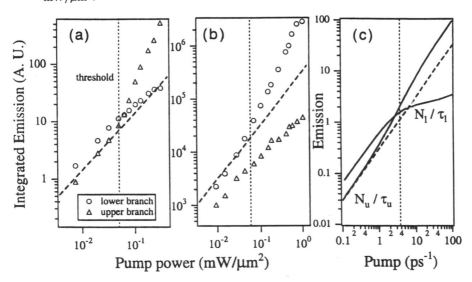

Fig. 55. (a) Observed emission intensities of the upper (photon-like) and lower (exciton-like) polaritons vs. pump power. (b) Observed emission intensities of the upper (exciton-like) and lower (photon-like) polaritons vs. pump power. (c) Theoretical emission intensities of the upper (photon-like) and lower (exciton-like) exciton-polaritons vs. pump rate using parameters: $\tau_r 100$ ps, $\tau_r = 5$ ps, $\tau_1 = 2$ ps, $C_u = C_1 = 0.1$ ps^{-1}, $n = 2$ and $D = 0.002$ ps^{-1}. Vertical dashed line marks location of threshold.

The boser threshold condition is calculated by neglecting exciton-polariton scattering effects in (24)-(26) as

$$\frac{1}{\tau_{u(\ell)}} = C_{u(\ell)}(N_{0,\text{th}} - n_{u(\ell)}) \quad , \tag{27}$$

or

$$P_{\text{th}} = C_{u(\ell)}\left(1 + \frac{1}{\tau_0 C_{u(\ell)}}\right)\left(n_{u(\ell)} + \frac{1}{\tau_{u(\ell)} C_{u(\ell)}}\right) \quad . \tag{28}$$

To obtain a net stimulated emission gain of the exciton polaritons, the hot exciton population N_0 must exceed the phonon population $n_{u(\ell)}$. This is a "population inversion condition" for an exciton polariton boser.[107] The threshold pump power increases with increasing a temperature because of the temperature dependent phonon population $n_{u(\ell)}$.

Above the threshold, the lasing exciton polariton also establishes a macroscopic phase due to onset of phase coherent stimulated emission, which results in the decreased emission linewidth. As shown in Fig. 56 (a) and (b), the emission linewidth of the lasing photon-like upper (or lower) polariton decreases from the original value of ~ 0.5 meV inversely proportional to pump power above the threshold. At higher pump powers, the linewidth reduction saturates or the linewidth even increases. The emission linewidth of the non-lasing polariton continuously increases from the original value of ~ 1 meV with a pump rate. These results are due to the exciton-polariton scattering effect, which is responsible for the attenuation of the exciton polariton in high excitation regimes shown in Fig. 53. The theoretical curves (Fig. 56 (c)) calculated by (24)-(26) well describes such linewidth behaviours.

Observation of the final state stimulation and the boser phase transition of a microcavity exciton polariton suggests a new physical process, "coherent matter wave (exciton) optics in solids." A direct electrical pumping as well as non-resonant optical pumping can generate such a coherent exciton-polariton population via the boser phase transition. It has been demonstrated recently that a QW hole-assisted electron resonant tunneling create directly a QW Wannier exciton.[108] This is a spontaneous tunneling process. If there is enough exciton polariton population in the system, electron resonant tunneling must be enhanced by the final state stimulation. It is expected that two fermion particles (a QW hole and n-type semiconductor bulk electron) directly form a boson quasi-particle (exciton polariton) by a stimulated tunneling process.

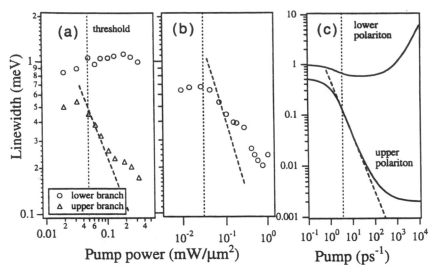

Fig. 56. (a) Observed spectral linewidths of the upper (photon-like) and lower (exciton-like) exciton-polaritons vs. pump power density (inverse power dependence of the linewidth is represented by a dashed line). (b) Observed spectral linewidth of the lower (photon-like) exciton-polariton vs. pump power density. (c) Theoretical spectral linewidths of the upper (photon-like) and lower (exciton-like) exciton-polaritons vs. pump rate, using same numerical parameters as in Fig. 55 except for D coefficient. ($D = 0$ dashed, $D = 0.002$ ps^{-1} solid).

5.5. NONLINEAR QUANTUM DYNAMICS OF DRESSED EXCITONS

When a microcavity exciton polariton in a highly inhomogeneous (MOCVD grown) sample is resonantly excited by a strong (200 psec) pump pulse and the exciton-polariton density exceeds a certain critical value, a bi-exciton effect emerges before the excitons are dissociated into electron-hole plasma and modifies a simple dressed boson picture. A lowest-order nonlinear effect is an attractive interaction between two excitons with opposite spins due to the bi-exciton effect.[109] The Hamiltonian of such a system is given by

$$\mathcal{H} = \sum_{\sigma} \left[\hbar\omega \, a_\sigma^\dagger a_\sigma + \hbar\omega \, b_\sigma^\dagger b_\sigma + \hbar\Omega(a_\sigma b_\sigma^\dagger + a_\sigma^\dagger b_\sigma) \right] + \hbar \sum_{\sigma,\sigma'} A_{\sigma\sigma'} \, b_\sigma^\dagger b_{\sigma'}^\dagger b_\sigma b_{\sigma'} \quad . \quad (29)$$

Here σ stands for two circular polarizations for a microcavity field a_σ and two orbital angular momenta (spins) for a QW Wannier exciton b_σ, and the exciton-exciton scattering coefficient $A_{\sigma\sigma'}$ is given by $-\frac{|E_b|}{2\hbar}$ for $\sigma \neq \sigma'$ (opposite spins) where E_b is a bi-exciton binding energy (~ 1 meV for GaAs), while $A_{\sigma\sigma'} = +O$ (weak repulsive interaction) for $\sigma = \sigma'$ (same spins). As shown in Fig. 57, not only two central transition lines but also all the other transition lines (two additional lines from the $N = 2$ to the $N = 1$ excitation manifolds and four additional lines from the $N = 3$ to the $N = 2$ excitation manifolds) have a non-zero matrix element due to the nonlinear term $A_{\sigma\sigma'}$. The emission intensities of the two central peaks are no more equal but the low-energy central peak has a higher intensity. These modifications can be readily seen by calculating the new dressed states in terms of the bare states, for instance, the new dressed states for the $N = 2$ excitation manifold are now $|2, +\rangle = \frac{1}{2}\left(1 - \frac{3}{8}\frac{A}{\Omega}\right)|2\rangle_e|0\rangle_f + \frac{1}{\sqrt{2}}\left(1 - \frac{1}{8}\frac{A}{\Omega}\right)|1\rangle_e|1\rangle_f + \frac{1}{2}\left(1 + \frac{5}{8}\frac{A}{\Omega}\right)|0\rangle_e|2\rangle_f$, $|2, 0\rangle = -\frac{1}{\sqrt{2}}|2\rangle_e|0\rangle_f - \frac{1}{2\sqrt{2}}\frac{A}{\Omega}|1\rangle_e|1\rangle_f + \frac{1}{\sqrt{2}}|0\rangle_e|2\rangle_f$ and $|2, -\rangle = \frac{1}{2}\left(1 + \frac{3}{8}\frac{A}{\Omega}\right)|2\rangle_e|0\rangle_f - \frac{1}{\sqrt{2}}\left(1 + \frac{1}{8}\frac{A}{\Omega}\right)|1\rangle_e|0\rangle_f + \frac{1}{2}\left(1 - \frac{5}{8}\frac{A}{\Omega}\right)|0\rangle_e|2\rangle_f$. The eigenfrequencies of these dressed states of the $N = 2$ excitation manifold are "redshifted" to $2\omega_0 + 2\Omega + \frac{A}{2}$, $2\omega_0 + A$ and $2\omega_0 - 2\Omega + \frac{A}{2}$, respectively. Thus, the emission spectrum is also red-shifted as shown in Fig. 57.

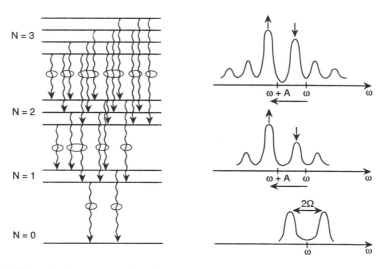

Fig. 57. The emission spectra of nonlinear quantum dressed excitons from the $N = 3$ to $N = 2$ manifold and from the $N = 2$ to $N = 1$ manifold.

Figure 58 shows the observed emission spectra from the MOCVD grown sample under weak, intermediate and strong resonant excitations with linearly polarized (200 fsec) pump pulse. In the weak excitation case (Fig. 58 (a)), the two peaks separated by 0.35 meV are central at 813.35 nm. In the intermediate excitation case (Fig. 58 (b)), two additional side peaks appear and longer-wavelength (lower-energy) central peak has a higher intensity. The center wavelength of this four peak emission spectrum is 813.6 nm, which corresponds to a red-shift of ~ 0.25 nm (~ 0.5 meV). This is in agreement with the bi-exciton binding energy of GaAs ($E_b = 2\hbar A \simeq 1$ meV). In the strong excitation case (Fig. 58 (c)), four additional peaks appear and the emission spectrum has six peaks in total.

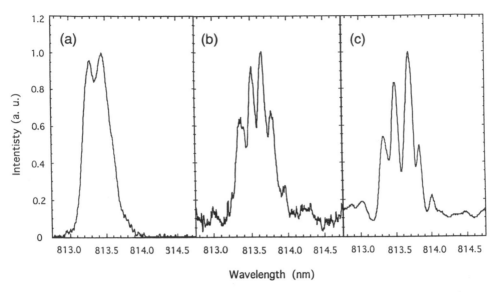

Fig. 58. Observed emission spectra of the MOCVD grown sample under on resonance (200 fsec) pulse excitation. (a) weak excitation, (b) intermediate excitation and (c) strong excitation.

Figure 59 compares the two emission spectra under strong resonant excitation with circular polarization and linear polarization. When the quantum well microcavity is excited by a circularly polarized pump light, QW Wannier excitons with only one spin are excited and thus the above nonlinear behavior can be switched off. The experimental results qualitatively support this prediction. Quantitative comparison between the experimental and theoretical results indicate that there are small populations of opposite spin excitons even under circular polarization excitation (maybe, due to exciton spin relaxation effect).

The exciton-exciton scattering Hamiltonian in (29) expresses a cross-phase-modulation between two excitons with opposite spins. This means that the two input optical fields with orthogonal (circular) polarizations are cross-phase-modulated inside the microcavity and output after a certain interaction time. This is an ideal nonlinear cross-phase-modulation process, in which the field excites resonantly a nonlinear particle (the exciton) and extracts the phase modulated excitation back from the exciton after an appropriate interaction time. The quasi-particle (exciton polariton) has a reduced scattering rate due to its small effective mass and so a long interaction time is available without suffering from dissipation. The applications of such efficient cross-phase-modulation include the quantum nondemolition measurement of photon number,[110] classical Fredkin gate operating with single photon state control signal $|1\rangle_c$,[111] quantum Fredkin gate operating with a linear superposition state control signal, $\frac{1}{\sqrt{2}}(|0\rangle + |1\rangle)_c$[112] and photonic de Broglie wave interferometer.[113]

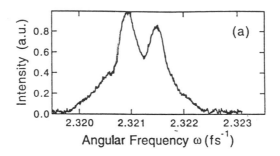

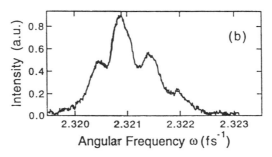

Fig. 59. Measured emission spectra of the MOCVD grown sample for linear (a) and circular (b) polarization pump with the same intensity.

However, the bi-excitonic nonlinear spectrum usually cannot be seen in MBE grown samples. This is partly because the exciton in a homogeneous QW is easily scattered by other excitons as compared to the weakly localized excitons in an inhomogeneous QW. This is also because the side peak intensity is proportional to $(A/\Omega)^2$ and so the larger normal mode splitting results in the smaller side peak intensity.

5.6. DRESSED QUON

Particles with small violations of Fermi or Bose statistics have considerable attention recently. A quon is a mathematical model for the particle field which smoothly extrapolates the fermion and boson as two extreme cases.[114] A quon operator satisfy the following q-deformed commutator bracket (q-mutator):

$$[c, c^\dagger]_q = cc^\dagger - qc^\dagger c = 1 \quad , \tag{30}$$

where a real number q goes from -1 to 1. If $q = 1$, we have a boson and, if $q = -1$, we have a fermion. A Fock-like Hilbert space is defined by the recursion relations: $c|n\rangle = \sqrt{1 + q + \cdots + q^{n-1}}|n - 1\rangle$, $c^\dagger|n\rangle = \sqrt{1 + q + \cdots + q^n}|n + 1\rangle$ and $c^\dagger c|n\rangle = (1 + q + \cdots + q^{n-1})|n\rangle$. The total system Hamiltonian for a coupled boson-quon system, $\mathcal{H} = \hbar\omega\, a^\dagger a + \hbar\omega\, c^\dagger c + \hbar\Omega(ac^\dagger + a^\dagger c)$, is expressed in the

$N = 1$, $N = 2$ and $N = 3$ excitation manifolds in the bare state representation:

$$\begin{pmatrix} \omega & \Omega \\ \Omega & \omega \end{pmatrix} \begin{array}{l} |1,0\rangle \\ |0,1\rangle \end{array}$$

$$\begin{pmatrix} 2\omega & \sqrt{2}\Omega & 0 \\ \sqrt{2}\Omega & 2\omega & \sqrt{1+q^2}\Omega \\ 0 & \sqrt{1+q^2}\Omega & 2\omega \end{pmatrix} \begin{array}{l} |2,0\rangle \\ |1,1\rangle \\ |0,2\rangle \end{array} \qquad . \quad (31)$$

$$\begin{pmatrix} 3\omega & \sqrt{3}\Omega & 0 & 0 \\ \sqrt{3}\Omega & 3\omega & \sqrt{2(1+q^2)}\Omega & 0 \\ 0 & \sqrt{2(1+q^2)}\Omega & \omega(2+q) & \sqrt{1+q+q^2}\Omega \\ 0 & 0 & \sqrt{1+q+q^2}\Omega & (1+q+q^2)\omega \end{pmatrix} \begin{array}{l} |3,0\rangle \\ |2,1\rangle \\ |1,2\rangle \\ |0,3\rangle \end{array}$$

The eigen-frequencies of dressed quons can be calculated by diagonalizing (31) and are shown in Fig. 44.

If the dressed quon is only slightly deviated from the dressed boson ($\varepsilon = 1-q \ll 1$), such a dressed quon resembles a nonlinear dressed exciton system. This can be seen by writing the Heisenberg equation of motion for the quon operator,

$$i\frac{d}{dt}c = \omega(1 - \varepsilon c^\dagger c)c + \Omega(1 - \varepsilon c^\dagger c)a \quad . \tag{32}$$

The first term in the right hand side (R.H.S) of this equation indicates a "red-shift" of the center frequency, which is identical to a bi-exciton effect. The second term in R.H.S indicates a reduced quon-boson coupling constant, which corresponds to a reduced exciton-photon coupling constant due to phase-space-filling effect. For GaAs, these two nonlinear effects do not have identical coefficients[109] and so the analogy is not perfect. Nevertheless, the mathematical dressed quon model qualitatively describes a quantum nonlinear dressed exciton. Figure 60 shows the emission spectra of the $N = 1$ to the $N = 0$, the $N = 2$ to the $N = 1$ and the $N = 3$ to the $N = 2$ transitions for $\varepsilon = 1-q = 0.002$. Note that similarity between the observed nonlinear dressed exciton spectra (Fig. 58) and the calculated dressed quon spectra (Fig. 60).

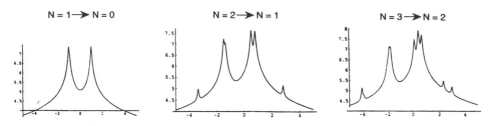

Fig. 60. Theoretical emission spectra of dressed quons with $\varepsilon = 0.002$ for the $N = 1$ to the $N = 0$, the $N = 2$ to the $N = 1$ and the $N = 3$ to the $N = 2$ transitions.

5.7. DIRECT CREATION OF EXCITON-POLARITON BY RESONANT ELECTRON TUNNELING

It has been demonstrated that QW excitons can be directly created in a pn tunnel junction by QW hole-assisted resonant electron tunneling.[108] In this process, the free electrons in a n-type semiconductor tunnel across the potential barrier and recombine with QW holes to create QW excitons. No photon emission process is required to create $k_{//} \simeq 0$ excitons, but not only optically active excitons but also optically inactive excitons are created.[115]

If such a pn tunnel junction is embedded in a high-Q microcavity, the resonant electron tunneling can directly excite lower-branch microcavity exciton polaritons rather the QW bare excitons. Only optically active excitons are selectively excited in this resonant tunneling process, so the excitation energy can be extracted out of the microcavity in a cavity photon lifetime.

Acknowledgement

The author wishes to thank S. Machida, O. Nilsson, G. Björk, A. Imamōglu, W.H. Richardson, S. Inoue, J. M. Jacobson, S. Pau, H. Cao, S. Kasapi, S. Lathi, R. Liu, J. Kim, H. Heitmann and F. M. Matinaga for their helpful discussions, and M. Shioda for her careful typing the manuscript.

References

[1] Schrödinger, E. (1927) *Naturwiss,* **14**, 644.

[2] Takahasi, H. (1965) in A. V. Balakrishnan (ed.), *Adv. Commun. Syst.,* Academic Press, New York, **1**, p. 227.

[3] Stoler, D. (1970) *Phys. Rev. D* **1**, 3217; ibid, (1971) *D* **4**, 1925.

[4] Yuen, H. P. (1976) *Phys. Rev. A* **13**, 2226.

[5] Glauber, R. J. (1963) *Phys. Rev.* **131**, 2766; (1965) in C. M. Dewitt, A. Blandin and C. Cohen-Tannoudji (eds.), *Quantum Optics and Electronics,* Gordon & Breach, New York, p. 138.

[6] Susskind, L. and Glogower, J. (1964) *Physics* (N.Y.) **1**, 49; Carruthers, P. and Neito, M. M. (1965) *Phys. Rev. Lett.* **14**, 387.

[7] Jackiw, R. (1968) *J. Math. Phys.* **9**, 339; Yamamoto, Y., Machida, S., Imoto, N., Kitagawa, M., and Björk, G. (1987) *J. Opt. Soc. Am.* **4**, 1645.

[8] Some of the early refernces are: Slusher, R. E., Hollberg, L. W., Yurke, B., Mertz, J. C., and Valey, J. F. (1985) *Phys. Rev. Lett.* **55**, 2409; Shelby, R. M., Levenson. M. D., Perlmutter, S. H., DeVoe R. G., and Walls, D. F. (1986) *Phys. Rev. Lett.* **57**, 691; Wu, L. A., Kimble, H. J., Hall, J. L., and Wu, H. (1986) *Phys. Rev. Lett.* **57**, 2520; Maeda, M. W., Kumar, P., and Shapiro, J. H. (1987) *Opt. Lett.* **12**, 161. For a recent review article: Kimble, H. J. (1992) in J. Dalibard, J. M. Raimond and J. Zinn-Justin (eds.), *Fundamental Systems in Quantum Optics, Les Houches Session 53,* Elsevier Science Publishers, Amsterdam, p. 549.

[9] Kitagawa, M. and Yamamoto, Y. (1986) *Phys. Rev. A* **34**, 3974.

[10] Bergman, K., Boen, C., Haus, H. A., and Shirasaki, M. (1993) *Opt. Lett.,* **18**, 643.

[11] Watanabe, K. and Yamamoto, Y. (1988) *Phys. Rev. A* **38**, 3556.

[12] Björk, G. and Yamamoto, Y. (1988) *Phys. Rev. A* **37**, 4229.

[13] Reynaud, S., Heidmann, A., Giacobino, E., and Fabre, C. (1992) in E. Wolf (ed.), *Progress in Optics,* North-Holland, Amsterdam, **30**.

[14] Yamamoto, Y. et al. (1990) in E. Wolf (ed.), *Progress in Optics,* North-Holland, Amsterdam, **28**, p. 89; Yamamoto, Y., Machida, S., and Richardson, W. H. (1992) *Science* **255**, 1219; Yamamoto, Y., Inoue, S., Richardson, W. H., and Machida, S. (1993) *Int. J. Mod. Phys.* **B7**, 1577.

[15] Yamamoto, Y., Machida, S., and Nilsson, O. (1986) *Phys. Rev. A* **34**, 4025.

[16] Imamōglu, A. and Yamamoto, Y. (1993) *Phys. Rev. Lett.* **70**, 3327.

[17] Yamamoto, Y. and Machida, S. (1987) *Phys. Rev. A* **35**, 5114.

[18] Kim, J., Kan, H., and Yamamoto, Y. (1995) *Phys. Rev. B,* **52**, 2008.

[19] Machida, S. and Yamamoto, Y. (1989) *Opt. Lett.* **14**, 1045.

[20] Richardson, W. H., Machida, S., and Yamamoto, Y. (1991) *Phys. Rev. Lett.* **66**, 2867.

[21] Machida, S. and Yamamoto, Y. (1988) *Phys. Rev. Lett.* **60**, 792.

[22] Richardson, W. H. and Yamamoto, Y. (1991) *Phys. Rev. Lett.* **66**, 1963.

[23] Ito, T., Machida, S., Nawata, K., and Ikegami, T. (1977) *IEEE J. Quantum Electron.* **QE-13**, 574; Liu, P. L., (1991) in Y. Yamamoto (ed.) *Coherence, Amplification and Quantum Effects in Semiconductor Lasers,* John Wiley & Sons, New York, p. 411.

[24] Inoue, S., Ohzu, H., Machida, S., and Yamamoto, Y. (1992) *Phys. Rev. A* **46**, 2757.

[25] Machida, S., Yamamoto, Y., and Itaya, Y. (1987) *Phys. Rev. Lett.* **58**, 1000.

[26] Freeman, M. J., Wang, H., Steel, D. G., Craig, R., and Scifres, D. R. (1993) *Opt. Lett.* **18**, 2141.

[27] Kitching, J., Yariv, A., and Shevy, Y. (1995) *Phys. Rev. Lett.* **74**, 3372.

[28] Yamamoto, Y. and Haus, H. A. (1984) *Phys. Rev. A* **29**, 1261.

[29] Gillner, L., Björk G., and Yamamoto, Y. (1990) *Phys. Rev. A* **41**, 5053.

[30] Inoue, S., Ohzu, H., Machida, S., and Yamamoto, Y. (1993) *Phys. Rev. A* **48**, 2230.

[31] Wang, H., Freeman, M. J., and Steel, D. G. (1993) *Phys. Rev. Lett.* **71**, 3951.

[32] Polzik, E. S., Carri, J. C., and Kimble, H. J. (1992) *Phys. Rev. Lett.* **68**, 3020.

[33] Xiao, M., Wu, L. A., and Kimble, H. J. (1987) *Phys. Rev. Lett.* **59**, 278; Grangier, P., Slusher, R. E., Yurke, B., and LaPorta, A. (1987) *Phys. Rev. Lett.* **59**, 2153.

[34] Kasapi, S., Lathi, S., and Yamamoto, Y. (1995) in *Proc. SPIE Photonics West,* San Jose, CA; ibid, (1995) J. H. Eberly et al. (eds.), *Coherence and Quantum Optics VII,* Plenum Press, NY.

[35] Kobayashi, S. (1991) in Y. Yamamoto (ed.), *Coherence, Amplification and Quantum Effects in Semiconductor Lasers,* John Wiley & Sons, New York; Saito, S., Nilsson, O., and Yamamoto, Y. (1982) *IEEE J. Quantum Electron.* **QE-18**, 961.

[36] Björklund, G. C., Levenson, M. D., Lenth, W., and Oritz, C. (1983) *Appl. Phys. B* **32**, 145.

[37] Inoue, S., Björk, G., and Yamamoto, Y. (1995) in *Proc. SPIE Photonics West*, San Jose, CA.

[38] Caves, C. M. (1980) *Phys. Rev. Lett.* **45**, 75.

[39] Yurke, B. (1986) *Phys. Rev. Lett.* **56**, 1515.

[40] Loudon, R. (1989) in J. H. Eberly, L. Mandel and E. Wolf (eds.), *Coherence and Quantum Optics VI*, Plenum Press, New York, p. 703.

[41] Hong, C. K., Ou, Z. Y., and Mandel, L. (1987) *Phys. Rev. Lett.* **59**, 2044.

[42] Arrechi, F. T. et al., (1972) *Phys. Rev. A* **6**, 2211.

[43] Liu, R. and Yamamoto, Y. (1994) *Phys. Rev. B* **49**, 10520.

[44] Landauer, R. (1970) *Philos. Mag.* **21**, 863; Buettiker, M. (1986) *Phys. Rev. Lett.* **57**, 1761.

[45] Johnson, M. B. (1927) *Nature* **119**, 50; Johnson, M. J. (1928) *Phys. Rev.* **32**, 97; Nyquist, H. (1928) *Phys. Rev.* **32**, 110.

[46] Callen, H. B. and Welton, T. A. (1951) *Phys. Rev.* **83**, 34.

[47] Landauer, R. (1989) *Physica D* **38**, 226.

[48] Liu, R. and Yamamoto, Y. (1994) *Phys. Rev. B* **50**, 17411.

[49] Buettiker, M. (1986) *Phys. Rev. B* **33**, 3020; Beenakker, C. W. J. and Buettiker, M. (1992) *Phys. Rev. B* **46**, 1889.

[50] Liu, R and Yamamoto, Y. (1995) *Physica B* **210**, 37.

[51] Liu, R., Eastman, P., Odom, B., and Yamamoto, Y. (1995) *March Meeting of the American Physical Society*, San Jose, CA I p. 1711.

[52] Carmichael, H. J. (1993) *Phys. Rev. Lett.* **70**, 2273.

[53] Dalibard, J., Castin, Y., and Molmer, K. (1992) *Phys. Rev. Lett.* **68**, 580.

[54] Dum, R., Zoller, P., and Ritsu, H. (1992) *Phys. Rev. A* **45**, 4879.

[55] Imamōglu, A. and Yamamoto, Y. (1994) *Phys. Lett. A* **191**, 425.

[56] Imamōglu, A., Yamamoto, Y., and Solomon, P. (1992) *Phys. Rev. B* **46**, 9555.

[57] Imamōglu, A. and Yamamoto, Y. (1992) *Phys. Rev. B* **46**, 15982.

[58] Imamōglu, A. and Yamamoto, Y. (1994) *Phys. Rev. Lett.* **72**, 210.

278

[59] Purcell, E. M. (1946) *Phys. Rev.* **69**, 681.

[60] Drexhage, K. H. (1974) in E. Wolf (ed.), *Progress in Optics*, North-Holland, New York **12**.

[61] Hulet, R. G., Hilfer, E. S., and Kleppner, D. (1985) *Phys. Rev. Lett.* **55**, 2137.

[62] Haroche, S. and Raimond, J. M. (1984) *Adv. Atomic and Molecular Physics* **20**, 347.

[63] Gabrielse, G. and Dehmelt, H. (1985) *Phys. Rev. Lett.* **55**, 67.

[64] Yamamoto, Y., Machida, S., Igeta, K., and Horikoshi, Y. (1989) in J. H. Eberly, L. Mandel, and E. Wolf (eds.), *Coherence and Quantum Optics VI*, Plenum Press, New York, p. 1249.

[65] DeMartini, F., Innocenti, G., Jacobovitz, G. R., and Mataloni, P. (1987) *Phys. Rev. Lett.* **59**, 2955.

[66] Yokoyama, H. and Brorson, S. D. (1989) *J. Appl. Phys.* **66**, 4801; Yokoyama, H. et al., (1992) *Opt. Quantum Electron.* **24**, S245.

[67] Dicke, R. H. (1954) *Phys. Rev.* **93**, 99.

[68] Björk, G., Pau, S., Jacobson, J., and Yamamoto, Y. (1994) *Phys. Rev. B* **50**, 17336.

[69] Raizen, M. G., Thomason, R. J., Brecha, R. J., Kimble, H. J., and Carmichael, H. J. (1989) *Phys. Rev. Lett.* **63**, 240.

[70] Kaluzny, Y., Goy, P., Gross, M., Raimond, J. M., and Haroche, S. (1983) *Phys. Rev. Lett.* **51**, 1175.

[71] Zhu, Y., Gauthier, D. J., Morin, S. E., Wu, Q., Carmichael, H. J., and Mossberg, T. W. (1990) *Phys. Rev. Lett.* **64**, 2499.

[72] Weisbuch, C., Nishioka, M., Ishikawa, A., and Arakawa, Y., (1992) *Phys. Rev. Lett.* **69**, 3314.

[73] (1994) in P. R. Berman (ed.) *Cavity Quantum Electrodynamics*, Academic Press, Boston; (1995) in E. Burstein and C. Weisbuch (eds.), *Confined Electrons and Photons*, Plenum Press, New York; (1995) in H. Yokoyama and K. Ujihara (eds.), *Spontaneous Emission and Laser Oscillation in Microcavities*, CRC Press, Boca Raton.

[74] Arrechi, F. T., Courtens, E., Gilmore, R., and Thomas, H. (1972) *Phys. Rev. A* **6**, 2211.

[75] Suematsu, Y. and Furuya, K. (1977) *Trans. IECE Japan* **E-60**, 467.

[76] Björk, G. and Yamamoto, Y. (1991) *IEEE J. Quantum Electron.* **QE-27**, 2386; Yamamoto, Y., Machida, S., and Björk, G. (1992) *Opt. Quantum Electron.* **24**, S-215.

[77] Einstein, A. (1917) *Z. Phys.* **18**, 121.

[78] Rice, P. R. and Carmichael, H. J. (1994) *Phys. Rev. A* **50**, 4318.

[79] Yamamoto, Y. and Björk, G (1991) *Jpn. J. Appl. Phys.* **30**, 2039.

[80] Heinzen, D. J., Childs, J. J., Thomas, J. E., and Feld, M. S. (1987) *Phys. Rev. Lett.* **58**, 1320.

[81] Yamamoto, Y., Björk, G., Karlsson, A., Heitmann, H., and Matinaga, F. M. (1993) *Int. J. Mod. Phys.* **B7**, 1653.

[82] Yamamoto, Y., Machida, S., Igeta, K., and Björk, G. (1991) in Y. Yamamoto (ed.), *Coherence, Amplification and Quantum Effects in Semiconductor Lasers*, John Wiley & Sons, New York, p. 561; Björk, G., Machida, S., Yamamoto, Y., Igeta, K., (1991) *Phys. Rev. A* **44**, 669.

[83] Horowicz, R. J., Heitmann, H., Kadota, Y., and Yamamoto, Y. (1992) *Appl. Phys. Lett.* **61**, 393.

[84] Heitmann, H., Kadota, Y., Kawakami, T., and Yamamoto, Y. (1993) *Jpn. J. Appl. Phys.* **32**, L1141.

[85] Björk, G., Heitmann, H., and Yamamoto, Y. (1993) *Phys. Rev. A* **47**, 4451.

[86] Baba, T., Hamano, T., Koyama, F., and Iga, K. (1991) *IEEE J. Quantum Electron.* **27**, 1347; ibid, (1992) **28**, 1310.

[87] Matinaga, F. M., Karlsson, A., Machida, S., Yamamoto, Y., Suzuki, T., Kadota, Y., and Ikeda, M. (1993) *Appl. Phys. Lett.* **62**, 443.

[88] McCall, S. L., Levi, A. F., Slusher, R. E., Pearton, S. J., and Logan, R. A. (1992) *Appl. Phys. Lett.* **60**, 289.

[89] Björk, G., Karlsson, A., and Yamamoto, Y. (1994) *Phys. Rev. A* **50**, 1675.

[90] Hanamura, E. (1988) *Phys. Rev. B* **38**, 1228.

[91] Andreani, L. C. et al., (1991) *Solid State Commun.* **77**, 641.

[92] Citrin, D. S. (1993) *Phys. Rev. B* **47**, 3832.

[93] Björk, G., Pau, S., Jacobson, J., and Yamamoto, Y. *Phy. Rev. B.* (to be published).

[94] Andreani, L. C. (1994) in E. Burnstein and C. Weisbuch (eds.), *Confined Electrons and Photons: New Physics and Devices*, Plenum Press, New York.

[95] Savona, V., Andreani, L. C., Schwendimann, P., and Quattropani, A. (1995) *Solid State Commun.* **93**, 733.

[96] Pau, S., Björk, G., Jacobson, J., Cao, H., and Yamamoto, Y. (1995) *Phys. Rev. B* **51**, 14437.

[97] Houdre, R., Weisbuch, C., Stanley, R. P., Oesterle, U., Pellandini, P., and Ilegems, M. (1994) *Phys. Rev. Lett.* **73**, 2043.

[98] Yamamoto, Y., Jacobson, J., Pau, S., Cao, H., and Björk, G. (1994) in H. Sakaki and H. Noge (eds.), *Nanostructures and Quantum Effects*, Springer-Verlag, Berlin, p. 157; Jacobson, J., Pau, S., Cao, H., Björk, G., and Yamamoto, Y. (1995) *Phys. Rev. A* **51**, 2542.

[99] Cao, H., Jacobson, J., Björk, G., Pau, S., and Yamamoto, Y. (1995) *Appl. Phys. Lett.* **66**, 1107.

[100] Norris, T., et al., (1994) *Phys. Rev. B* **50**, 14663.

[101] Schmitt-Rink, S., Chemla, D. S., and Miller, D. A. B. (1985) *Phys. Rev. B* **32**, 6101.

[102] Pau, S., Björk, G., Jacobson, J., Cao, H., and Yamamoto, Y. (1995) *Phys. Rev. B* **51**, 7090.

[103] Griffin, A., et al., (1995) *Bose-Einstein Condensation*, Cambridge, New York.

[104] Lin, J. L. and Wolfe, J. P. (1993) *Phys. Rev. Lett.* **71**, 1222; Fortin, E. et al., (1993) ibid, **70**, 3951.

[105] Butov, L. V., et al., (1994) *Phys. Rev. Lett.* **73**, 304.

[106] Pau, S., Björk, G., Jacobson, J., and Yamamoto, Y. (1995) in J. H. Eberly et al. (eds.), *Coherence and Quantum Optics VII*, Plenum Press, New York.

[107] Ram, R. J. and Imamoğlu, A. (1995) in J. H. Eberly et al. (eds.), *Coherence and Quantum Optics VII*, Plenum Press, New York.

[108] Cao, H., Klimovitch, G., Björk, G., and Yamamoto, Y. (1995) *Phys. Rev. Lett.* **75**, 1146.

[109] Hanamura, E. and Haug, H. (1977) *Phys. Rev.* **33**, 209.

[110] Imoto, N., Haus, H. A., and Yamamoto, Y. (1985) *Phys. Rev. A* **32**, 2287.

[111] Yamamoto, Y., Kitagawa, M., and Igeta, K. (1988) in C. N. Yang et al. (eds.), *Proc. 3rd Asia Pacific Physics Conference*, World Scientific.

[112] Chuang, I. and Yamamoto, Y. *Phys. Rev. A* (to be published).

[113] Jacobson, J. M., Björk, G., Chuang, I., and Yamamoto, Y. (1995) *Phys. Rev. Lett.* **74**, 4835.

[114] Greenberg, O. W. (1991) *Phys. Rev. D* **43**, 4111.

[115] Cao, H., Klimovitch, G., Björk, G., and Yamamoto, Y (Dec. 1995) *Phys. Rev. B.*

PHYSICS OF SEMICONDUCTOR LASERS

K. J. EBELING
Department of Optoelectronics
University of Ulm, Germany

1. Introduction

We are going to study optical processes in semiconductors and semiconductor heterostructures in some detail and describe fundamental properties of classical edge emitting and novel vertical cavity laser diodes. Topics to be discussed include generation of inversion by current injection in double heterostructures, characteristics of gain and absorption in direct semiconductors, basic laser structures and lasing threshold analysis, multi-mode rate equations and dynamic behavior as well as emission properties of vertical cavity laser diodes and two-dimensional laser arrays. Most of the fundamentals is based upon the comprehensive treatment in [1]. Details on edge emitting lasers can be found in various textbooks [1-4] and special issues of the IEEE Journal of Quantum Electronics [5-6] whereas on novel vertical cavity lasers there is not much review work [7] that one has to refer to various technical papers [8-11] to provide an overview of the state-of-the-art.

2. Interaction of Electrons and Photons in Semiconductors

2.1. ELECTRONS AND HOLES

In a semiconductor electrons are located in energy bands. Of particular interest are the conduction band and the valence band whose electrons determine the electrical and optical properties of a semiconductor, like GaAs. Fig. 1 illustrates the band structure, densities of states, distribution functions and spectral densities of free charge carriers in n- and p-type semiconductors. Near the band edge energy W_c of the conduction band the electron energy is described to a good approximation by an isotropic parabolic band structure

$$W(\vec{k}) = W_c + \frac{\hbar^2}{2m_e}(k_x^2 + k_y^2 + k_z^2) \quad , \tag{1}$$

where $\vec{k} = (k_x, k_y, k_z)$ denotes the electron wave vector, m_e is the effective mass of the electron and $h = 2\pi\hbar$ is Planck's constant. The density of states for electrons in the conduction band is then given by

$$D_c(W) = 2^{5/2}(2\pi)^{-2}\hbar^{-3}m_e^{3/2}(W - W_c)^{1/2} \quad , \tag{2}$$

283

M. Ducloy and D. Bloch (eds.), *Quantum Optics of Confined Systems*, 283-308.
© 1996 *Kluwer Academic Publishers. Printed in the Netherlands.*

284

where $D_c(W)dW$ represents the number of states per unit volume in the energy interval from W to $W + dW$. For the density of states of the holes, i.e. missing electrons in the valence band, a corresponding square root growth law holds for the density of states near the band edge W_v of the valence band

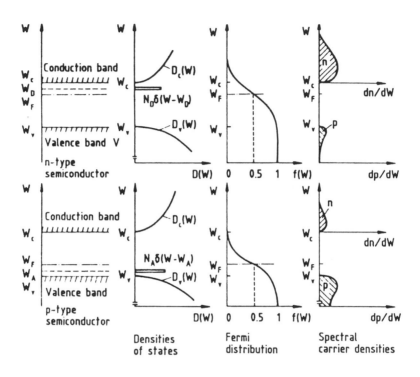

Figure 1. Band structure, densities of states, distribution functions and spectral densities of the free charge carriers in a doped semiconductor with donors (above) and acceptors (below). The shaded areas in the right-hand parts of the figure indicate the electron density n and the hole density p

$$D_v(W) = 2^{5/2}(2\pi)^{-2}\hbar^{-3}m_h^{3/2}(W_v - W)^{1/2} \quad , \tag{3}$$

where m_h denotes the effective mass of the hole. The occupation of states by electrons in the conduction band and holes in the valence band are governed by the Fermi distributions

$$f_{c,v}(W) = \left(1 + \exp\left\{\frac{W - W_{F_c,F_v}}{kT}\right\}\right)^{-1} \tag{4}$$

where W_{F_c} and W_{F_v} are quasi-Fermi levels of the conduction band and valence band, respectively, and kT is the thermal energy. In thermal equilibrium we have

$$W_{F_c} = W_{F_v} = W_F \quad . \tag{5}$$

The density dn of the electrons in the conduction band with an energy between W and $W + dW$ is determined by the density of the states $D_c(W)dW$ in this interval, multiplied by the probability that states of this energy are occupied by electrons

$$dn = \frac{dn}{dW}dW = f_c(W)D_c(W)dW \quad . \tag{6}$$

Integration of the spectral electron density dn/dW then gives the electron density in the conduction band

$$n = \int_{W_c}^{\infty} f_c(W)D_c(W)dW \quad , \tag{7}$$

where the upper limit of the integral has been extended to infinity without making a too large error. Similary, we obtain for the spectral density of holes

$$dp = (1 - f_v(W)) \, D_v(W)dW \quad , \tag{8}$$

since with holes one is concerned with the electron states not occupied. The density of holes in the valence band is then given by

$$p = \int_{-\infty}^{W_v} (1 - f_v(W)) \, D_v(W)dW \quad . \tag{9}$$

For non-degenerate semiconductors with $W_{F_v} - W_v \gg kT$ and $W_c - W_{F_c} \gg kT$ we obtain the simple approximate formulas

$$n = 2\left(\frac{2\pi m_e kT}{h^2}\right)^{3/2} \exp\left\{-\frac{W_c - W_{F_c}}{kT}\right\} \tag{10}$$

and

$$p = 2\left(\frac{2\pi m_h kT}{h^2}\right)^{3/2} \exp\left\{-\frac{W_{F_v} - W_v}{kT}\right\} \tag{11}$$

that relate electron and hole densities n and p to quasi-Fermi levels W_{Fc} and W_{Fv}, respectively.

In the semiconductor the density of electrons and holes can be changed by absorption of high energy photons or by injection of free carriers due to a current flowing across a pn-junction. Such a perturbation causes non-equilibrium densities

$$n = n_0 + \Delta n \quad , \quad p = p_0 + \Delta p \tag{12}$$

with n_0, p_0 denoting the equilibrium values and $\Delta n, \Delta p$ the excess densities. Not too large excess carrier densities are related to quasi-Fermi levels by

$$\Delta n = n_0 \left(\exp \left\{ \frac{W_{F_c} - W_F}{kT} \right\} - 1 \right) \tag{13}$$

and

$$\Delta p = p_0 \left(\exp \left\{ \frac{W_F - W_{F_v}}{kT} \right\} - 1 \right) \quad , \tag{14}$$

where W_F is the equilibrium Fermi energy. Fig. 2 compares equilibrium and perturbation of the equilibrium.

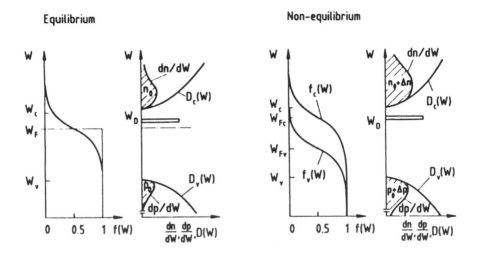

Figure 2. Occupation probability, density of states and spectral carrier density in an n-type semiconductor in equilibrium (left) and non-equilibrium (right)

In a direct semiconductor excess carriers preferably recombine via photon emission, where spontaneous as well as stimulated emission processes can occur. For simplicity we will assume that intrinsic equilibrium carrier densities n_0 and p_0 are vanishingly small throughout that the actual free carrier densities are well approximated by the corresponding excess carrier densities, i.e.

$$n \approx \Delta n \quad , \quad p \approx \Delta p \quad . \tag{15}$$

2.2. EXCESS CARRIERS IN DOUBLE HETEROSTRUCTURES

The model double heterostructure consists of an AlGaAs-GaAs-AlGaAs epitaxial layer triplett as indicated in Fig. 3.

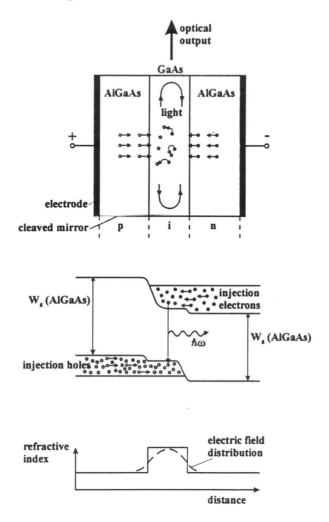

Figure 3. AlGaAs double heterostructure, injection of electrons and holes, and refractive index distribution with profile of guided wave (schematic)

The inner GaAs layer of $W_g = 1.42$ eV bandgap energy or $\lambda_g = 870$ nm bandgap wavelength is surrounded by AlGaAs claddings of larger bandgap energy. P- and n-type doping of the AlGaAs claddings and applying a forward current to the pn-junction results in an accumulation of free carriers in the intrinsic inner GaAs layer. Excess electrons and holes recombine under the emission of photons. This process is the basic mechanism of a light emitting diode and is also responsible for stimulated emission in a laser diode. As a side effect the optical refractive index of the active GaAs layer is larger than that of the surrounding AlGaAs cladding.

As a consequence light can be guided within the GaAs layer by total internal reflection. This means that the inner layer serves as a film waveguide for photons emitted in the direction of the plane of the layer. In a ray model guided photons bounce back and forth at the GaAs-AlGaAs interfaces on a zig-zag path. This confinement of photons is widely exploited for efficient optical amplification in laser diodes.

2.3. ELECTRONIC TRANSITIONS IN DIRECT SEMICONDUCTORS

In a semiconductor electronic transitions occur between valence band and conduction band. The wave functions of the electrons are Bloch waves of the form

$$\Psi_c(\vec{r}) = \frac{1}{\sqrt{V_K}} u_c\left(\vec{k}_c, \vec{r}\right) \exp\left\{i\vec{k}_c \cdot \vec{r}\right\} \tag{16}$$

for the conduction band and

$$\Psi_v(\vec{r}) = \frac{1}{\sqrt{V_K}} u_v\left(\vec{k}_v, \vec{r}\right) \exp\left\{i\vec{k}_v \cdot \vec{r}\right\} \tag{17}$$

for the valence band. V_K denotes the crystal volume and u_c and u_v are lattice-periodic functions. In direct semiconductors like GaAs or InP electronic transitions preferably occur under conservation of the crystal momentum implying

$$\vec{k}_v \approx \vec{k}_c \quad . \tag{18}$$

The transitions therefore take place almost in the vertical direction in the Wk-diagram of Fig. 4 where parabolic bands have been assumed.

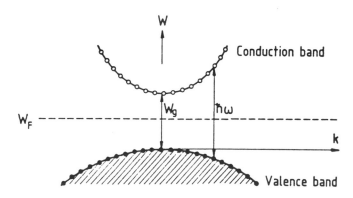

Figure 4. Direct transition of an electron between valence band and conduction band

In a transition of an electron a photon of energy

$$\hbar\omega = W_c(\vec{k}_c) - W_v(\vec{k}_v) \tag{19}$$

with crystal momentum $\vec{k} \approx \vec{k}_c \approx \vec{k}_v$ is created or annihilated. The photon delivers or emits the energy difference between the energy levels $W_c(\vec{k}_c)$ and $W_v(\vec{k}_v)$ taking part in the transition. One distinguishes three elementary processes: (stimulated) absorption, stimulated emission and spontaneous emission. These processes are illustrated in Fig. 5 .

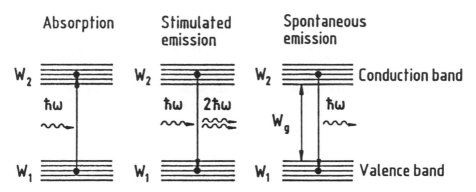

Figure 5. Absorption, stimulated emission and spontaneous emission in interband transitions of electrons

Spontaneous transition processes from the conduction band to the valence band tend to restore thermodynamic equilibrium even without external radiation. Photons are emitted at random phases and directions. If we are interested in the amplification or attenuation of an incident plane electromagnetic wave the fraction of random photons is usually neglected since its contribution in the direction of the incident plane wave is vanishingly small. Consequently it will be left out for calculating absorption and gain coefficients.

In absorption an electron transition takes place from the valence band to the conduction band, with extraction of a photon. The rate of transitions, i.e. the number of transitions per unit volume and time at photon energy $\hbar\omega$, depends on the matrix element $H_{cv}(\vec{k})$ of the transition and the probability that the initial state is occupied and the final state is empty. Thus, the transition rate for absorption is given by

$$P_{v \to c}(\hbar\omega) = \frac{1}{2\pi^2 \hbar} \int\!\!\!\int\!\!\!\int_{-\infty}^{\infty} |H_{cv}(\vec{k})|^2 f_v(1 - f_c)\, \delta\Big(W_c(\vec{k}) - W_v(\vec{k}) - \hbar\omega\Big)\, d^3k \quad , \tag{20}$$

where we have to sum over all possible states with energy difference $\hbar\omega$ and take into account $\vec{k}$-selection with the δ-function. For an incident linearly polarized plane wave of electric field amplitude $\hat{E}_y$ the modulus of the matrix element is calculated as

$$|H_{cv}(\vec{k})|^2 = \frac{q^2 |\hat{E}_y|^2}{m_0^2 \omega^2} \frac{1}{V_K^2} \left| \int\!\!\!\int\!\!\!\int_{V_K} u_c^*(\vec{k}, \vec{r}) \frac{\hbar}{i} \frac{\partial}{\partial y} u_v(\vec{k}, \vec{r}) d^3r \right|^2$$

$$= q^2|\hat{E}_y|^2|M_{cv}(\vec{k})|^2/(m_0^2\omega^2) \quad , \tag{21}$$

where q is the electronic charge, m_0 the electronic mass and ω the angular frequency of the light wave. The quantity $M_{cv}(\vec{k})$ defined in the second equation is called momentum matrix element.

In stimulated emission an electron transition takes place from the conduction band to the valence band, with a generation of a new photon in phase and same direction as the incident photon. Considering occupation probabilities of states and hermiticity $|H_{cv}| = |H_{vc}|$ of the matrix element the transition rate for stimulated emission becomes

$$P_{c\to v}(\hbar\omega) = \frac{1}{2\pi^2\hbar} \iiint\limits_{-\infty}^{\infty} |H_{cv}(\vec{k})|^2 f_c(1 - f_v)\,\delta(W_c(\vec{k}) - W_v(\vec{k}) - \hbar\omega)d^3k \quad . \tag{22}$$

Attenuation or amplification of an incident plane wave is determined by the net transition rate, given by the difference between the absorption and stimulated emission processes

$$\begin{aligned} P(\hbar\omega) &= P_{v\to c}(\hbar\omega) - P_{c\to v}(\hbar\omega) \\ &= \frac{1}{2\pi^2\hbar} \iiint\limits_{-\infty}^{\infty} |H_{cv}(\vec{k})|^2 (f_v - f_c)\,\delta(W_c(\vec{k}) - W_v(\vec{k}) - \hbar\omega)d^3k \quad . \end{aligned} \tag{23}$$

2.4. ABSORPTION AND GAIN COEFFICIENT

The net transition rate is simply related to the macroscopic absorption coefficient $\alpha=\alpha(\hbar\omega)$ of a plane wave propagating in z-direction with Poynting vector S varying as

$$S(z) = S_0 \exp\{-\alpha z\} \quad . \tag{24}$$

The relation is

$$\alpha(\hbar\omega) = \frac{P(\hbar\omega)}{S/\hbar\omega} \tag{25}$$

showing that the absorption coefficient is given by the fraction of incoming photons that are involved in an electronic transition. The expression for α is simply rewritten when the electric field amplitude $\hat{E}_y$ or the photon density N are introduced using

$$S = 2\bar{n}c\varepsilon_0|\hat{E}_y|^2 = N \cdot \hbar\omega \cdot c/\bar{n} \quad , \tag{26}$$

where $\bar{n}$ is the refractive index, c the vacuum velocity of light, and ε_0 the dielectric constant. For computing the absorption coefficient $\alpha(\hbar\omega)$ from (25) and (23) it is to be noticed that the quasi-Fermi distributions f_c and f_v are in general functions of the wave vector $\vec{k}$.

If, however, the dependence is only on the modulus of the wave vector, it follows from (23) with $m_r = m_e m_h/(m_e + m_h)$ that

$$P(\hbar\omega) = \frac{2}{\pi\hbar} \int\limits_0^\infty |H_{cv}(k)|^2 \left[f_v(W_v(k)) - f_c(W_c(k))\right] \delta\left(\frac{\hbar^2 k^2}{2m_r} + W_g - \hbar\omega\right) k^2 dk \quad, \quad (27)$$

where $W_c(k) = W_g + \hbar^2 k^2/(2m_e)$ and $W_v(k) = -\hbar^2 k^2/(2m_h)$. Further simplification is obtained for a matrix element $|H_{cv}|$ independent of k leading to

$$P(\hbar\omega) = \frac{2}{\pi\hbar}|H_{cv}|^2 \int\limits_0^\infty \left[f_v\left(-\frac{\hbar^2 k^2}{2m_h}\right) - f_c\left(\frac{\hbar^2 k^2}{2m_e} + W_g\right)\right] \delta\left(\frac{\hbar^2 k^2}{2m_r} + W_g - \hbar\omega\right) k^2 dk \; . \; (28)$$

The integration is readily performed and using (25) results in a simple expression for the absorption coefficient

$$\alpha(\hbar\omega) = \frac{q^2 |M_{cv}|^2 (2m_r)^{3/2}}{2\pi m_0^2 \hbar^3 \bar{n} c \varepsilon_0 \omega} \sqrt{\hbar\omega - W_g} \left[f_v\left(\frac{m_e W_g - m_e \hbar\omega}{m_e + m_h}\right) - f_c\left(\frac{m_e W_g + m_h \hbar\omega}{m_e + m_h}\right)\right] . (29)$$

Obviously, occupation probabilities f_c and f_v determine the amount of absorption. For $f_c > f_v$ one finds negative absorption, i.e. gain. We define the gain coefficient as

$$g(\hbar\omega) = -\alpha(\hbar\omega) \quad . \quad (30)$$

Strongest gain is obtained for $f_c = 1$ and $f_v = 0$. Fig. 6 illustrates the general behavior of gain and absorption coefficient.

If f_v and f_c are quasi-Fermi distribution functions, positive gain, i.e. inversion of the material, requires

$$W_{F_c} - W_{F_v} > \hbar\omega > W_g \quad . \quad (31)$$

For a quantitative evaluation of (29) the momentum matrix element has to be known. We note that a rough estimate is given by

$$|M_{cv}|^2 = \frac{m_0}{2} W_g \left(1 + \frac{m_0}{m_h}\right) \quad . \quad (32)$$

Also, in deriving (29) we have assumed strict $\vec{k}$-conservation in electron transitions which is expressed by the δ-functions in the integral for the transition rate. More realistic, the finite lifetime of the electrons leads to a deviation from the strict selection rule and to spectral broadening. This effect is taken into account by replacing the δ-function in (28) by a Lorentz-type broadening factor

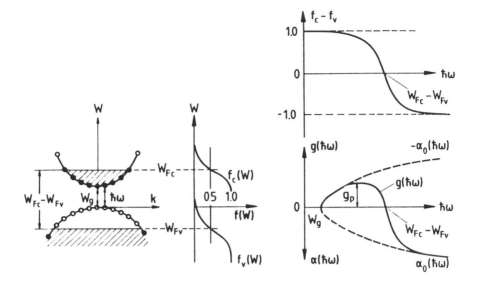

Figure 6. Occupation of bands; occupation probabilities; difference in the occupation probabilities and profiles of the absorption and gain coefficients (schematic)

$$L(W_c(\vec{k}) - W_v(\vec{k}) - \hbar\omega) = \frac{\hbar/(\tau_{in}\pi)}{(W_c(\vec{k}) - W_v(\vec{k}) - \hbar\omega)^2 + (\hbar/\tau_{in})^2} \quad , \tag{33}$$

which is characterized by the intraband relaxation time τ_{in}. For semiconductors like GaAs or InP τ_{in} is in the order of 0.1 ps. In the limit $\tau_{in} \to \infty$ the normalized Lorentzian in (33) tends to the δ-function $\delta(W_c(\vec{k}) - W_v(\vec{k}) - \hbar\omega)$.

Fig. 7 shows numerically calculated gain coefficients $g(\hbar\omega) = -\alpha(\hbar\omega)$ for GaAs as functions of the photon energy for various excess carrier densities $n \approx \Delta n$. For $n > 2.1 \cdot 10^{18}$ cm^{-3} amplification sets in. In the neighborhood of the gain peak the gain profile is approximated well by a parabolic function. At a carrier density $n = 3 \cdot 10^{18}$ cm^{-3} the peak gain coefficient is $g = 200$cm^{-1} and the half-width of the gain curve is 90 meV.

The peak gain coefficient g_p increases, as a shown in Fig. 8, about linearly with the carrier density. One can write

$$g_p = a(n - n_t) \quad , \tag{34}$$

where from Fig. 6 the differential gain coefficient is $a = \partial g_p/\partial n \approx 2.4 \cdot 10^{-16}$cm^2 and the transparency density $n_t \approx 2.1 \cdot 10^{18}$cm^{-3}. Indicated in Fig. 7 is the difference of the quasi-Fermi levels $W_{F_c} - W_{F_v}$ which for $n = 4 \cdot 10^{18}$cm^{-3} is about 100 meV larger than the bandgap energy W_g.

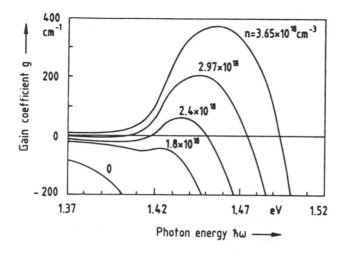

Figure 7. Gain coefficients as functions of the photon energy for GaAs at room temperature, calculated with an intraband relaxation time of $\tau_{in}=0.1$ ps. The bandgap energy is $W_g=1.42$ eV

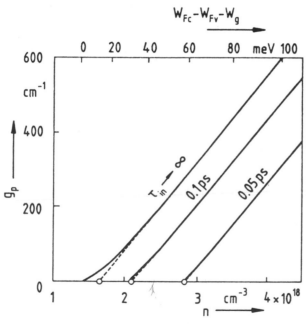

Figure 8. Peak gain coefficients g_p as functions of the injected carrier density n, calculated for different intraband relaxation times τ_{in}. The difference of the quasi-Fermi levels is shown at the upper axis

2.5. QUANTUM WELLS

A quantum well consists of a thin layer, a few nanometers thick, of a semiconductor with small bandgap (e.g. GaAs) in an environment with larger bandgap (e.g. AlGaAs). The motion of electrons is more or less constraint to the plane of the well. As a consequence the density of states is changed from parabolic to a step-like form as indicated in Fig. 9.

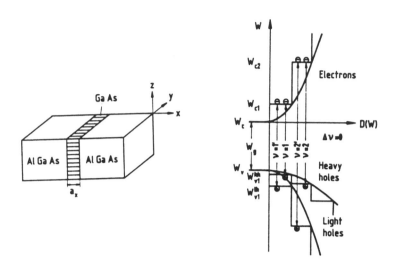

Figure 9. GaAs quantum well, with AlGaAs barriers densities of states and direct electronic transitions in a quantum well. The thickness of the GaAs layer is a few nanometers

Allowed electron transitions occur between subbands of equal order. The calculation of absorption and gain is similar to bulk material but, due to the modified density of states, results in a maximum gain that is generally larger than in the bulk. Fig. 10 shows calculated peak gains for various quantum well thicknesses. It is to be noted that for small gains the differential gain for thin quantum wells is larger than in bulk systems. This observation predicts higher modulation cut-off frequencies for quantum well laser diodes than for conventional semiconductor lasers.

2.6. CURRENT DEPENDENCE OF THE GAIN

Fig. 3 already indicates how excess carriers are accumulated in double heterostructures. Carriers injected with current density j recombine in the distance d of the inner layer. In the steady state the injection rate is just equal to the recombination rate. With carrier lifetime τ_s we can simply write ($n \gg n_0$)

$$\frac{j}{qd} = \frac{n}{\tau_s} \quad . \tag{35}$$

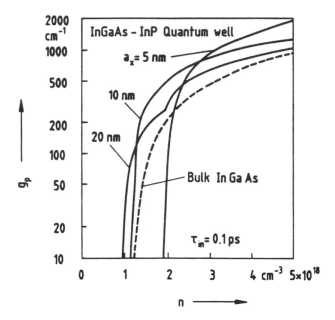

Figure 10. Peak gain coefficients as functions of the carrier density for InGaAs quantum wells of different thickness a_x at room temperature. The peak gain in bulk InGaAs is also shown for comparison

The lifetime τ_s decreases with increasing carrier density. For strong excitation ($\tau_s \propto 1/n$) one obtains approximately the square root behavior

$$n \propto \sqrt{\frac{j}{qd}} \quad . \tag{36}$$

Fig. 11 shows gain curves calculated according to a more refined theory. For the bulk material an injection depth of $d=150$ nm is assumed, whereas the quantum well has a thickness of 10 nm. Obviously, quantum wells favor high gain.

3. Edge Emitting Laser Diodes

3.1. LASER STRUCTURE AND THRESHOLD CONDITION

We have seen how light can be amplified in a semiconductor double heterostructure. Providing feedback of light, typically by using mirrors, self-sustained oscillation is achieved. In semiconductor lasers cleaved endfaces of the crystal serve as mirrors. Fig. 12 shows a broad area double heterostructure laser. Current is injected uniformly over the whole area. Amplification of light takes place approximately uniform in the middle GaAs layer. At the same time, because of its high refractive index, the GaAs layer acts as an optical film waveguide. Evanescent tails of the film wave extend into the adjacent AlGaAs cladding layer. The portion

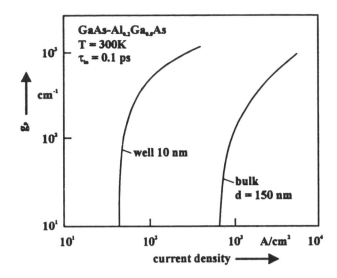

Figure 11. Peak gain coefficients as functions of the injection current for GaAs-Al$_{0.2}$Ga$_{0.8}$As quantum structures with a characteristic dimension of 10 nm and for bulk GaAs with an injection depth of $d=150$ nm

$$\Gamma = \int\limits_{-d/2}^{+d/2} |E(x)|^2 dx \Big/ \int\limits_{-\infty}^{\infty} |E(x)|^2 dx \qquad (37)$$

of the wave propagating in the active GaAs layer is amplified. The confinement factor Γ may vary from 0.5 for bulk lasers to less than 0.01 for quantum well lasers. Light traveling in z-direction is incident perpendicular to the (110) cleaving planes and is reflected back and forth between the laser mirrors constituting a Fabry-Perot resonator. If the reflection losses of the mirrors of reflectivity R are overcome by the gain, lasing occurs.

Considering one round trip of light in the cavity of length L leads to threshold condition

$$R^2 \exp\{2\Gamma g L\} = 1 \quad . \qquad (38)$$

The requirement for the threshold gain thus is

$$\Gamma g_{th} = \Gamma a(n_{th} - n_t) = -\frac{1}{L} \ln R \quad , \qquad (39)$$

where (34) with threshold carrier density n_{th} has been used. Except for losses to be overcome also constructive interference of the waves is necessary.

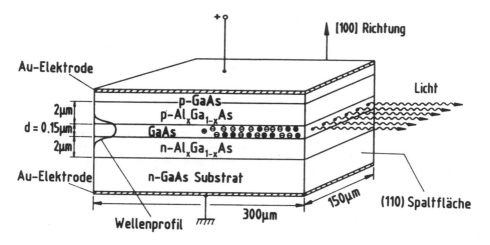

Figure 12. Broad area double heterostructure semiconductor laser

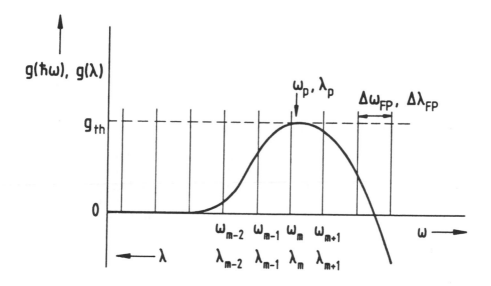

Figure 13. Wavelength dependence of the gain and position of the Fabry-Perot resonances ω_m, λ_m (schematic). The peak gain occurs at ω_p and λ_p

This phase condition leads to the selection of wavelengths λ_m of the possible modes according to

$$m\lambda_m = 2\bar{n}L \quad , \tag{40}$$

where m is an integer and $\bar{n}$ the effective refractive index of the waveguide structure. Amplitude and phase conditions (38) and (40), respectively, are illustrated schematically in Fig. 13.

Those modes will start lasing whose wavelengths λ_m are close to the spectral gain maximum at λ_p. The gain maximum is near the bandgap wavelength of $\lambda_g = hc/W_g \approx 870$nm for GaAs. For typical resonator lengths $L=300\mu$m and refractive indices $\bar{n} \approx 3.5$ we find that the integers m lie at about

$$m = \frac{2\bar{n}L}{\lambda_g} \approx 2400 \tag{41}$$

giving about 2400 half wavelengths along the resonator axis. Neglecting dispersion$(d\bar{n}/d\lambda = 0)$ the separation between two neighboring Fabry-Perot modes is

$$\Delta\lambda = \lambda_m - \lambda_{m+1} = \lambda^2/(2\bar{n}L) \quad . \tag{42}$$

Typical mode separations lie in the order of magnitude $\Delta\lambda \approx 0.3$ nm and therefore are small in comparison with the width of the gain curve.

3.2. STRIPE LASERS AND RATE EQUATIONS

To suppress higher order transverse mode oscillation a strip waveguide in the form of e.g. rib or buried geometry has to be incorporated in the laser diode structure. Fig. 14 shows a so-called buried heterostructure laser. The GaAs active stripe is surrounded by AlGaAs layers that have a lower refractive index than the GaAs. Thus an optical strip waveguide is formed. Also, on both sides of the active stripe a reverse biased pn-junction is incorporated to force current injection in the active stripe only. For sufficiently small width of the active stripe oscillation occurs in the fundamental transverse mode. However, several longitudinal modes fall in the gain spectrum and are amplified. Fig. 15 shows a typical light output characteristics as a function of the diode current and corresponding spectra. Above the lasing threshold the output power increases linearly with current. Slightly above threshold several modes appear in the emission spectrum whereas far above threshold the central mode clearly dominates. It should be noted that quantum wells may be included in the active stripe to improve the gain.

Lasing behavior is well described by rate equations for the electron density n and the photon density N_m in mode m. For simplicity we denote the central mode by $m = 0$ and neighboring modes by $m = \pm1, \pm2, \ldots$ The rate of change of the electron density is determined by current injection as $j/(qd)$, spontaneous emission n/τ_s and stimulated emission in the modes giving

$$\frac{dn}{dt} = \frac{j}{qd} - \frac{n}{\tau_s} - \frac{c}{\bar{n}} \sum_m g_m N_m \quad , \tag{43}$$

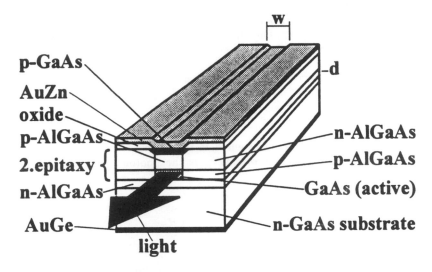

Figure 14. BH (buried heterostructure) laser diode in AlGaAs-GaAs as an example of an index-guided stripe laser. Typical length is 300μm, width w=3μm and active layer thickness d=100 nm

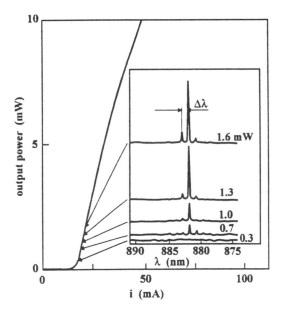

Figure 15. Typical output light power and emission spectra of an index-guided stripe laser diode as a function of the pumping current

where the modal gains $g_m = g(\lambda_m)$ decrease with increasing distance from the spectral gain peak as illustrated in Fig. 12. The photon density is altered by spontaneous emission into the mode of consideration, stimulated emission and resonator losses, e.g. by transmission through the mirrors.

The photon rate equation of mode m is thus given by

$$\frac{dN_m}{dt} = \Gamma \frac{\bar{\beta}n}{\tau_s} + \frac{c}{\bar{n}}\Gamma g_m N_m - \frac{N_m}{\tau_m} \quad , \tag{44}$$

where $\bar{\beta} \approx 10^{-5}$ is the spontaneous emission factor characterizing the fraction of spontaneous emission going into mode m and τ_m is the photon lifetime in mode m. If mirror losses dominate we have simply

$$\frac{1}{\tau_m} \approx -\frac{c}{\bar{n}}\frac{1}{L}\ln R_m \tag{45}$$

with R_m being the reflectivity for mode m.

For a simple evaluation of the rate equations we assume stationary conditions $dn/dt = dN_m/dt = 0$, a dominating central mode $N_0 \neq 0$ but $N_m = 0$ for $m \neq 0$, and negligible spontaneous emission into the mode $\bar{\beta} \approx 0$.

The result is

$$N_0 = \frac{\Gamma\tau_0}{qd}(j - j_{th}) \tag{46}$$

with threshold current density

$$j_{th} = \frac{qd}{\tau_s}\left(n_t + \frac{\bar{n}}{ca\Gamma\tau_0}\right) \quad . \tag{47}$$

The linear increase of photon density with current valid above threshold is in full accordance with the experimental finding in Fig. 12 since the output power is proportional to the photon density. A more rigorous discussion of the multi-mode rate equations again for stationary condition gives the current dependent modal output powers shown in Fig. 16. Side modes are saturated above threshold whereas the central mode takes over all the power far above threshold. This experimentally observed behavior is typical for homogenously broadened lasers.

3.3. MODE SELECTIVE LASER DIODES AND EMISSION LINEWIDTH

Laser diodes with Fabry-Perot resonator tend to oscillate on several longitudinal modes not too far above threshold since the modal reflectivities R_m are very much the same. Mode selection is simply achieved in laser structures with wavelength dependent feedback as in distributed Bragg reflector (DBR) laser diodes shown schematically in Fig. 17 . This device favors the mode determined by the Bragg reflection condition of the corrugated waveguide while suppressing all other modes. Other single-mode emitting laser diodes are distributed feedback (DFB) or coupled-cavity laser diodes.

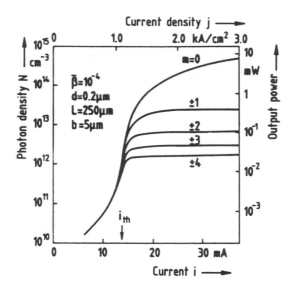

Figure 16. Calculated photon densities and output powers of the central mode ($m=0$) and neighboring modes $m= \pm 1, \pm 2, \pm 3, \pm 4$ as functions of the injection current

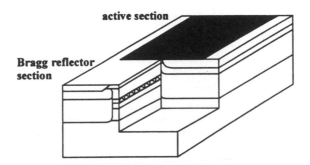

Figure 17. Laser diode with integrated Bragg reflector (schematic)

The emission linewidth of a single-mode laser diode is given by

$$\delta\nu = \frac{\bar{\beta}(1 + \alpha_H^2)j_{th}}{2\pi\tau_0(j - j_{th})} = \frac{\bar{\beta}(1 + \alpha_H^2)\Gamma j_{th}}{2\pi q d N_0} \tag{48}$$

showing that the lasing linewidth decreases with increasing modal photon density N_0 or light output power.

Formula (48) is very similar to that for other types of lasers except for the so-called linewidth broadening factor

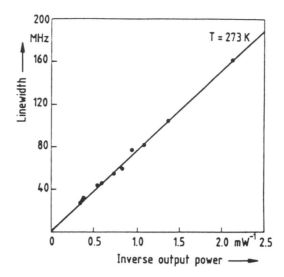

Figure 18. Linewidth of a single-mode AlGaAs laser as a function of the inverse output power

$$\alpha_H = \frac{4\pi}{\lambda} \frac{\partial \bar{n}/\partial n}{\partial g/\partial n} = \frac{4\pi c\tau_0 \Gamma}{\bar{n}\lambda} \frac{\partial \bar{n}}{\partial n} \left(\frac{j_{th}\tau_s}{qd} - n_t \right) \tag{49}$$

describing the coupling of amplitude and phase fluctuations. The factor α_H is not negligible in semiconductor lasers but in the order of $\alpha_H \approx -2 \ldots -7$ depending on laser structure and operating conditions. Fig. 18 shows the experimentally observed linear increase of the emission linewidth with increasing inverse output power. The residual linewidth observed for zero inverse output power may be attributed to carrier number fluctuations or mode competition with not fully suppressed side modes.

4. Vertical cavity laser diodes

4.1. PLANAR PROTON IMPLANTED STRUCTURES

In edge emitting laser diodes light propagates parallel to the plane of the pn-junction and emission occurs perpendicularly from the cleaving planes of the chip. Vertical cavity lasers, on the other hand, radiate perpendiculary from the wafer surface like in a light emitting diode. The resonator axis is perpendicular to the plane of the pn-junction. The active length of the resonator is extremely short ($< 1\mu$m) and highly reflective mirrors are necessary in order to achieve reasonably low threshold currents.

Fig. 19 shows a planar proton implanted vertical cavity laser. The active layer contains three typically 8 nm thick strained layer InGaAs quantum wells embedded in GaAs barriers and AlGaAs claddings. Upper and lower mirrors consist of AlAs-GaAs Bragg reflector stacks with quarter wavelength thick epitaxially grown layers. The emission wavelength of the InGaAs quantum well is larger than the bandgap wavelength of GaAs and therefore the

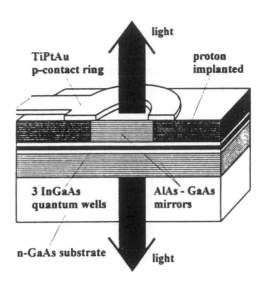

Figure 19. Planar proton implanted vertical cavity surface emitting laser diode (VCSEL)

surrounding layers and the substrate are transmissive for the lasing light. The reflectivity of the Bragg mirrors has to be above 99% requiring about 20 layer pairs in each of the p-doped top and n-doped bottom reflector. Current is supplied through both reflectors and modulation and δ-doping is applied in the Bragg stacks to reduce the series resistance for electron and hole transport across the heterojunctions.

From the longitudinal field distribution in Fig. 20 it is seen that the effective length of the vertical cavity laser is in the order of just one micrometer. It is important to place the quantum wells exactly in an antinode of the standing wave pattern to attain a good coupling of carriers and photons. The incomplete longitudinal overlap of mode and gain has to be taken into account by an appropriate confinement factor in the photon rate equations (44) and also in the threshold condition (38). Fig. 20 also indicates that step grading of the AlAs-GaAs interfaces in the mirrors is applied to further reduce the series resistance.

Fig. 21 shows light output and current voltage characteristics of a planar proton implanted vertical cavity laser of 25 μm diameter. Also included is the electrical to optical power conversion efficiency. Lasing threshold current density is about $600A/cm^2$, maximum output power 15 mW and maximum conversion efficiency 18% . The operating voltage remains below 2.5 V. Roll-over of the light output characteristics is due to device heating. Due to the short cavity length only modes of lowest longitudinal order can oscillate. At threshold the fundamental transverse mode starts lasing and depending on the diameter of the device higher order transverse modes can start to oscillate at higher current levels. For the device of Fig. 21 emission remains single-mode up to 2.5 mW output power.

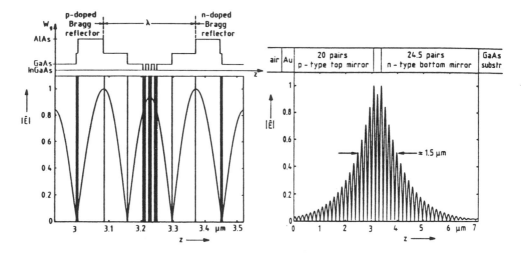

Figure 20. Electric field distribution and band structure of vertical cavity surface emitting laser diode

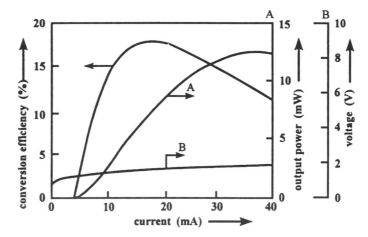

Figure 21. Output power, driving voltage, and electrical to optical conversion efficiency as a function of the driving current of a 25μm diameter vertical cavity laser diode

4.2. TRANSVERSE MODES AND FIBER COUPLING

Due to current heating the refractive index of planar proton implanted vertical cavity lasers is increased on the symmetry axis. The lateral dependence of the refractive index can be well approximated by the parabolic profile

$$\bar{n}(r) = \bar{n}_0 - \bar{n}_2 r^2/2 \quad . \tag{50}$$

Assuming that the index profile is parabolic throughout the cavity the transverse eigenmodes are Laguerre-Gaussian functions. Multi-mode lasing is generally described by the rate equations (43) and (44).

For laser diameter of less than about $10\mu m$ emission predominantly occurs in the lowest order transverse mode whose intensity distribution of Gaussian form

$$I(r,z) = I_0 \exp\left\{-\pi\sqrt{\bar{n}_2\bar{n}_0}r^2/\lambda\right\} \tag{51}$$

is observed as near field pattern on the laser endface. The corresponding far field pattern is also circularly symmetric and of Gaussian shape.

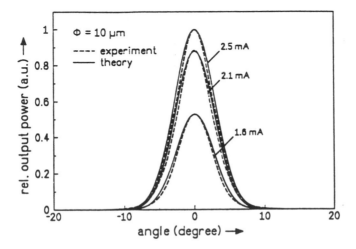

Figure 22. Gaussian far field pattern of a $10\mu m$ diameter single-mode vertical cavity laser diode for various driving currents

Fig. 22 shows that Gaussian far field patterns are observed experimentally. For $10\mu m$ diameter laser diodes the full width half-maximum divergence angle is about five degrees. This is very much in contrast to edge emitting lasers where the radiation cone is elliptical with typical angles of $10°$ in the plane of the pn-junction and $40°$ normal to it resulting in an astigmatic beam. The Gaussian mode patterns (51) of small size vertical cavity lasers can be well matched to the mode profile of single-mode optical fibers. Simple butt-coupling of laser and plane-cut fiber then easily results in coupling efficiencies of larger than 90 % for launching light into the fiber.

The radiation of single-mode vertical cavity lasers is linearly polarized. The minimum emission linewidth observed is 30 MHz and the residual linewidth is found to be in the same order of magnitude. The comparatively large residual linewidth is attributed to mode competition noise from interaction with the mode of orthogonal polarization which due to residual mechanical stress in the device is never completely degenerate.

4.3. MODULATION BEHAVIOR AND BIT RATE TRANSMISSION

The output power of semiconductor lasers can be simply modulated by varying the driving current. The rate equations describe the dynamic behavior. For an arbitrary driving current $j = j(t)$ one has to reside to a numerical solution of the nonlinear system of differential equations. A small signal analysis based on linearized differential equations can already provide insight in the general behavior. In particular, this analysis gives the frequency transfer characteristics

$$H(\nu) = \frac{N_0(\nu)}{N_0(\nu = 0)} = \frac{P(\nu)}{P(\nu = 0)} = \frac{\nu_r^2}{\nu_r^2 - \nu^2 + i\gamma\nu/2\pi} \tag{52}$$

which describes the relative output power fluctuations at Fourier component $P(\nu)$ when the driving current is sinusoidally modulated at frequency ν about the constant bias current $\hat{j}$, i.e. $j(t) = \hat{j} + \Delta j \exp\{2\pi i\nu t\}$. The resonance frequency

$$\nu_r = \frac{1}{2\pi}\sqrt{\frac{ac\,\hat{N}_0}{\bar{n}\,\tau_0}} \tag{53}$$

and the damping constant

$$\gamma = 1/\tau_s + a\hat{N}_0 c/\bar{n} \tag{54}$$

both increase with increasing bias current $\hat{j}$ since according to (46) the average photon density $\hat{N}$ is given by

$$\hat{N}_0 = \frac{\Gamma\tau_0}{qd}\left(\hat{j} - j_{th}\right) \tag{55}$$

assuming oscillation on the dominating central mode.

Fig. 23 shows typical transfer characteristics of a single-mode vertical cavity laser for various bias conditions. It is seen that the 3 dB frequency limit is pushed to above 5 GHz already at comparatively low average output power levels of 240μW. From this behavior it is not surprising that vertical cavity laser diodes can be used as transceivers in high bit rate fiber transmission. Fig. 24 demonstrates transmission at 3 Gbit/s with bit error rates down to 10^{-11}. This result together with the high coupling efficiency to single-mode fibers shows that vertical cavity laser diodes are extremely attractive sources for optical communication systems.

4.4. TWO-DIMENSIONAL ARRAYS

Edge emitting stripe lasers can be arranged in one-dimensional arrays, for instance, to support parallel transmission in fiber ribbons or to produce high power output. Vertical cavity laser diodes, on the other hand, allow simple fabrication of densely packed two-dimensional arrays as schematically indicated in Fig. 25.

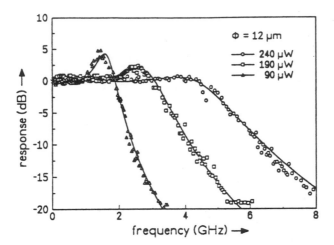

Figure 23. Modulation response of a single-mode vertical cavity laser diode for various average output powers

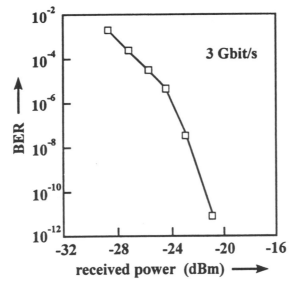

Figure 24. Bit error rate (BER) as a function of received power in 3 Gbit/s data transmission with vertical cavity laser diode source

This property makes vertical cavity laser diodes ideally suited for optical signal processing in 3D systems and also for display applications if visible vertical emitters become available. Aside from wafer scale manufacturability without cleaving and on-wafer testing the possibility of two-dimensional array formation is considered a major advantage of vertical cavity laser diodes compared to edge emitters.

light

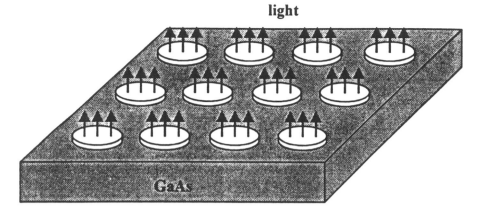

Figure 25. Two-dimensional vertical cavity laser diode array (schematic)

5. Conclusion

We have discussed some fundamental classical theory of semiconductor lasers and outlined some important properties of edge emitting and vertical cavity laser diodes like output power, emission spectrum, beam profile or dynamic behavior. In a more extensive treatment effects of strained quantum wells, waveguiding by separate confinement layers, index guided vertical cavity structures or noise behavior of the emission should be addressed in some more detail. Also, physical aspects of semiconductor laser applications in optical communications and optical interconnects, compact disk storage and bar code scanning or high power and narrow band emission should find some more attention.

References

1. K.J. Ebeling, Integrated Optoelectronics. Springer Verlag, Berlin 1993.
2. K. Iga, Fundamentals of Laser Optics. Plenum Press, New York 1994.
3. G.P. Agrawal, N.K. Dutta, Long-Wavelength Semiconductor Lasers. Van Nostrand Reinhold, New York 1986.
4. L.A. Coldren, S.W. Corzine, Diode Lasers and Photonic Integrated Circuits, John Wiley, New York 1995.
5. Special Issue on Semiconductor Lasers, IEEE J.QE, 27, June 1993.
6. Special Issue on Semiconductor Lasers, IEEE J.QE, 29, June 1995.
7. K. Iga, Vertical cavity surface emitting lasers, Short Course, IEEE Int. Semiconductor Laser Conference, Hawaii 1994.
8. K.L. Lear, K.D. Choquette, R.P. Schneider, S.P. Kilcoyne, K.M. Geib, Selectively oxidized vertical cavity surface emitting lasers, Electr. Lett. 31 (1995) 208-209.
9. B. Möller, E. Zeeb, T. Hackbarth, K.J. Ebeling, High speed performance of 2-D vertical cavity laser diode arrays, IEEE Phot. Techn. Lett. 6 (1994) 1056-1058.
10. U. Fiedler, E. Zeeb, G. Reiner, K.J. Ebeling, 1 Gbit/s single-mode fiber transmission with individually addressable elements of 10 x 10 VCSEL arrays, Proceedings Conference on Optical Fiber Communications 1995, pp. 105-106.
11. R.P. Schneider, K.D. Choquette, J.A. Lott, K.L. Lear, Efficient room temperature continuous-wave AlGaInP/AlGaAs visible (670nm) vertical-cavity surface emitting laser diodes, Phot. Techn. Lett. 6 (1994) 313-316.

NEAR FIELD OPTICS AND SCANNING NEAR FIELD OPTICAL MICROSCOPY

U.C. FISCHER, J. KOGLIN, A. NABER,
A. RASCHEWSKI, R. TIEMANN AND H. FUCHS
Physikalisches Institut,
Westfälische Wilhelms- Universität,
Wilhelm Klemmstr. 10,
48149 Münster, Germany

ABSTRACT By near field optics the diffraction limit of light microscopy can be avoided. Contact imaging by energy transfer is a simple scheme to achieve this goal. Scanning Near Field Optical Microscopy (SNOM) using a tapered metal coated fibre with an aperture at the tip seems to reach a resolution limit at 30 nm. Alternative probes offer the opportunity to extend the resolution to the nm range. The concept of the tetrahedral tip as an efficient SNOM probe and its performance in a hybrid SNOM/STM (Scanning Tunneling Microscope) mode is described, where optical images of 6 nm resolution can be obtained.

1. Introduction

1.1 LIMITS OF CLASSICAL LIGHT MICROSCOPY

Figure 1

Light micrograph of 10 nm thick triangular aluminium patches of a center to center distance of 0.26 µm, taken with a Leica Confocal microscope using an oil immersion objective lens of a n.a. of 1.3. By Courtesy of U. Kubitschek, Institute of Medical Physics and Biophysics, University of Münster

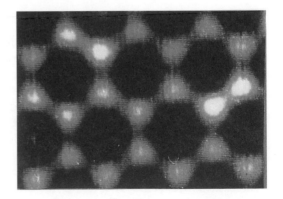

Light microscopy, which was invented more than 300 years ago still is a very important technique in various fields of science, especially in biology. With the recent development of confocal microscopy, the 3-D imaging of fluorescent labels in

309

M. Ducloy and D. Bloch (eds.), Quantum Optics of Confined Systems, 309-326.

biological cell preparations has become possible at a resolution of about 0.2 μm (1). The implementation of sensitive optoelectronic detectors such as photomultiplier tubes, photoavalanche diodes and CCD cameras has improved the sensitivity of detection to an extent, that it is possible to detect single molecules in video rate imaging with remarkable contrast. Yanagida et al (2) reported recently about the measurement of the kinetics of dissociation of ATP molecules from a single myosin molecule by recording the fluorescence of single molecules. It is thus now possible to measure important elementary biochemical events on the level of single molecules with unprecedented sensitivity by using a light microscope. The diffraction limited lateral resolution of light microscopy in the visible spectral range using oil immersion optics is about 0.2 μm as shown in fig. 1.

There are schemes to extend this limit of far field microscopy by nonlinear techniques such as two photon luminescence microscopy (3) and 4π confocal microscopy (4).

Near field microscopy (5-9), on the other hand, is a technique which principally is not limited by diffraction and it offers an opportunity to decrease the resolution of light microscopy substantially below 100 nm and possibly down to molecular resolution. Although other microscopic techniques with atomic resolution such as electron microscopy and scanning probe microscopies exist, Scanning Near Field Optical Microscopy (SNOM) has the potential to combine the resolution of scanning probe microscopy with spectroscopic optical contrast.

2. Near field imaging

2.1 CONTACT IMAGING BY ENERGY TRANSFER

Kuhn proposed to exploit near field effects for imaging purposes down to molecular resolution in his scheme of contact imaging by energy transfer (10) as described in fig. 2 in the form in which it was implemented experimentally by Fischer and Zingsheim (11). A monomolecular film of a dye serves as light sensitive film and a very thin only partially absorbing metal film which is embedded into the surface of a pliable polymer film serves as a conformal mask. For contact imaging the film is brought into contact with the mask and is then irradiated with light. As a result of irradiation the pattern is transferred from the mask to the monomolecular film as a pattern of areas where the dye is bleached and where it is not bleached. After release of contact, the areas of unbleached dye correspond to the areas where there was contact to the slightly absorbing metal pattern. The resolution of this pattern transfer is not limited by the wavelength of light but by the distance between the mask and the dye layer (11). If the distance can be made comparable to the molecular dimension of the dye molecule, molecular resolution should therefore be obtainable. Experimentally resolution of 70 nm was demonstrated in contact copies, which still is far from molecular resolution.

Conformal Mask:
Planar metal pattern embedded into the surface
of a pliable plastic film

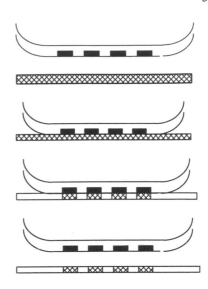

Lightsensitive film:
monomolecular layer of a dye

1.) Contacting

2.) Photochemical reaction during contact
Result:
Pattern transferred
to dye layer

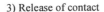

3) Release of contact

Figure 2. Scheme of contact imaging by energy transfer

2.2. SNOM WITH APERTURES

A different approach of using near field effects for imaging is based on the idea to illuminate an aperture of a diameter small compared to the wavelength of light in a thin and opaque metal screen and to use the local emission from this aperture as a source of light of subwavelength dimensions (5-9). This source of light is scanned at a distance equal to or less than the dimensions of the source over the surface of an object to be investigated. The modulation of the emission of this source by near field interaction with the surface of the object is used as the near field optical signal for image formation. With the advent of scanning tunneling microscopy it became apparent, that it would be advantageous to use a source at the end of a tip as a probe and Pohl (7) et al and Dürig et al (12) demonstrated near field microscopy with an apeture at the apex of a metal coated tip serving as a source. This technique was substantially improved 1) by Betzig et al. (13) by introducing a convenient method to fabricate such tips and 2) by Betzig (14) et al. and by Vaez Iravani et al. (15) by implementing shear force microscopy as a tool to control the distance between the tip and the object independently of the near field optical signal. By shear force microscopy the topography of the sample can be obtained simultaneously whereas the near field optical signal displays specific optical contrast. These improvements allowed a more routinely investigation of surfaces and for the first time it was possible not only to obtain images of test objects but also to address more interesting problems by SNOM:

Imaging of single fluorescent molecules by SNOM (16); the measurement of fluorescence decay times of single molecules (17, 18); magnetooptic reading and

312

presumably thermal writing of magnetooptic material at a resolution of 50 nm (19). Spatially resolved low temperature luminescence spectroscopy of semiconductor materials (20). Photolithography by SNOM at a lateral resolution of 80 nm (21,22). The localisation of fluorescent labels on chromosomes (23).

The lateral resolution of SNOM with the fibre probe is typically about 100 nm and seems to reach a principal limit at 30 nm. The limitations of the method are related to the complex structure of the probe and to the inefficiency of light transmission through a tapered metal coated fibre.

A specific advantage of this method as compared to high resolution light microscopy using oil immersion optics and also to other SNOM configurations is the simultaneous recording of surface topography and in some cases also the good accessibility of the fibre tip to the sample, e.g. for low temperature applications.

3. The characteristic components of a near field optical microscope

The optical components of a SNOM are shown schematically in fig.3.:

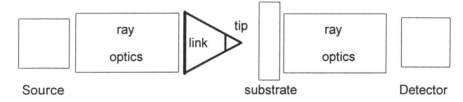

Figure 3. Optical components of a SNOM

Light from a macroscopic light source is directed by ray optical components to the probe. A probing tip serves as a nanoscopic emitter or detector of light. In the case of Aperture SNOM this tip has the form of a nanoscopic aperture in a metal film. The probing tip is connected to the ray optical components of the SNOM by a link which serves to channel light efficiently from the macroscopic or microscopic dimensions of a free or guided light beam to the nanoscopic dimensions of the probing tip. Further ray optical components serve to transmit light from probe to the detector.

The probing tip and the link are the characteristic near-field optical components of a SNOM.

3.1. THE RAY OPTICS OF A SNOM

It is quite useful to classify different types of SNOMs by differences in their ray optical components. As shown schematically in fig. 4, we can differentiate between three

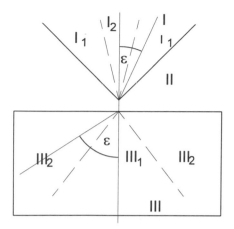

Figure 4. Ray optics of a SNOM

different regions where the rays propagate; the body of the probe I, the outside II and the substrate III of the object. Regions I and III will have in general a higher refractive index than the outside region II. Angular domains of rays III_2 and I_2 exist, where rays are totally reflected within the respective domain, and into which no rays are refracted from the outside II. In the complementary angular domains III_1a nd I $_1$ rays are refracted into the outside and rays from the outside can be refracted into these domains.

For a transparent substrate III of refractive index 1.5, corresponding to a critical angle of total reflection $\varepsilon_T = 41.5^o$, the domain is defined as $III_2 = (90 > \varepsilon > \varepsilon_T$ or $-90 < \varepsilon < -\varepsilon_T)$. This domain is sometimes called the domain of forbidden light (47). For the case of a 2-D analogon of a tip, e.g. a rectangular transparent wedge of refractive index n = 1.5 we obtain correspondingly $I_2 = (-3.5 < e < +3.5)$.

In summary, in regions I and III, angular domains may exist, from which no rays are refracted into the outside II and vice versa, no rays are refracted into these domains from the outside II.

Different types of SNOM have been realized, which can be classified according to the rays of the excitation pathway and the ones of the detection pathway propagating in the different respective domains as listed in table 1. The different types of SNOM according to this classification have different advantages. The types marked with the symbol o in table 1 can be used for opaque samples. The types marked with the symbol * allow opaque bodies of the tip to be used, e.g. simple etched metal tips. The symbol + indicates the advantage, that at a large distance between tip and sample no signal from the tip is obtained and they show a pronounced increase of the signal, when the tip is approached to the sample.

TABLE 1 Ray Optical Classification of different SNOM types

Excitation	Detection	Type	properties	Reference
I	I	Internal reflection	o	24, 45
I_2	I_1	internal reflection	o	9
I	II	external reflection	o	24, 46
I	III_2	inverse PSTM;	+	36, 47
I	III_1	Transmission mode		7, 14
III_2	I	PSTM	+	48, 49
III_1	I	Collection mode		50
III_2	III_2	SPNM	*	28
III_2	III_1		*	31
III_1	III_1	apertureless SNOM	*	25

Ideally no light is leaking from the ray optics of the illumination pathway to the ray optics of the detection pathway. For a very efficient SNOM, light should be coupled into the detection pathway only by the nanoscopic probing tip. This condition is however never strictly reached and the way in which the near field optical components, the link and the tip itself are made decide whether this condition can be fulfilled. The design of the link and the tip itself are therefore the clue to a realisation of an efficient SNOM. At the same time the confinement of light to nanoscopic dimension is a physical problem which has not yet been satisfactorily solved. Models exist such as the coaxial tip discussed below, which are based on the assumption of ideally conducting materials which can be met quite well at microwave frequencies but not at optical frequencies.

Leakage of light causes artifacts in SNOM images and reduces contrast but these effects can be partly overcome by modulation techniques as shown e.g. by Spajer et al. (24) and by Wickramasinghe et al (25,26) .Wickramasinghe et al. did not even use a "link" at all but irrradiated the tip directly with a focussed beam of light.

Even if the effect of leakage on the SNOM signal can be reduced, leakage should be reduced as much as possible to allow e.g. optical nanolithography with a SNOM and local photochemistry in general, and in order to avoid heating and photochemical damage by unnecessary exposure of the object which is also very important for fluorescence studies, where the photodestruction limits the sensitivity of detection. Aperture SNOM up to now was the method with the least leakage problems. This probably is the reason why with this SNOM method nanolithography and fluorescence studies could be successfully performed.

3.2 THE LINK BETWEEN RAY OPTICS AND THE NANOSCOPIC TIP

As mentioned above we consider the link between the ray optical components of a SNOM and the tip as an important component of a SNOM.

For the fibre SNOM the link is the tapered part of the fiber, which connects the waveguide modes of the fibre to the tip. The link should serve the function to channel light efficiently from the ray optical component to the tip. The tapered metal clad monomode fiber can be considered as a tapered cylindrical hollow metal waveguide. Propagation of light is attenuated over - exponentially for the section of the tip which has a diameter smaller of than half the wavelength (27) . Therefore the tapered metal coated fibre cannot be considered to be an effective link. Instead concepts of transmission lines without cutoff can be used for the design of a link which does not suffer from the limitation of cutoff.

3.3 THE TIP

The nanoscopic tip itself is the second characteristic component of the SNOM. It serves as an antenna of nanoscopic dimensions which interacts with the object. The radiation from this antenna is used as a signal for SNOM. The near field distribution of this antenna determines also its radiative signal. As the intensity of the radiation of such an antenna scales with the 6th power of its dimensions, this radiative signal will become very small for nanoscopic tips. Specht et al (28), Vaez Iravani et al (29) Wickramasinghe et al (26) Ash and Nicholls (6) and Fee et al (30) point out, that this limitation can be reduced by interference techniques, where the additive overlap of the amplitudes of the exciting beam and of the field scattered from the tip yield a multiplicative term in the intensity which scale with the third power of the dimensions of the probing tip. The interaction of the antenna with the object is determined at optical frequencies mainly by the electric component of the near field, and therefore it is important that e.g. an electric dipole type of exitation with a high local electric field rather than a magnetic dipole type of exitation with a high local magnetic field is generated on this antenna (6,31). In the case of an aperture NSOM, the aperture serves as the antenna. An aperture cannot be made much smaller than 50 nm and the spread of the near field of an aperture is determined not only by its size but also by the skindepth of the surrounding metal. In order to obtain a resolution significantly below 50 nm it is therefore essential to use a more simple structure such as a small metal grain as an antenna.

4. New probe concepts

We consider the development of alternative probe concepts as a decisive step in obtaining SNOM with a resolution significantly below 50 nm. Two such probe concepts are considered, the coaxial tip and the tetrahedral tip. Both concepts are derived from the concept of the transmission line as a waveguide without cutoff.

4.1. THE COAXIAL TIP

The concept of the coaxial tip as a probe for SNOM (30,32,33) is derived from the consideration, that the well known coaxial cable is a device by which - unlike by the hollow waveguide - electromagnetic energy can be transported through a constriction which is negligibly small compared to the wavelength. This property is due to the TEM_0 transmission line mode of the coaxial cable which has no cutoff. Moreover, the coaxial tip - a conically tapered coaxlial cable - has in theory the interesting property, that the electric and magnetic fields of its TEM_0 mode increase as $1/r$ with decreasing distance from the tip. This means that this concept allows us to increase the field intensities in the tip as compared to the intensity by which the tip is irradiated. This focussing effect of the coaxial tip (41) should be of considerable interest for it to be used as a probe for SNOM, where a high local intensity of the nanoscopic probe is desirable. These properties of the coaxial cable rely on the metal to be a perfect conductor. Therefore the concept can only be considered as a crude model for optical frequencies. A realization of this concept for a SNOM in the infrared region is being attempted by Keilmann (34). Fischer and Zapletal (33) tried to realize this concept by a fabrication scheme for the coaxial tip based on the fabrication of Taylor wires - metal filled glas capillaries- but the fabrication scheme was not yet fully worked out.

4.2 THE TETRAHEDRAL TIP

The idea to use the concept of transmission lines without cutoff for the design of a probe of SNOM with the property of an effective link for confining electromagnetic energy from macroscopic to nanoscopic dimensions is not limited to the coaxial line as was already pointed out by Fee et al (30). A full scheme for a realisation of such a probe was suggested by Fischer (25) and Koglin et al (36) on the basis of the concept of the tetrahedral tip.

4.2.1 FABRICATION OF THE TETRAHEDRAL TIP

For the practice of near field microscopy it is essential to find a simple and reliable fabrication scheme for the tips, because tips have to be replaced frequently. Either a simple laboratory scale fabrication scheme or a method of microfabrication is suitable. A laboratory scale fabrication scheme for the body of the tetrahedral tip was developed on the basis of simple triangular glass fragments. Three faces and three sharp edges - as defined by the fracture process - yield into a well defined corner of a radius of curvature in the nm range. In two succesive steps of metal evaporation as described by Koglin et al (36) the three faces are coated with a thin film (50 nm) of gold such that also two edges and the corner itself are coated with metal, whereas the third edge remains uncoated or coated with less metal. Fig.8a shows schematically such a glass fragment. In order to be able to irradiate the tip with light from within, the coated fragment is glued onto a glass prism as shown in fig. 8b.

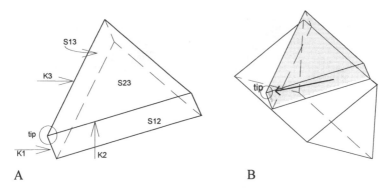

A   B

Figure 5. a) glass fragment and b) coated glass fragment, mounted onto a prism, so that the tip can be irradiated with light

Fig. 6 shows the overall scheme relevant for the function of the tetrahedral tip. The body consists of a glass prism with three faces and edges converging to a common corner which serves as a nanoscopic light emitting tip. The faces are coated with a thin film of metal, preferentially gold or silver. One of the edges is assumed not be coated with metal. A tiny metal grain is also deposited on the tip.

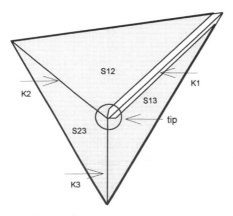

Figure 6. Scheme of the tetrahedral tip

4.2.2 *Optical properties of tetrahedral tip*

4.2.2.1 *Ray optics and interferences in the tetrahedral tip.* We want to use the tetrahedral tip as a nanoscopic source of light for SNOM. In order to get an understanding of how light might get confined to the region of the tip one has to trace a beam of light falling into this tip. A beam of light which is directed into the corner of a triangular glass fragment will be reflected within the fragment and forms there a complicated interference pattern. An example of such an interference pattern is shown in fig. 7a. for a triangular glass fragment of an angle of 75^0.

318

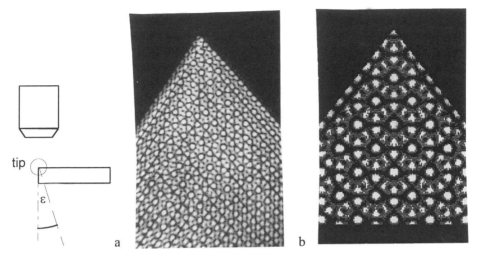

Figure 7 a) Light micrograph of interference pattern on a triangular glass fragment of an angle of 75°, which is irradiated at an angle δ = 25°, as observed with a light microscope. b) calculated pattern.

This pattern was observed with a light microscope directed onto the triangular face of a glass fragment when a beam of light was directed at an angle $\varepsilon = 31^0$ relative to the edge into the fragment as shown in fig. 7a. The interference pattern is due to an interference of - in this case - five beams which are generated by multiple total reflections of the incoming beam. This result can be reproduced as shown in fig. 7c by simple numerical ray tracing calculations (40). For this case of a shallow angle ε of incidence smaller than the angle of total reflection $\varepsilon_T = 41.5^0$ the five beams, giving rise to the interference pattern are refracted out of the tip. If e.g. by prism coupling to the glass fragment ε is chosen to be larger than ε_T, no beams are refracted out of the upper triangular face in the direction of the objective lens.

If in addition the angle of the triangular fragment is chosen to be larger than 2x $\varepsilon_T = 83^0$, e.g. 90°, rays entering the tip from a limited 3-D angular domain will not be refracted to the outside of the tip as was shown for the 2-D analogon in fig. 4.

4.2.2.2 Optical properties of the nanoslot. Whereas these properties of the light beam being reflected within and refracted out of the tip are mainly determined by the glass body of the tetrahedral tip, the confinement of light to the edges and finally to the tip will be determined by the metal coating of the tip. Such a confinement to an edge was investigated experimentally in the 2-dimensional configuration as shown in fig.8. A beam of light of an opening angle of approximately +/-5° was directed into the edge of the metal coated nearly rectangular wedge similar as it is formed also by the edge K_1 of the tetrahedral tip shown in fig. 6. Using a light microscope, light emission from the edge, masked by a slot in the image plane of the microscope, was detected by a photomultiplier. The degree of polarisation P = I_p / I_s as defined by the ratio of transmitted intensities of light polarized perpendicular (I_p) and parallel (I_s) to the edge is shown for different metal coatings and for different wavelengths in fig.8. The behaviour of aluminium coated edges can be understood on the basis of the theory of light transmission through a slot of subwavelength width of 40 nm in a metal screen (see fig. 9b). For a slot of a width in the order of $1/10 \times \lambda$ the intensity I_p is expected to be nearly independent of the wavelength λ, whereas I_s is expected to vary as $1/\lambda^3$ assuming the model of a slot in an infinitely thin perfectly conducting screen (41). In the wavelength range of 544 to 633 nm P should increase by a factor of 1.6 which comes close to the observed value of 1.5. However the absolute experimental values are much smaller than those expected for the ideally conducting screen. Novotny et al. (42) however calculated the transmission of a slot in a wedge coated with aluminum and arrived at an absolute value of P of approximately 10, which comes close to the observed value. Our experimental results therefore suggest, that the experimental conditions correspond to the ones of a slot of a width of a few tens of nm in a screen of aluminum. This result implies, that the field intensity in the near field of the slot is largely enhanced as compared to the intensity of

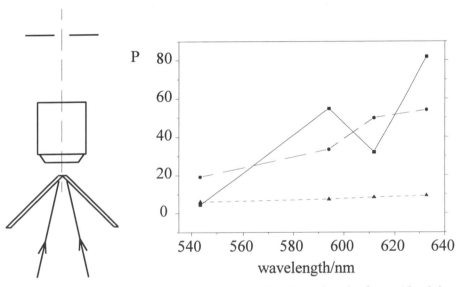

Figure 8. Degree of polarisation of light transmitted through a slot for ▲ Aluminium, ● silver, and ■ gold

the incoming beam for the incident beam being polarized perpendicular to the slot. In the case of silver and gold the increase of P with the wavelength is much stronger than for the case of aluminum. This indicates, that the emission from the edge cannot be considered to be due to light transmission through the slot. We think - however- that this behaviour is consistent with the model of fig. 9a of an exitation of surface plasmons by the irradiating beam which lead to a propagation of surface plasmons towards the edge where they are transformed to radiation. For this case a large value of P is expected, because only p-polarized light can exite surface plasmons. The strong wavelength dependence of P can be explained qualitatively by the decrease of the internal damping of surface plasmons with decreasing wavelength. Surface plasmons are then excited from a larger fraction of the irradiated area. For silver the corresponding decay length L rises from 24.4 μm to 45.4 μm in the wavelength range of 540 nm to 640 nm nm whereas for gold it increases from 0.6 μm to 8 μm as calculated from the complex dielectric constants according to Raether (37). The ratio R of the values L for the different values R_L=L(640 nm)/L(540 nm) is nearly equal to the observed ratios of the corresponding ratio R_P =P(640 nm)/P540 nm). The exitation of surface plasmons on the metal films and their convergence towards the edge should lead to a much larger field enhancements as compared to the case of aluminum. We believe, that the results give evidence for a strong field enhancement at the edge.

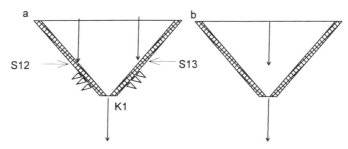

Figure 9. A beam of light directed perpendicular to the edge K_1 from the inside of the tetrahedral may a) excite surface plasmons on the faces S_{12} and S_{13} which travel to the edge or b) it may reach the edge from within and be partially transmitted through the slot

4.2.3 *The concept of the tetrahedral tip as a probe for SNOM*

From the optical properties described above we deduce a hypothetical model for the function of the tetrahedral tip as a probe for SNOM. Such a tetrahedral tip contains several elements which may support electromagnetic surface waves which allow us to confine and concentrate electromagnetic energy at the tip as illustrated schematically in fig. 10 in three steps by "compression" in the transition

1) of an incoming 3 dimensional beam to a 2 dimensional surface wave on metalized faces of the prism

2) of these 2-dimensional surface waves to a one dimensional transmission line mode along the edge and

3) of this 1-dimensional mode to a localised excitation of the particle at the end.

If in the first step the incoming beam of light has a component directed perpendicularly to the edge K_1, it can excite surface plasmons as shown in fig. 9a on the metal films on fthe faces S_{12} and S_{13} by a Kretschmann type of configuration (37). These surface waves are localized on the outside of the metal films and they leak out to a distance of a fraction of the wavelength. These surface waves as excited by an incoming beam which is directed obliquely with respect to the edge K_1, travel towards this edge and may excite there a linear wave travelling along the gap between the metal films. This gap should have similar properties as a parallel wire transmission line which was described for ideal (and to some extent also for non ideal metal wires) by Sommerfeld (38). Such parallel wire transmission line modes can be strongly localized to the gap between the

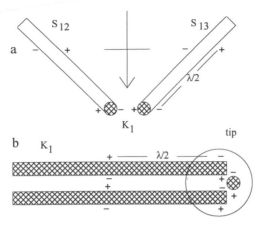

Figure 10. Schematic view of light confinement towards the tip as described in the text

"wires", with their localization not being determined by the wavelength of this traveling wave - as is the case for the two dimensional surface wave - but rather by the width of the gap between the "wires" of a few tens of nanometers. Fig. 10b illustrates the case of a mode consisting of two synchronized polarisation waves traveling in phase along the two "wires". In the third step this mode excites a localized mode on a metal grain at the end of the "double wire" line, leading to a strongly localized near field. We intend to use such a local exitation as a nanoscopic source of light for SNOM.

It should be possible to test rigorously the validity of such a concept of light confinement by numerical calculations using methods which have been worked out by Dereux et al (39).

5. Scanning Near field optical microscopy with the tetrahedral tip

We used the tetrahedral tip as a probe for scanning near field optical microscopy. With this approach it is possible to extend the resolution limit of SNOM from 50 nm, a value which is typical for aperture based SNOM, to the nm range. In first experiments an inverted PSTM type of operation was used, allowing an all optical mode of operation of the tip where no independent control of distance between tip and object is needed. There we obtained a resolution of approximately 30 nm (36). In this mode of operation the optical signal is used to control the distance while scanning. Since, however, the optical signal does not only depend on distance but also on the optical properties of the object and to some extent also on the non perfect inverted PSTM configuration, it is likely that the resolution limit cannot be tested in this configuration. In the experiments the resolution obtained with the same tip varied quite strongly which we believed to be due to a varying distance between tip and object. In order to test the resolution limit of our SNOM probe we therefore used a simultaneous SNOM/STM mode of operation, where the STM signal is used to control the distance while scanning and the transmitted light is used to obtain the SNOM image. This hybrid SNOM/STM mode was realized in a simple transmission SNOM configuration as outlined in fig.11. The glass fragment is mounted in this case onto a specially prepared prism of a size of 2 - 3 mm, such that light can be coupled in from the side.

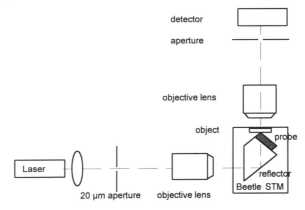

Figure 11. Scheme of the hybrid SNOM/STM mode

The prism is attached to an adaptor which fits into the socket which normally holds the STM tip in our beetle type STM (43). Light emission from the tip is observed through the semitransparent sample by a light microscope and the light signal is detected by a photodiode. There is an electrical connection from the gold coating of the tetrahedral tip to the socket, which is necessary for the STM mode.

Fig. 12 shows first results which we obtained in the hybrid STM/SNOM mode on thin films of evaporated silver (44). The STM image shows the typical appearance of aggregated islands of an evaporated metal film. The simultaneaosly obtained SNOM image shows the same grains, larger grains showing a stronger absorption of light. The resolution in the SNOM image as determined between two minima or two maxima in the absorption is determined to a value of 6 and 8 nm respectively in fig. 12 C. Fig.12 D shows another example where two grains of a distance of 12 nm are clearly resolvedin the lower right part of the fgure.

6. Conclusion

In conclusion it is shown, that by SNOM with the tetrahedral tip a resolution in the nm range can be obtained. This opens the possibility of near field microscopy at molecular resolution similar as was also shown by Wickramasinghe et al. (26). In our case however, the total load of light on the sample is very low. The light emitted from the tip had an intensity in the order of a nW, corresponding to a flux of 10^4 W/cm^2 through an apparent aperture of 10 nm as determined by the obtained resolution of our SNOM. On the other hand the light flux of the irradiating light coupled into the tip was in the order of 500 W/cm$^{2\cdot}$. This conservative estimate indicates, that the tetrahedral tip is a very efficient SNOM probe which has a very high field enhancement. This property opens the possibility of light induced modifications of a surface at a nm scale by local irradiation with very high field intensities.

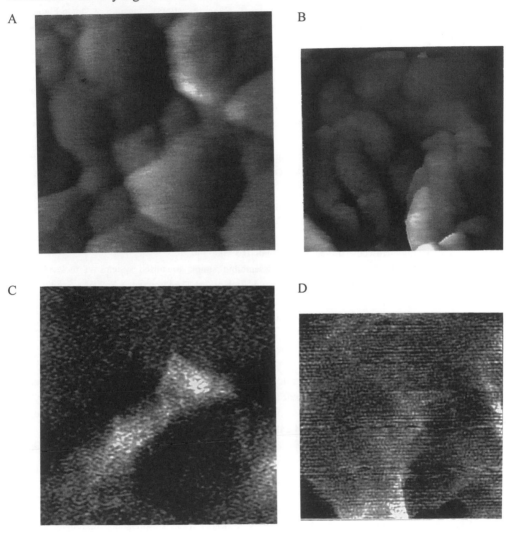

Figure 12. Differential STM and simultaneous SNOM images of a 50nm thick evaporated silver film. A) STM, total xy Scan ranges 62.5 nm. C) SNOM image of the same area as A. B) STM, total scan range 120 nm . D) SNOM image of the same area as B.

Acknowledgement

Part of this work was supported by the German Science Foundation (J.Koglin), by the Volkswagen Foundation and by the Carl Zeiss Jena GmbH.

References

1   Wilson, T, C.J.R. Sheppard. (1984)."Theory and Practice of Scanning Optical Microscopy." Academic Press, London.

2   Funatsu, T., Y. Harada, M. Tokunaga, K. Saito, T. Yanagida. (1995). "Imaging of Single Fluorescent Molecules and Individual ATP Turnovers by Single Myosin Molecules in Aqueous Solution." Nature 374, 558.

3   Denk, W. J.H. Strickler, W.W. Webb. (1990). "Two Photon Laser Scanning Fluorescence Microscopy." Science, 248, 73-75.

4   Hell, S.W., S. Lindek, E.H.K. Steltzer. (1994). "Enhancing the Axial Resolution: Two Photon Excitation 4Pi-confiocal Microscopy." Mod. Opt. 41, 675-681.

5   Synge, E.H. (1928). "A suggested Method for Extending the Resolution into the Ultr-Microscopic Region." Phil. Mag. 6, 356-362.

6   Ash, E.A., G. Nichols (1972). "Super Resolution Aperture >Scanning Microscope". Nature 237, 510-516

7   Pohl, D.W., W. Denk, M. Lanz (1984). "Optical Stethoscopy: image recording with resolution λ /20". Appl. Phys. Lett. 44, 651-653

8   Lewis, A., M. Isaacson, A. Harootunian, A. Muray. (1984). "Development of a 500 A Resolution Light Microscope. " Ultramicroscopy 13, 227-232.

9   Fischer, U.C. (1985). "Optical Characteristics of 0.1 μm Apertures in a Metal Film as Light Sources for Scanning Ultramicroscopy." J. Vac. Sci. Technol. B 3, 386-390.

10 Kuhn, H. (1968). "On possible ways of assembling simple organised systems of molecules". In: "Structural Chemistry and Molecular Biology". A. Rich and N. Davidson (eds.), Freeman, San Francisco, 566-571.

11 Fischer U. Ch., H. P. Zingsheim (1982). "Submicroscopic Contact Imaging with Visible Light by Energy Transfer". Appl. Phys. Lett. 40, 195.

12 Dürig, U., D.W. Pohl, F. Rohner (1986). Near-field optical-scanning microscopy. J. Appl. Phys. 59 (10), 3318-3327.

13 Betzig E. , J.K. Trautman, T.D. Harris, J.S. Weiner, R.L. Kostelak, (1991). "Breaking the diffraction barrier: optical microscopy at a nanometric scale." Science 251, 1468-1470 .

14 Betzig, E., P.L. Finn, J.S. Weiner. (1992). "Combined Shear Force and Near-Field scanning optical microscopy." Appl. Phys. Lett. 60, 2484-2486.

15 Yang, P.C., Y. Chen, M. Vaez-Iravani. (1992). " Attractive Mode Force Microscopy with Optical Detection in an Orthogonal Tip/Sample Configuration".J. Appl. Phys. 71, 2499-2503.

16 Betzig, E., J. Chichester, (1993), " Single Molecules Observed by Near- Field Scanning Optical Microscopy". Science 262, 1422-1425.

17 Xie, X.S., R.C. Dunn. (1994) "Probing Single Molecule Dynamics". Science 265, 361-364.

18 Ambrose, W.P., P.M. Goodwin, J.C. Martin, R.A. Keller. (1994) "Alterations of Single Molecule Lifetimes in Near-Field Optical Microscopy." Science 264, 364-367.

19 Betzig, E., J.K. Trautman, R. Wolfe, E.M. Gyorgy, P.L. Finn, M.H. Kryder, C.H. Chang. (1992). " Near - Field Magneto-Optics and High Density Data Storage." Appl. Phys. Lett. 61, 142-144.

20 Hess, H.F., E. Betzig, T.D. Harris, L.N. Pfeiffer, K.W. West. (1994). "Near-Field Spectroscopy of the Quantum Constituents of a Luminescent System." Science 264, 1740-1745.

21 Betzig, E., J.K. Trautman. (1992) "Near Field Optics: Microscopy, Spectroscopy, and Surface Modification beyond the Diffraction Limit." Science, 257, 189-195.

22 Krausch, G., S. Wegscheider, A. Kirsch, H. Bielefeldt, J.C. Meiners, J. Mlynek. (1995). "Near Field Microscopy and Lithography with Uncoated Fiber Tips: A Comparison." Optics. Comm. 119, 283

23 Moers, M.H.P. (1995). "Near -Field Optical Microscopy." PhD Thesis: Proefschrift Universiteit Twente Enschede. ISBN 90-9008593-9.

24 Spajer, M., D. Courjon, K. Sarayeddine, A. Jalocha, J.M. Vigoureux. (1991). J. Phys. III 1, 1.

25 Zenhausern, F., Y. Martin, H.K. Wickramasinghe. (1995). "Scanning Interferometric Apertureless Microscopy: Optical Imaging at 10 Angstrom Resolution." Science 269, 1083-1085.

26 Zenhäusern F. , M. P. O´Boyle, H. K. Wickramasinghe (1994). Appl. Phys.
Lett. 65, 1623 - 1625.

27 Roberts A. (1991) 'Field Detection by Subwavelength Aperture probes.´ SPIE 1556 Scanning Microscopy Instrumentation, 11-18.

28 Specht M. , J.D. Pedarnig, W.M. Heckl, T.W. Hänsch. (1992) Phys Rev. Lett.
68, 476

29 Vaez Iranani, M., R. Toledo - Crow (1993). "Amplitude, Phase Contrast, and Polarization Imaging in Near-Field Scanning Optical Microscopy". In : "Near Field Optics".D.W. Pohl and D. Courjon eds. Kluwer Academic Publ. Netherland, 25-34.

30 Fee M., S. Chu, T.W. Hänsch (1989). " Scanning Electromagnetic Transmission Line Microscope with Sub- wavelength Resolution." Optics Communications 69, 219 - 224.

31 Fischer U. Ch. (1990). "Resolution and Contrast Generation in Scanning Near Field Optical Microscopy". In: R. J. Behm et al. (eds.), "Scanning Tunneling Microscopy and Related Methods". Kluwer Academic Publishers, Netherlands, 475-496.

32. Keilmann, F. (1988) German Patent DE 3837389 C1.

33 Fischer U.C., M. Zapletal (1991). "The Concept of the Coaxial Tip as a Probe for Scanning Near Field Optical Microscopy and Steps towards a Realisation." Ultramicroscopy 42-44, 393-398.

34 Keilmann F., R. Merz (1993) "Far -Infrared Near-Field Spectroscopy of Two-Dimensional Electron Systems". in: "Near Field Optics". D.W. Pohl and D. Courjon eds. Kluwer Academic Publ. Netherlands; 317-327.

35 Fischer, U.C. (1993). "The Tetrahedral Tip as a Probe for Scanning Near-Field Optical Microscopy". In : "Near Field Optics".D.W. Pohl and D. Courjon eds. Kluwer Academic Publ. Netherland, 255-262.

36 Fischer, U.C., J. Koglin, H. Fuchs. (1994), "The Tetrahedral Tip as a Probe for Scanning Near-Field Optical Microscopy at 30 nm Resolution." J. Microscopy 176, 281-286.

37 Raether H. (1988). "Surface Plasmons on Smooth and Rough Surfaces and on Gratings". in G. Höhler Ed. : Springer Tracts in Modern Physics 111. Springer, Berlin.

38 Sommerfeld A. (1948). Vorlesungen über Theoretische Physik. Dietrich´sche Verlagsbuchhandlung, Wiesbaden.

39 Martin, O.J.F., C. Girard, A. Dereux (1995). "Generalized Field Propagator for Electromagnetic Scattering and Light Confinement". Phys. Rev. Lett. 74, 526-529.

40 Raschewski, A, U.C. Fischer, H. Fuchs, to be published

41 Harrington, R.F. (1961). "Time Harmonic Electromagnetic Fields". Mc Graw Hill.

42 Novotny, L. D.W. Pohl, P. Regli (1994). "Light Propagation through nanometer sized structures: The Two Dimensional Aperture Scanning Near-Field Microscope." J. Opt. Soc. Am.A 11, 1768.

43 Besocke. K. (1987). Surf. Sci. 181, 145.

44 Koglin, J. U.C. Fischer, H. Fuchs. submitted for publication.

45 Bielefeldt, H., J. Hörsch, G. Krausch, J. Mlynek, O. Marti. (1994) Appl. Phys. A 59, 103

46 Cline, J.A., H. Barshatzky, M. Isaacson (1991). "Scanned Tip Refleection Mode Near-Field Scanning Optical Microscope." Ultramicroscopy 38, 299-304.

326

47 Hecht, B., H. Heinzelmann, D.W. Pohl (1993). "Combined Aperture SNOM/PSTM: Best of both worlds?" Proceedings of the 2nd Conference on Near Field Optics, Rayleigh NC. USA, Oct. 20-22 (Ultramicroscopy, in print)

48 Courjon, D., K. Sarayeddine, M Spajer. (1989). "Scanning Tunneling Optical Microscopy".Opt. Commun., 71, 23-28.

49 Reddick, R.C., R.J.Warmack, T.L. Ferrell. (1989)."New Form of Scanning Optical Microscopy". Phys. Rev.B, 39, 767-770.

50 Betzig, E., M. Isaacson, A. Lewis. (1987). "Collection Mode Near Field Optical Microscopy." Appl. Phys. Lett. 51, 2088- 2090.

51 Fischer U.C., D.W. Pohl (1989). "Observation of Single Particle Plasmons by Near-Field Optical Microscopy". Phys. Rev. Lett. 62, 458-461.

LASER RESONANCE PHOTOELECTRON SPECTROMICROSCOPY with a SUBWAVELENGTH SPATIAL RESOLUTION

V.S.LETOKHOV

Institute of Spectroscopy, Russian Academy of Sciences, Troitzk, Moscow Region, 142092, RUSSIA
and
Universite Paris-Nord, Institut Galilee, Avenue J.-B. Clement - F93430 Villetaneouse, FRANCE

ABSTRACT

This lecture discusses a new trend in microscopy, based on the synthesis of wave and corpuscular microscopies, namely, the so-called resonance photoelectron microscopy. This type of microscopy enables one to attain in a natural way a high (subwavelength) spatial resolution combined with a high spectral resolution governed by the spectral line width of absorbing centers on a surface. At the root of photoelectron spectromicroscopy lies the effect of laser-induced resonance stepwise photoionization of absorbing centers on surface.

1. WAVE and CORPUSCULAR MICROSCOPY

There are two well-known types of microscopy, namely, *wave* (optical) and *corpuscular* (electron and ion) microscopy. In optical microscopy, a good

327

M. Ducloy and D. Bloch (eds.), Quantum Optics of Confined Systems, 327-340.
© 1996 *Kluwer Academic Publishers. Printed in the Netherlands.*

spectral resolution can be obtained since the resonant absorption of light is responsive to small changes in the photon energy ($\Delta E \approx 0.001\text{-}0.01$ eV). This is quite understandable since the photon energy is usually commensurate with the energy of electronic transitions and molecular bonds. However, the spatial resolution of optical microscopy is low ($\Delta x \approx 10^3\text{-}10^4$ $\overset{\circ}{A}$). It depends on the wavelength λ of light and the aperture of the light beam 2α. This diffraction limit of spatial resolution can be overcome in optical near-field microscopy [1,2] discussed in one of the preceding lectures. In this way they have already achieved a subwavelength spatial resolution of $\lambda/20$ while retaining a good spectral resolution. This is now being successfully used, for example, in experiments on the fluorescence observation of single molecules [3]. Of course, there is the possibility of x-ray microscopy with a potentially high spatial resolution, though attained at a sacrifice in spectral selectivity. An exception is the case of operation on wavelengths close the edge of characteristic absorption of various elements. Such element-sensitive x-ray microscopy is being considered one of the most important potential applications of x-ray lasers.

In corpuscular microscopy, a high spatial resolution can be achieved. For example, with an electron energy of 1 MeV, a deBroglie wavelength of $\lambda_{dBr} = 10^{-2}$ $\overset{\circ}{A}$, and an angular aperture of $2\alpha = 10^{-3}$ rad, the spatial resolution of electron microscopy may be as high as $\Delta x = 1$ $\overset{\circ}{A}$. However, the spectral resolution of such an electron microscopy will not be very high since a 1-MeV electron is almost insensitive to any type of chemical bond. The fact that the electron interaction cross section depends on the charge of the atomic nuclei makes it possible to form an image with a low contrast. Therefore, to attain an angstrom resolution with a high contrast, it is necessary that heavy atoms be incorporated into the molecules under study. Field-ion microscopy features an extremely high spatial resolution, but it is not very sensitive to the type of ion being produced. In combination with mass spectrometry, it can also provide information about the mass of the ion being produced [4]. Another very efficient approach to the observation of individual atomic and molecular particles on surface is electron scanning tunneling microscopy (STM) and atomic force microscopy (AFS) [5], but to achieve any spectral resolution in these cases is very difficult, if at all possible.

Laser photoelectron (photoion) spectromicroscopy discussed in this lecture potentially combines the merits of two types of microscopy, namely, the high spatial resolution of corpuscular (electron and field-ion) microscopy and the high spectral (energy) resolution of optical spectroscopy. The graph of spectral resolution versus spatial resolution of Fig.1 shows the areas occupied by the various types of microscopy. Resonance photoelectron (photoion) microscopy can be applied to the direct spatial localization of atoms and molecular bonds with an Å – or nm-high spatial resolution. The idea of using resonance photoionization for the spatial localization of molecular bonds or absorbing centers is based on combining the spectrally selective photoionization of a chosen bond or chromophore on a surface with electron or ion microscopy [7].

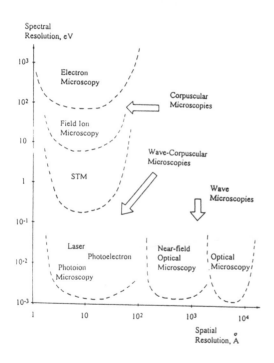

Figure 1. Spatial and spectral resolution of various types of microscopy (wave and corpuscular) and the position of photoelectron spectromicroscopy considered in this lecture, based on the synthesis of wave spectroscopy with corpuscular microscopy.

2. PRINCIPLE of the LASER PHOTOELECTRON (PHOTOION) SPECTROMICROSCOPY

The photoionization of an atom or a molecule gives rise to charge particles at the electron ejection site. In principle, it is quite possible to localize the site where a photoelectron (photoion) is ejected with an accuracy much better than the wavelength of the photoionizing radiation. For this purpose, use can be made of (1) the electron (ion) beam focusing technique or (2) the electron (ion) projection microscopy technique. As a result, it will be possible to observe to ejection site of a charged particle with a high magnification ($X10^5$-$X10^6$.

Figure 2 presents a schematic diagram of a laser photoelectron micrsocope. This system uses one of the possible image construction versions, namely, the projection microscopy technique. It is exactly this version that has recently been implemented experimentally in [8] where it helped to achieve a subwavelength spatial resolution and realize the

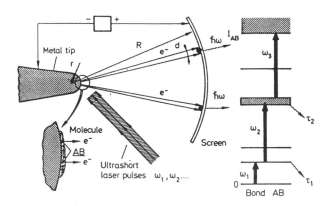

Figure 2. (a) Schematic diagram of a laser photoelectron projection microscope and (b) spectrally-selective multistep photoionization scheme for absorbing centers (color centers, molecular chromophores, etc.)

visualization of individual absorbing centers on a surface [9]. The hemispherical needle tip with a radius of curvature of r is used as a cathode, and a confocal screen with a radius of curvature of R serves as an anode. An absorbing center with a size of $a << r$ is located on the surface of the cathode. A laser is used to effect the resonance photoionization of the absorbing center. The laser irradiation conditions depend on the type of the absorbing center (a doped ion, a chromophore in a molecule, etc.). Most universal is irradiation with a series of ultrashort laser pulses differing in frequency ($\omega_1, \omega_2, \ldots$), which causes a multistep excitation of quantum levels in the desired absorbing center, followed by ionization, i.e., the ejection of an electron. A strong electric field near the cathode causes the photoelectrons to move along radial lines toward the screen. As in the field-ionization electron microscopy (Müller's microscope), the screen displays a magnified image of that region of the cathode which emits electrons. The difference is that the electric field only serves to transfer the photoelectrons to the anode rather than extract them out of the cathode. Such a "soft" and "photoselective" action allows the chosen type of absorbing center to be ionized on the surface of the cathode without involving "nonselective" cathode electrons.

The spatial resolution of the laser photoelectron (photoion) microscope is limited by the following two fundamental factors: the uncertainty principle and the presence of a nonzero tangential velocity component in the emitted photoelectron (photoion). The same factors restrict the spatial resolution of the field-ionization electron (ion) microscope [4].

The uncertainty principle causes a spread of the transversal velocity of the photoelectron (photoion) $\Delta\upsilon_1 = \hbar / 2ma_1$, if its coordinate is accurately determined to be a_1, where m is the mass of the photoelectron (photoion). This spread, $\Delta\upsilon_1$, results in a circle of electron (ion) scattering on the screen with a diameter $d_1 = 2T\upsilon_1$, where $T = R[m/(2eU)]^{1/2}$ is the flight time of the particle from the tip to the screen, and R is the radius of the screen i.e., the distance between the tip and the screen (see Fig.2.). On the other hand, the circle of scattering on the screen, d_1, is also related to the uncertainty of the coordinate at the point of extraction a_1 ($a_1 = d_1/M$), where M is the projector magnification coefficient. As a result, the spatial resolution is limited by the uncertainty ratio at the level,

$$a_1 = (2\hbar T / mM)^{1/2} = [\chi \hbar r (meU / 2)^{-1/2}]^{1/2}. \qquad (1)$$

In Eq.1 the relationship between the magnification coefficient M and the radii of the needle tip r and the screen R, i.e., $M=R/(\chi r)$, was taken into account. Here χ is a numerical factor ranging from 1.5 to 2.0 [4]. The maximum potential difference is limited by the value of the destructive field U near the tip of the needle (equal to 0.5 V/$\overset{o}{A}$). For molecular bonds, it may be assumed that $F_{max}=eU/r<0.2eV/\overset{o}{A}$. Thus, a simple estimate of the ultimate resolution can be made

$$a_1 = [\chi \hbar (2r / mF_{max})^{1/2}]^{1/2}. \qquad (2)$$

When the radius of the needle tip $r=6 \cdot 10^3 \overset{o}{A}, a_1 = 30 \overset{o}{A}$ for the photoelectron microscope. For the photodetachment of H^+, $a_1 = 3 \overset{o}{A}$. When a molecular ion with a molecular weight A=100 photodetaches, the indeterminacy ratio limits the spatial resolution to 1 $\overset{o}{A}$.

An emitted photoelectron (photoion) can have a nonzero transversal velocity, $\Delta \upsilon_1$, if a part of the excitation energy is converted to kinetic energy during the photodetachment from the surface. It is assumed that the average kinetic energy of the transverse motion of a photoelectron (photoion) equals ε_{kin}. Then, in a way similar to the above procedure, it is possible to estimate the spatial resolution due to this effect as

$$a_2 = 2\chi (rE_{kin} / F_{max})^{1/2}. \qquad (3)$$

Let the photoelectron (photoion) have thermal energy, i.e., under normal conditions the energy of transversal motion $E_{kin}=0.01$ eV. Then, all other parameters being the same ($\chi=1.5-2$, $r=6 \cdot 10^3 \overset{o}{A}$, and $F_{max}=0.2 eV/\overset{o}{A}$). $a_2=30$ $\overset{o}{A}$. First, it is essential that the limit of the spatial resolution due to the nonzero transverse velocity of the emitted electrons (ions) be independent of the mass of the particle. Second, the temperature of the needle and the adsorbed molecules must be reduced to 4-10 K to reach a resolution of about several angstroms.

These simple estimates show that with photoelectron (photoion) microscope it is possible to reach resolutions of several nanometers,

3. LASER RESONANCE PHOTOIONIZATION of ABSORBING CENTERS on a SURFACE

The key step for the development of photoelectron spectromicroscopy is the possibility of the photoselective ionization of the absorbing centers of interest on a surface. From the standpoint of the reproducibility of experimental results, most suitable for this purpose have proved wide-gap dielectric crystals doped with rare-earth or transition-metal ions, as well as color centers wherein a number of individual energy levels emerge between the ground state of a system and its ionization threshold. These levels belong to impurity ions or color centers, respectively. Consequently, resonant multistep excitation and ionization of such doped ions or color centers can be performed using photons with a sufficiently low energy that are not absorbed by the crystal host itself. The external photoeffect observed in these conditions will result from selective photoionization of absorbing impurity centers, i.e., the observation of emitted electrons with high spatial resolution will allow visualization of the locations of these centers in the host. At a resolution of 10 nm, one can resolve the arrangement of absorbing centers in hosts at their concentrations up to 10^{18} cm^{-3}. The mobility of impurity ions or centers, at least at temperatures that are not exceedingly high, is low, and the cross sections of the transitions used here ($\sigma=10^{-20}$-10^{-16} cm^2) and the excited level lifetimes (which are in the range of 10^{-9}-10^{-3} s) are sufficiently large, considerably simplifying the organization of the corresponding experiment.

In the process of search investigations of the laser resonant multistep photoeffect, we studied various samples and conditions of laser irradiation: YAG:Nd$^+$ and neodimium silicate glass, $CaMoO_4$-Nd^{3+}, ZrO_2-Nd^{3+}, CaF_2-RE$^{2+,3+}$ (where RE$^{2+,3+}$ are bi- and trivalent impurity ions of rare-earth metals), and LiF crystals with various color centers. The crystals under study were irradiated by means of various nanosecond laser pulses (harmonics of a Nd:YAG laser, a copper vapor laser, tunable dye and solid-state lasers), as well as CW radiation of an argon laser in the case of CaF_2-RE^{2+} and LiF crystals.

The most descriptive and easily explained experiments demonstrating the laser resonant stepwise external photoeffect are our experiments on photoemission from LiF crystals with aggregate F_2-color centers [10,11].

Crystal lattice defects of various types having optical absorption in IR, visible and near-UV regions are termed color centers in alkali-halide crystals. The ground energy state of such defects lies much higher than the ground state of the crystal itself, and correspondingly, photons with much smaller energy are needed to ionize color centers. Aggregate F_2-color centers in LiF crystals are paired anion vacancies in the LiF crystal lattice capturing various numbers of electrons: one for positively charged F_2^+ – centers, two for neutral F_2-centers, and three for negatively charged F_2^- – colors centers. LiF crystals with aggregate color centers are sufficiently stable at room temperatures.

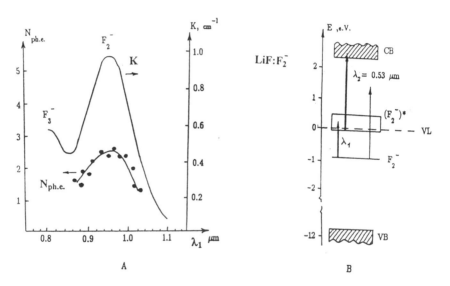

Figure 3. (a) Absorption spectra and the number of emitted photoelectrons as a function of the laser wavelength and (b) energy levels and ransitions involved in the two-step photoionization scheme for the LiF:F_2^- crystal (borrowed from [10,11]).

F_2^- – colors centers resonantly absorb light in a range of 0.85-1.1 μm (Fig.3a) with a cross section $\sigma = 2 \cdot 10^{-17}$ cm^2 at a wavelength of 1.06 μm. As a result of this absorption, they come to a relatively long-lived excited state

$(F_2^-)^*$ with a lifetime of about 60 ns. The laser radiation energy density required to saturate this transition is about 10^{-2} J/cm^2 at a wavelength of 1.06 µm and still lower at the absorption maximum of the transition $F_2^- \rightarrow (F_2^-)^*$.

Figure 3b. presents a two step ionization scheme for the F_2^--centers in the LiF crystal. In experiment, where an area of a sample is irradiated simultaneously by frequency-doubled Nd:YAG laser pulses (λ_2=0.53 µm wavelength) and tunable (λ_1) on color centers, noticeable photoelectron emission is observed. Under optimum conditions, this emission is more than an order of magnitude stronger than that observed when the same area is irradiated by frequency-doubled Nd:YAG laser pulses only. Spectral tuning of IR laser radiation reveals a good correlation between the photoemission signal and F_2^- — color center absorption (Fig.3a). The photoemission signal coincides in time with the laser pulse and is a near-linear function of both the IR laser radiation power and the frequency-doubled Nd:YAG laser power.

These experimental results give strong evidence for the observation of a new physical effect, selective laser stepwise resonant external photoeffect. Light absorption and impurity center ionization are shown to be responsible for photoelectron emission from samples irradiated by suitable chosen laser radiation in a number of cases, for instance, for aggregate F$_2$-color centers in LiF crystals. The observation of this effect opens the way to the realization laser photoelectron spectromicroscopy.

4. LASER PHOTOELECTRON PROJECTION MICROSCOPE

Figure 4 presents a schematic diagram of a laser photoelectron projection microscope. The needles were fabricated from LiF:F$_2$ and LiF:F_2^- crystal fragments by etching. They were a few millimeters long, and their pointed end had an almost conical shape with the radius of curvature of the tip, r, as small as 1 µm and less. The experiments were conducted by V.Konopskii and S.Sekatskii [8,9]. The needle was fastened to an electrode with a voltage of U ranging between 0 and 2.5 kV was applied to it. The vacuum in the microscope chamber amounted to $3 \cdot 10^{-7}$ Torr. The needle tip was irradiated with the 488- and/or 514-nm lines of a CW Ar$^+$ laser. The photoelectrons emitted from the needle tip were directed by an electric field to the input of a microchannel plate (MCP) combined with a fluorescent screen at a distance of R = 10 cm from the tip. The optical image formed on the screen was registered with a CCD camera combined with an Argus-50 Model computer image processor (Hamamatsu Photonics K.K., Japan). The photoelectron collection solid angle was determined by the size of the working area of the MCP (diameter 32 mm) and amounted to some $20°$. Therefore, it was only a small part of the needle tip that was displayed on the screen. The cleaning of

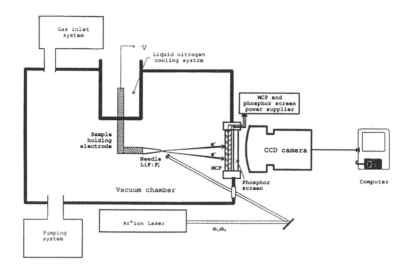

Figure 4. Schematic diagram of the laser photoelectron projection microscopy experiment.

the needle tip that was displayed on the screen. The cleaning of the needle surface from adsorbed molecules was effected on account of photostimulated ion desorption by applying to the needle a stepped-up (up to 20 kV) voltage of the opposite polarity. Immediately following this cleaning the polarity of the voltage applied to the needle was reversed and the photoelectron images of the needle tip were registered, these images being stable and repeatable.

A typical photoelectron image (PEI) of a LiF:F_2^- needle tip is presented in Fig.5. The laser intensity was varied in the range (4 to $8 \cdot 10^3$ W/cm^2) determined by the dynamic range of the registration system used. Since the photoemitted electrons possess some nonzero transverse motion kinetic energy E_{kin}, each photoelectron must be imaged as a spot with a diameter defined by the following expression similar to (3)

$$a \approx 2\chi r (E_{kin} / eU)^{1/2}. \qquad (4)$$

Putting $E_{kin} = 1$ eV and $U = 2.5$ kV, we get $a \approx 30$ nm, which agrees quite well with the size of individual bright spots in Fig.5.

The photoemission of electrons from LiF crystals under the effect of visible light is due to the photoionization of defects in the surface region of the crystals, and the main defect in the specimens under study are the F_2-centers (adjacent anionic vacancies in the LiF crystal lattice that have captured two electrons). It is therefore only natural to associate the bright spots observed in photoelectron images with the visualization of individual F_2-centers. With the electron escape length l_{esc} in the LiF crystal being equal

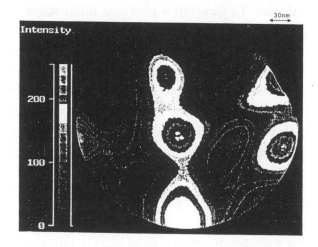

Figure 5. Photoelectron image of the surface of a LiF F_2 needle tip with a radius of a curvature of $r \approx 1$ μm at a $\times 10^5$ magnification. Individual F_2-centers are visible.

to some 3-10 nm and the F_2-centers concentration $n \approx 10^{16}$ cm^{-3}, the average distance between the photoelectron images of these centers must be $\ell \approx (h_{esc}n)^{-1/2} = 100$-170 nm. This agrees with the experimentally observed distances between the bright spots in Fig.5.

The magnification attended in the experiment with the photoelectron microscope was $M \approx \times 10^5$ and the spatial resolution, around 30 nm, which proved sufficient for the visualization of individual color centers in a LiF crystal with the concentration of such centers less than 10^{17} cm^{-3}. The results obtained in [8,9] may be considered the first successful implementation of laser resonance photoelectron microscopy possessing both subwavelength spatial resolution and chemical selectivity (spectral resolution). It will be necessary to increase the spatial resolution of the technique by approximately an order of magnitude and substantially improve its spectral resolution by effecting resonance multistep photoionization by means of tunable ultrashort laser pulses.

5. TOWARDS LASER PHOTOION MICROSCOPY

Especially interesting are the prospects for the development of laser photoion spectromicroscopy that potentially possesses a spatial resolution of a few angstroms, i.e., potentially allows visualizing absorbing centers in large molecules, biomolecules in particular. To develop a photoion microscope, it is necessary to learn how to detach with a laser pulse a chromophore from a large molecule adsorbed on the projector tip surface. The formation of photoions upon irradiation of a surface with a powerful UV laser pulse was demonstrated [12] that the production of photoions resulted from the multiphoton ionization (MPI) of neutral molecules undergoing laser-induced thermal desorption. To avoid this effect which leads to a loss of spatial resolution, the laser pulse duration must satisfy the following obvious condition

$$\tau_p \ll \tau_{det} \approx a_{mol} / \upsilon_0 ; \tau_{transf} \tag{5}$$

where τ_{det} is the time it takes for a photoion with a size of a_{mol} to move at an average velocity of $\upsilon_0 \approx 10^4$ cm/s for a distance of the order of a_{mol} from the tip surface and τ_{transf} is the time of excitation transfer from the excited chromophore to the other parts of the molecule. It is obvious that to realize the MPI of a molecular chromophore directly on the tip surface necessitates the use of femtosecond laser pulses.

   The experiments [13] on the femtosecond MPI of molecules on surface have shown that under femtosecond-pulse irradiation conditions the formation threshold of molecular ions is reduced by a factor of around 10 compared to that in the case of irradiation with picosecond pulses. This is evidence that the MPI of the chromophores under the effect of femtosecond pulses occurs directly on the tip surface and prior to their desorption.

To avoid too fast an energy transfer and enhance the selectivity of the laser excitation of the chosen sites in large biomolecules, it seems very promising to use their chemical labeling. This is especially important in the mapping of the sequences of DNA nucleotide bases having very close spectral properties. The first experiments on the MPI of dye-labeled nucleic acid bases were quite a success [14].

The next natural steps are the study of the mechanisms of photodetachment of dye chromophore ions from molecules adsorbed on metal and dielectric surfaces and the search for conditions necessary to effect preferably the MPI of the labeled chromophores while causing no laser-induced desorption of intact molecules. This is necessary in order to make it possible to irradiate a large molecule on a surface repeatedly and thus accumulate information on the location of the chromophores in its various parts.

In conclusion, the author expresses his gratitude to his colleagues at the Institute of Spectroscopy for joint work at various stages of this project.

REFERENCES

1. Betzig, E. and Trautman, J.K. (1992) Near-Field Optics: Microscopy, Spectroscopy, and Surface Modification Beyond the Diffraction Limit. *Science* 257, 189-195.

2. Pohl, D.W. and Courjon, D. (1993) *Near Field Optics*, Kluwer Academic Publishers, Dordrecht.

3. Trautman, J.K., Macklin, J.J., Brus, L.E. and Betzig, E. (1994) Near-field Spectroscopy of Single Molecules at Room Temperature. *Nature*, 369, 40-42.

4. Tzong, T.T. (1990) *Atom-Probe Field Ion Microscopy*, Cambridge Univ. Press, Cambridge.

5. Binning, G. and Rohrer, H. (1982) Scanning Tunneling Microscopy. *Helv.Phys.Acta*, 55, 726-735.

6. Letokhov, V.S. (1975) The Use of Laser Radiation in Autoelectron and Autoion Microscopy for the Observation of Biomolecules. *Sov.J. Quant. Electr.* 5, 506-510; Possible Laser Modification of Field Ion Microscopy, *Phys.Lett.* A51, 231-232.

7. V.S.Letokhov. *Laser Photoionization Spectroscopy.* (Academic Press, Orlando, 1987), Ch.12.

8. Konopskii, V.N., Sekatskii, S.K. and Letokhov, V.S. (1995) Single Color Centers on the Surface Observed by the Laser Resonance Photoelectron Microscopy (in press).

10. Sekatskii, S.K., Letokhov, V.S., Basiev, T.T., and Ter-Mikirtychev, V.V. (1994) Laser Resonance External Photoelectric Effect in F_2^- Color Centers of LiF Crystals. *Appl.Phys.* A58, 467-470.

11. Letokhov, V.S. and Sekatskii, S.K. (1994) Laser Resonance Photo-ionization of Absorbing Centers on Surface. *Optics and Spectroscopy*, 76, 271-277.

12. Egorov, S.E., Letokhov, V.S. and Shibanov, A.N. (1984). Mechanism of Ion Formation upon Irradiation of Molecular Crystal Surfaces with a Pulsed Laser Radiation. *Sov.J.Quant.Electron.*, 14, 940-946.

13. Chekalin, S.V., Golovlev, V.V., Kozlov, A.A., Matveets, Yu.A., Yartsev, A.P., Letokhov. V.S. (1988) Femtosecond Laser Photoioni-zation Mass Spectrometry of Tryptophan-Containing Proteins. *Journ. Phys.Chem.*, 92, 6855-6858; in *Ultrashort Phenomena VI*, ed. by Yajima, T., Yoshihara, K., Harris, C.B., Springer, Berlin, 414-419.

14. Chekalin, S.V., Golovlev, V.V., Matveets, Yu.A., Yartsev, A.P., Letokhov, V.S., Greulich, K.-O., Wolfrum, J. (1990). *IEEE Journ. Quant.Electr.*, 26, 2158-2161.

REFLECTION OF LIGHT FROM VAPOR BOUNDARIES

G.NIENHUIS

Huygens Laboratorium, Rijksuniversiteit Leiden
Postbus 9504, 2300 RA Leiden, The Netherlands

1. Introduction

Reflection spectroscopy of gases is to a large degree analogous to standard transmission spectroscopy. Standard techniques, such as saturation spectroscopy and magneto-optical methods, can be applied in both cases. Nevertheless, the spectral signatures display striking differences. For instance, absorption and dispersion curves are often interchanged, and the two types of signals contain different information. Moreover, reflection spectroscopy has several characteristic advantages. First, it can be applied to optically thick systems. Second, it is particularly sensitive to the behavior of the gas particles in the boundary region, where the interaction with surfaces can be probed [1]. For dilute gases, this boundary region gives rise to sub-Doppler structures, even for a single incident light beam at low intensity [2,3]. This provides an additional technique of Doppler-free spectroscopy.

In this contribution we discuss a few typical features of the theoretical description of selective reflection of light from the interface between a dielectric and an atomic vapor.

2. Emission by plane layers of dielectric polarization

We consider the interface between a transparent dielectric, and a gaseous medium. A monochromatic light beam is sent through the dielectric to the interface. We are interested in the internally

341

M. Ducloy and D. Bloch (eds.), Quantum Optics of Confined Systems, 341-353.
© 1996 *Kluwer Academic Publishers. Printed in the Netherlands.*

reflected light. It is then convenient to regard the dipole polarization in the vapor as a source of light. Since Maxwell's equations are linear in the sources, we can formally separate the field in the contribution that is present when the vapor is absent, and the field contribution emitted by the dielectric polarization of the vapor. A similar separation also lies at the basis of the microscopic description of long-range two-particle correlations in polar fluids [4]. When the dielectric susceptibility of the vapor is small, this polarization can be evaluated as the response to the refracted light ield that would be present in the absence of the vapor. Hence, the evaluation of the reflection breaks up into three distinct steps. First, we omit the vapor, and evaluate the field refracted into the vacuum. Next, we calculate the dipole polarization of the vapor induced by this refracted field. Finally, we calculate the field emitted by the polarized vapor, and refracted back into the dielectric.

We consider the field components at the frequency ω. Then the time dependence can be omitted, by indicating the polarization $\vec{P}$ and the electric fields $\vec{E}$ in the form

$$2 \text{ Re } \vec{P}(\vec{r}) \, e^{-i\omega t} \qquad , \qquad 2 \text{ Re } \vec{E}(\vec{r}) \, e^{-i\omega t} \qquad (1)$$

In this contribution, we restrict ourselves to normal incidence, so that all fields are invariant for translations parallel to the interface. We select the z-direction normal to this interface. The dipole polarization in the vapor can then be put in the form

$$\vec{P}(\vec{r}) = \vec{P}_o(z) \, e^{ikz} \qquad (2)$$

where $k = \omega/c$. The z-dependence of $\vec{P}_o$ reflects a possible effect of the boundaries.

2.1. FIELD EMITTED BY DIPOLE POLARIZATION

The radiation field emitted by the polarization (2) can be expressed in the Green's function corresponding to the one-dimensional Helmholtz equation

$$\frac{d^2}{dz^2}\phi + k^2 \phi = \delta(z - z_o) \tag{3}$$

which has the solution

$$\phi(z) = \frac{1}{2ik} \exp(ik|z - z_o|) \tag{4}$$

Likewise, the electric field emitted by the polarization $\vec{P}(\vec{r})$ solves the Maxwell equation

$$\vec{\nabla}^2\vec{E} - \vec{\nabla}(\vec{\nabla}.\vec{E}) + k^2\vec{E} = -k^2\vec{P}/\varepsilon_o \tag{5}$$

This shows that the radiation emitted by a layer of polarization

$$\vec{P}(\vec{r}) = \vec{P}_o \delta(z - z_o) \tag{6}$$

is given by the transverse electric field

$$\vec{E}(\vec{r}) = \frac{ik}{2\varepsilon_o} \exp(ik|z - z_o|)\vec{P}_o \tag{7}$$

provided that $\vec{P}_o$ is parallel to the xy-plane [5]. A possible polarization component normal to the layer cannot radiate.

We consider a layer of polarization of the form (2) between z = 0 and z = L. According to (7), this layer emits radiation in both directions normal to the layer. The emission in the negative z direction has the amplitude at z = 0

$$\vec{E}_r = \frac{ik}{2\varepsilon_o} \int_0^L dz\, e^{2ikz} \vec{P}_o(z) \tag{8}$$

which is simply an integral over the contribution from delta-layers of polarization. This field is then refracted back into the dielectric, where it attains the modified amplitude $2\vec{E}_r/(n+1)$. This is then the contribution to the reflected field from the dipole polarization, which must be added to the reflection from a dielectric-vacuum interface [5]. All the physics of selective reflection spectroscopy is contained in the amplitude (8), which represents the emission from the vapor polarization towards the dielectric. We shall simply refer to this amplitude as the field reflected from the vapor. In the same way, the field emitted by the vapor into the positive z direction may be called the transmitted field, with amplitude at z = L

$$\vec{E}_t = \frac{ik}{2\varepsilon_o} e^{ikL} \int_0^L dz\, \vec{P}_o(z) \tag{9}$$

The phase factor in the integrand in (8) reflects that the contribution to the reflected field at z = 0 from the polarization at the position z has twice a phase lag kz, once due to the propagation from the incident field, and once from the propagation backwards of the emitted field. In contrast, the contribution to the transmitted field at z = L has the same phase lag kL, for all positions z of the polarization.

In the simple case of a uniform medium without spatial dispersion, the polarization $\vec{P}_o$ to first order in the density is independent of z. In this case, the integral in (9) is proportional to L, which diverges in the limit of a thick layer. This divergence of the first-order term results from the fact that the medium changes the wavelength, which gives rise to a phase shift proportional to L. Only when this phase shift is small, is linearization in the vapor density justified. In contrast, the integral in (8) remains bounded also in the limit of infinite L.

2.2. EVALUATION OF THE DIPOLE POLARIZATION

The dipole polarization is determined by the average dipole moment of the vapor atoms. We denote by $\rho(z,v,t)$ the density matrix for the internal state of the atoms, as a function of their position and velocity. In the present case of normal incidence, only the z components of the position and velocity enter the description. We make a transformation to a rotating frame, by writing the optical coherence in the form

$$\rho_{eg}(z,v,t) = e^{ikz-i\omega t} \sigma_{eg}(z,v,t) \qquad (10)$$

where e and g indicate the excited and the ground state of the driven transition. For illustration, we consider the simplest possible case of atoms. Then the dielectric polarization of the vapor is

$$\vec{P}_o(z) = N \vec{\mu}_{ge} \int dv \, W(v) \, \sigma_{eg}(z,v) \qquad (11)$$

with N the density of active atoms, $\vec{\mu}_{ge}$ the transition dipole, and W the one-dimensional velocity distribution.

For a dilute vapor at non-saturating intensities, the steady-state value of the optical coherence $\sigma_{eg}(z,v)$ to first order in the field is determined by the equation

$$v \frac{\partial}{\partial z} \sigma_{eg} = - \Lambda_+ \sigma_{eg} + \frac{1}{2} i\Omega \qquad (12)$$

where we used that the excited-state population is of second order, and where the atoms are assumed to follow straight paths. This latter assumption is justified when the kinetic mean free path is larger than a few wavelengths. Inclusion of velocity-changing collisions greatly complicates the description. This is only necessary for dense vapors. We used the notation

$$\Lambda_+ = \gamma - i(\Delta - kv) \qquad (13)$$

with γ the homogeneous linewidth, $\Delta = \omega - \omega_0$ the detuning of the light frequency ω from resonance, and $\Omega = 2\vec{\mu}_{eg} \cdot \vec{E}_1 / \hbar$ the Rabi frequency. [Note that $\vec{E}_1$ describes the field in the vapor cell.] The optical coherence σ_{eg} is determined by (12) with the proper boundary condition. A most reasonable assumption is that at a collision with the surface, an atom is deexcited, so that the optical coherence vanishes. This leads to the simple integrals

$$\sigma_{eg}(z,v) = \frac{i\Omega}{2v} \int_0^z dz' \; e^{-\Lambda_+(z-z')/v} \qquad \text{for } v > 0$$

$$\sigma_{eg}(z,v) = - \frac{i\Omega}{2v} \int_z^L dz' \; e^{-\Lambda_+(z-z')/v} \qquad \text{for } v < 0 \qquad (14)$$

One obtains an explicit expression for $\vec{P}$ after substituting (14) into (11), which in turn leads to expressions for the low-intensity reflected and transmitted fields $\vec{E}_r$ and $\vec{E}_t$ when using (8) and (9). In the resulting expressions, one may view the integration variable z' as the position where an atom absorbs a photon, which is reemitted into the reflected or the transmitted field at the position z.

3. Selective reflection spectroscopy

Substitution of (11) and (14) into (8) leads to an expression for the reflected-field amplitude in terms of a double spatial integral, which can easily be performed. A remarkable feature of the result is that the contribution from atoms with a positive velocity v is identical to the contribution from the atoms with the negative velocity - v. This is characteristic for non-saturating fields. It has been noticed before, both for a half-infinite vapor (L → ∞), [7,5], and for a finite layer [8]. At first sight, this may seem surprising, since one would expect that the Doppler effect discriminates between these two opposite velocity groups. In fact, the Doppler effect on the absorption is just the opposite from that in emission. This can be traced back to the fact that the atoms absorbing at z' and emitting at z give the same contribution as the atoms with opposite velocity, which absorb at z and reemit at z'. These two histories are related by time reversal symmetry.

The explicit expression for the reflected-field amplitude may be cast in the form

$$\vec{E}_r = \vec{E}_1 \, A \, G_r(\Delta, L) \tag{15}$$

with

$$A = \frac{N|\vec{\mu}_{eg}|^2}{2\varepsilon_o \hbar} \tag{16}$$

a measure of the strength of the linear response of the vapor. The function G_r is given by the average

$$G_r(\Delta, L) = \int dv \, W(v) \, g_r(\Delta, v, L) \tag{17}$$

with W(v) the normalized Maxwell distribution. For positive values of v, the integrand is determined by the expression

$$g_r(\Delta, v, L) = \frac{1}{i}(\frac{1}{\Lambda_-} - \frac{1}{\Lambda_+} e^{2ikL}) - \frac{2kv}{\Lambda_+\Lambda_-} e^{-\Lambda_- L/v} \qquad \text{for } v > 0,$$

$$g_r(\Delta, v, L) = \frac{1}{i}(\frac{1}{\Lambda_+} - \frac{1}{\Lambda_-} e^{2ikL}) + \frac{2kv}{\Lambda_+\Lambda_-} e^{\Lambda_+ L/v} \qquad \text{for } v < 0, \tag{18}$$

with

$$\Lambda_- = \gamma - i(\Delta + kv) = \Lambda_+ - 2ikv \tag{19}$$

Time-reversal symmetry between opposite velocity groups makes f_r an even function of v, so that

$$g_r(\Delta, v, L) = g_r(\Delta, -v, L) \tag{20}$$

Note that the reflected field can also be expressed as a one-sided average over the Maxwell distribution for positive (or negative) velocities only. The modified reflection due to the presence of the vapor is found by adding $\vec{E}_r$ to the field reflected from the dielectric-vacuum interface, which has the amplitude $4n\vec{E}_1/(n-1)^2$. The resulting signal is proportional to the real part of G_r.

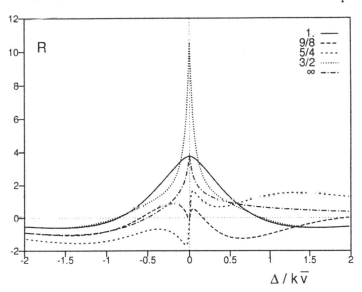

Fig.1. Reflection parameter R, defined in the text, as a function of the detuning Δ, for various values of L/λ, with λ the wavelength. The homogeneous width γ in this and the other figures is taken equal to $k\bar{v}/100$.

The oscillating terms proportional to $\exp(2ikL)$ in (18) apparently do not decay as a function of the layer thickness L. This indicates that the reflection signal strongly varies with L even for a very thick layer. This is an artefact of the linear approximation, which assumes that the driving field is not affected by the vapor. In fact, the driving field decreases with z when the thickness becomes comparable to the absorption length. This is accounted for by allowing k to have a small imaginary part. In the case of a half-infinite vapor, we find $g_r = 1/i\Lambda_-$ for $v > 0$, and $g_r = 1/i\Lambda_+$ for $v < 0$ [9,7,5]. Then the reflected field is simply a one-sided Maxwell average of a dispersion curve. The resulting convolution of a one-sided Doppler profile with a

Lorentzian dispersion curve gives rise to the sub-Doppler structure. For a thin layer, the reflected field has recently been studied in Ref. 7. Fig. 1 displays the reflectivity of the layer, as given by the dimensionless quantity $R = k\bar{v}$ Re G_r, for various values of the thickness L, with $\bar{v}$ the mean velocity. The strong dependence of G_r on L indicates that the reflected field can be viewed as a coherent superposition of the contribution from layers at various positions z.

4. Transmission spectroscopy

The amplitude of the transmitted field is likewise obtained after substituting (11) and (14) into (9). The result is

$$\vec{E}_t = \vec{E}_1 \, A \, e^{ikL} \, G_t(\Delta, L) \tag{21}$$

where

$$G_t(\Delta, L) = \int dv \, W(v) \, g_t(\Delta, v, L) \tag{22}$$

The function g_t is given by the expression

$$g_t(\Delta, v, L) = - \frac{k}{\Lambda_+}[L - \frac{v}{\Lambda_+}(1 - e^{-\Lambda_+ L/v})] \quad \text{for } v > 0$$

$$g_t(\Delta, v, L) = - \frac{k}{\Lambda_+}[L + \frac{v}{\Lambda_+}(1 - e^{\Lambda_+ L/v})] \quad \text{for } v < 0 \tag{23}$$

For large thickness L, the first term in (23) is obviously the dominant one. The expression for this term is the same for positive and negative velocities, and therefore it does not give rise to sub-Doppler structures. The remaining terms result from the transient behavior of the vapor atom near the boundaries. They make the expression for the integrand different for the two signs of the velocity, so they can give rise to sub-Doppler structures. These boundary effects are relatively important for a layer with a thickness of a few wavelengths.

The measured transmitted signal is found by mixing the field (20) due to the vapor polarization with the transmitted field in the absence of the vapor. The result is proportional to the real part of G_t. The dominant part is an absorption dip, which is the convolution of a Lorentzian with the Doppler profile. The transmission signal of the layer is illustrated by the plots in Fig. 2, which give $T = \bar{v}\, \text{Re}\, G_t/L$ for various values of L/λ. The net absorption per unit length is seen to vary strongly with L.

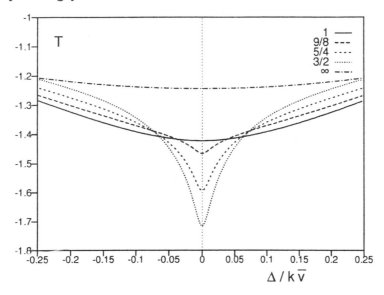

Figure 2. Plot of the transmission parameter T, defined in the text, for the indicated values of L/λ.

5. Magneto-optical effects in reflection

When light propagates through a vapor in the presence of a magnetic field parallel to the propagation direction, the susceptibility for right-hand and left-hand circular polarization can be different due to the Zeeman effect. When the light polarization is linear, the difference in the refractive indices for the two circular polarizations will cause a Faraday rotation of the polarization plane. Moreover, the differential absorption of the two circular polarization components will make the polarization slightly elliptical. This effect is called circular dichroism. The Zeeman shift of the magnetic sublevels simply causes opposite shifts of the susceptibility for the two circular polarizations. To first order in the magnetic field, this explains that in transmission spectroscopy the Faraday effect has the spectral signature of the derivative of a dispersion profile, whereas the circular dichroism is mainly proportional to the derivative of an absorption profile.

In a recent experiment [10] these two effects have been measured for a caesium vapor in reflection. Here we point out that these effects are also affected by the transient behavior of the atoms near the surface. Explicit expressions for these magneto-optical effects are simply obtained in the absence of saturation. We take the x direction along the linearly polarized driving field in the vapor, which is expressed as

$$\vec{E}_1 = E_1 \frac{1}{\sqrt{2}} (\vec{u}_+ + \vec{u}_-) \tag{24}$$

with $\vec{u}_\pm = (\hat{x} \pm i\hat{y})/\sqrt{2}$ the circular polarization vectors. It is sufficient to notice that the response of the vapor to the polarized light field is simply the sum of the responses to the two circularly polarized components. The effect of the magnetic field can be summarized by the simple statement that the polarization $\vec{u}_\pm$ drives a transition with $\Delta M = \pm 1$, so that the Zeeman effect enhances or decreases the transition frequency by the Larmor precession frequency ω_L. This implies that the detuning Δ is effectively replaced by $\Delta \mp \omega_L$, as compared with the case with zero magnetic field. Therefore the reflected field corresponding to the driving field (24) is

$$\vec{E}_r = E_1 A \frac{1}{\sqrt{2}} [\vec{u}_+ G_r(\Delta - \omega_L, L) + \vec{u}_- G_r(\Delta + \omega_L, L)] \tag{25}$$

Of course, this must be supplemented with the reflected field in the absence of the vapor, which is linearly polarized along the x direction. When we calculate the rotation angle ϕ of the plane of linear polarization to first order in the magnetic field, we obtain the result

$$\phi = \frac{4n}{n^2 - 1} A\omega_L \frac{\partial}{\partial \Delta} \mathrm{Im}\, G_r(\Delta, L) \tag{26}$$

In the limit of a thick layer, Im G_r is simply proportional to the convolution of a Lorentzian absorption profile with a one-sided Doppler profile. The Faraday rotation of the reflected polarization is just the frequency derivative of this convolution, which gives a sub-Doppler structure near resonance. For thin layers, the Faraday rotation displays wild variations with L. Fig. 3 displays the spectral signature of the Faraday rotation. The curves represent the dimensionless quantity

$$F = (k\bar{v})^2 \frac{\partial}{\partial\Delta} \operatorname{Im} G_r(\Delta,L) \tag{27}$$

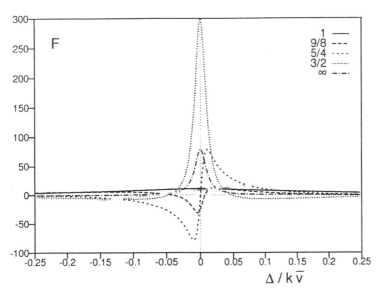

Figure 3. Behavior of the parameter F, measuring the Faraday effect in reflection.

The circular dichroism is defined as

$$\alpha = (I_+ - I_-)/(I_+ + I_-) \tag{28}$$

with $I_\pm$ the intensities of the two circular polarization components. To the same order, we find

$$\alpha = -\frac{4n}{n^2 - 1} \, 2A\omega_L \frac{\partial}{\partial\Delta} \operatorname{Re} G_r(\Delta,L) \tag{29}$$

Hence the circular dichroism is proportional to the frequency derivative of the selective-reflection signal (15), for all values of L. The behavior of α is illustrated in Fig. 4, which displays the quantity

$$D = (k\bar{v})^2 \frac{\partial}{\partial\Delta} \operatorname{Re} G_r(\Delta,L) \tag{30}$$

352

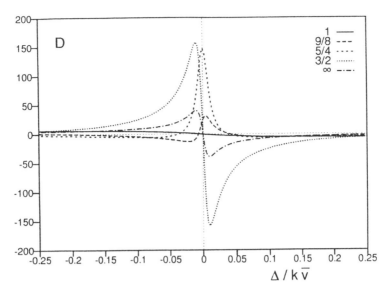

Figure 4. Plots of the parameter D, which determines the circular dichroism in reflection, for different values of L/λ.

6. Final remarks

It is noteworthy that for a thick layer, dispersive and absorptive signals are interchanged when one goes from a reflection to a transmission signal. The transmitted intensity has the shape of an absorption profile, and the intensity of the reflected signal ias governed by a dispersion profile. Also, whereas the Faraday rotation in transmission is basically the frequency derivative of a dispersion curve, the Faraday rotation in reflection is the frequency derivative of an absorption curve. The opposite is true for the circular dichroism: the signal is the frequency derivative of an absorption curve in the case of transmission, and it is the frequency derivative of a dispersion curve in the reflection case. This is related to the fact that the in-phase quadrature of the transmitted signal arises from the imaginary part of the susceptibility, whereas the real part of the susceptibility determines the in-phase quadrature of the reflected signal. Due to spatial dispersion, the signals in reflection are convoluted with a one-sided Doppler profile.

These statements are drastically modified when we consider the reflection from a vapor with a density that is not constant over a wavelength. This we have shown for a thin layer. Moreover, the expressions in terms of the real or imaginary parts of G_r become mixed if we consider the case of a periodic structure of stratified layers of a vapor. Then each layer gives the same complex contribution to $\vec{E}_r$, multiplied with a phase factor and, possibly, an absorption factor. The total reflection is simply the contribution from a single layer, multiplied with a complex factor. The reflection signals are then mixtures of the real and imaginary part of $\vec{E}_r$.

We have shown that even for a vapor of two-state atoms, the reflection and transmission signals display a rich structure, which can be explicitly treated. More complex situations, such as three-state systems [11,12], or vapors driven by several fields [13] have also been described.

1. Oria, M., Chevrollier, M., Bloch, D., Fichet, M. and Ducloy, M. (1991) *Europhys. Lett.* **14**, 527.

2. Cojan, J.L. (1954) *Ann. Phys.* (Paris) **9**, 385.

3. Woerdman, J.P. and Schuurmans, M.F.H. (1975) *Opt. Commun.* **16**, 248.

4. Nienhuis, G. and Deutch, J.M. (1971) *J. Chem. Phys.* **55**, 4213.

5. Nienhuis, G., Schuller, F. and Ducloy, M., (1988) *Phys. Rev.* *A*38, 5197.

6. Guo, J, Cooper, J., Gallagher, A. and Lewenstein, M. (1994) *Opt. Commun.* **110**, 732.

7. Schuurmans, M.F.H. (1980) *Contemp. Phys.* **21**, 463.

8. Vartanyan, D.A. and Lin, D.L. (1995) *Phys. Rev.* **A 51**, 1959.

9. Schuurmans, M.F.H. (1976) *J. Phys. France* **37**, 469.

10. Weis, A., Sautenkov, V.A. and Hänsch, T.W. (1993) *J. Phys. II France* **3**, 263.

11. Schuller, F., Gorceix, O. and Ducloy, M. (1993) *Phys. Rev.* **A 47**, 519.

12. Nienhuis, G. and Schuller, F. (1994) *Phys. Rev.* **A 50**, 1586.

13. Schuller, F., Nienhuis, G. and Ducloy, M., (1991) *Phys. Rev.* **A 43**, 443.

LIGHT INDUCED ATOM DESORPTION

A photo-atomic effect

E.MARIOTTI, L.MOI

INFM- Dipartimento di Fisica Universita' di Siena

via Banchi di Sotto 55 - 53100 Siena - Italy

Abstract

The effect of non resonant and weak light on the adsorption-desorption processes of alkali atoms on organic coated surfaces is discussed and a new phenomenon, the light-induced atom desorption, is presented. It is demonstrated that light can control the adsorption-desorption as well as the diffusion of the atoms inside silane coatings.

1. Introduction

In this paper we report an experimental study of the effects induced by light on adsorption-desorption processes of alkali atoms on dielectric surfaces. We show that light modifies the adsorption-desorption equilibrium in silane coated surfaces. Upon ilumination and depending on the molecular structure of the organic molecules, either huge amounts of atoms are desorbed or atoms are not adsorbed at all. Light, in other words, behaves as a "key" able to open or close the atom "cages" present or formed in the organic layer. This new effect has been observed by A. Gozzini et al. [1], and by M.Meucci et al., who named it LIAD (Light-Induced Atom Desorption) [2]. LIAD is completely different with respect to all the other light induced surface desorption

M. Ducloy and D. Bloch (eds.), Quantum Optics of Confined Systems, 355-365.
© 1996 *Kluwer Academic Publishers. Printed in the Netherlands.*

effects, where the atom ejection is essentially due to surface heating and where much higher laser power densities are used. LIAD seems to be a sort of photo-atomic effect [1,3], where light supports the atom for the energy needed to overcome the potential barrier present at the surface. Further tests are in progress in order to have a complete check of this hypothesis and to verify the similarities with the photoelectric effect.

2. Gas-surface interaction: alkali atoms on organic coatings

The interaction between atoms and surfaces is quite complicated and its discussion imposes some preliminary definitions. The general picture given about the atom-surface collision is the following: when the atom gets near the surface, its motion is determined by the van der Waals attractive force and the electrostatic repulsive force. Either the atom bounches back from the surface elastically or the atom sticks to the surface for a certain time, and then flies away, but in a direction unrelated to that from which it came. The second situation is what is called adsorption. Adsorption is carachterized by the adsorption energy σ. If n atoms strike a unit area of a surface per unit time and remain there for an average time τ_S, then σ atoms per unit area will stay at the surface, where σ is given by

$$\sigma = n\,\tau_S \tag{1}$$

From the kinetic theory of gases we have

$$n = \frac{Nu_m}{4} \tag{2}$$

where u_m is the mean velocity and N is the gas density. When the atom bounches elastically τ_S is very short and it is equal to the collision time $\tau_S = \tau_0 \cong 10^{-13}$ s. When it does not, the atom can reach thermal equilibrium with the surface, and it can be shown that

$$\tau_S = \tau_0\, e^{Ea/kT} \tag{3}$$

where τ_0 is the high-temperature limit of τ_S (i.e. an elastic bounching time τ_0). By combining eqs. 2 and 3 we get

$$\sigma = \frac{Nu_m}{4}\, \tau_0\, e^{Ea/kT} \tag{4}.$$

When the surface is exposed to the gas, σ is usually reached almost instantaneously. If this is not found to be the case, we must seek the cause in transport problems either in the gas phase or in the diffusion of the adsorbed gas from the outer surface to deeper parts of the "porous" structure forming the surface. This last point is important in the LIAD effect and it will be stressed later on.

The time an atom spends at the surface ends by the re-evaporation of the atom. In absence of light, this happens when it picks up so much energy from the thermal fluctuations, that it can overcome the forces which try to keep it at the surface. The adsorbed atoms can either migrate on the surface or diffuse inside it. In fact, E_a is not uniform but it varies on the atomic scale by giving origin to potential wells. If one calls ΔE_a the mean difference E_a takes between neighboring sites, one sees that the atom will jump from place to place whenever its kinetic energy is of the order of ΔE_a.

The surface behaviour therefore depends essentially on the two correlated parameters: the sticking time τ_S and the adsorption energy E_a. As a consequence, it is important to measure them and classical measurements are based on gas flow through a capillary [4]. This same approach can be nowadays implemented by using both laser spectroscopy and the recently observed light induced kinetic effects [5]. The Light-Induced Drift effect (LID), proposed few years ago by Gel'mukhanov and Shalagin [6], gives the possibility of pushing or pulling the gas along a capillary and of studying its diffusion back to equilibrium. The effect is present when a velocity selective excitation is combined with the collisions with a buffer gas, where the collision cross-sections are state dependent. LID has been used to measure E_a of sodium on the alkali resistant Gehlenite glass [7] and on pyrex [8]. The measured values are $E_a \cong 1$ eV and $E_a=(0.71\pm0.02)$eV respectively. Recent measurements, where laser spectroscopy is used to monitor the low vapor densities, have been made by Bonch-Bruevich et al. [9] for sodium on sapphire [$E_a = (0.75\pm0.25)$ eV] and by Stephens et al. [10] for cesium both on pyrex [$E_a=(0.53\pm0.03)$eV] and on sapphire [$E_a=0.4\pm0.1$eV].

From these measurements it derives that the alkali-dielectric surface interaction is in general not negligeable and, according to eq.4, it brings to the consequence that alkali atom layers at the dielectric surface are formed.

When the atom-surface interaction considerably affects the experiment, the problem of its minimization becomes urgent. A classical example is given by optical pumping experiments where special surfaces have been used in order to minimize the depolarization rate in the atom-wall collisions. It is worth noticing that these same solutions have been adopted, since that time, in many other experiments ([5] and references therein).

A small E_a value (of the order of 0.1 eV) can be obtained by choosing a surface that minimizes the van der Waals force. This means to have a surface that is not very polarizable or has a low dielectric constant [11]. Many surfaces have been tested in optical pumping experiments through the relaxation rate of spin polarization (see for example ref.12). In some cases the interaction is so weak that the atom can make more than 1000 collisions before being depolarized.

The two organic compounds, that we used in LIAD experiments, show a weak interaction with the colliding atoms. They are the octamethyl-cyclotetrasiloxane (OCT) and the poly-(dimethylsiloxane) (PDS). PDS has been checked both in optical pumping experiments where, for example, surface reflection of spin polarized sodium atoms has been directly observed [13], and in LID experiments with sodium, where high drift velocities have been measured [14]. OCT has been studied by Xu [3] who observed LIAD with sodium.

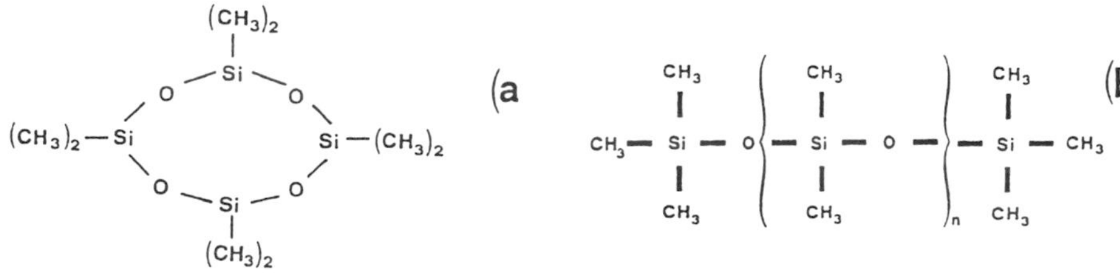

Figure 1. Sketch of the molecular structures of OCT (a) and PDS (b)

The OCT molecule has a crown structure and PDS is a polymer whose chain length may change by many orders of magnitude. In fig.1 their molecular structures are sketched. These compounds "work" because terminated or surrounded by methyl groups. In fact, surfaces terminated by methyl groups have low dielectric constants as these groups are polarizable enough to interact with their neighbors to form a smooth surface, yet they are not so polarizable that they form an image charge for the approaching atom. It is important to have enough layers of methyl groups perpendicular to the wall to effectively shield the underlying surface.

Coating procedure consists in rinsing the cell with an ether solution of the selected silane and then in placing it in an oven at T=250 ^{0}C for few hours. A uniform and perfectly transparent layer is in that way deposited on the cell surface [14].

3. Light Induced Atom Desorption (LIAD)

Although PDS has been used for many years in many experiments and by many groups, only recently a surprising observation has been made [1,2]: when PDS is weakly illuminated it desorbs huge amounts of atoms. The first evidence has been obtained by Gozzini et al. [1] with sodium. A PDS coated cell, placed in front of a dye laser tuned to resonace, showed a bright fluorescence at room temperature, while usually the cell has to be heated to more than 160 ^{0}C, in order to get the same fluorescence intensity. This same result is obtained by illuminating the cell with laser light out of resonance.

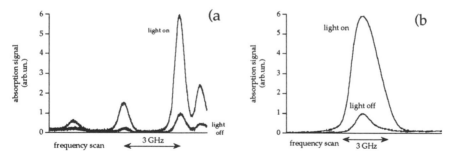

Figure 2. Absorption spectra of rubidium (fig. 2a) and cesium (fig. 2b) with desorbing light off and on. The spectra are obtained at room temperature in PDS coated cells. The desorbing light is a hot wire lamp for Rb and a high pressure mercury lamp for Cs.

Similar observations have been made by Meucci et al. [2] with rubidium and by Mariotti et al. [15] with cesium. In this case the vapor density is detected by using resonant diode lasers and the desorbing light is, besides the available He-Ne and diode lasers, either a thermal source (a hot wire lamp) or a high pressure mercury lamp. Huge amounts of rubidium and cesium atoms are desorbed and the vapor densities are increased by one and two orders of magnitude. In fig.2 the rubidium and cesium absorption spectra are shown with desorbing light off and on. The huge absorption signal increase corresponds, in both cases, to a strong increasing of the vapor densities inside the cell.

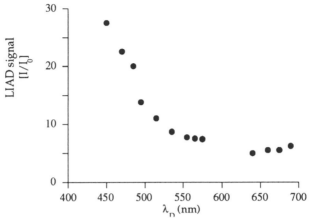

Fig. 3. Relative increasing of rubidium vapor absorption as a function of λ_D. The light power density is kept constant. The coating is PDS.

In fig.3 LIAD dependence on the desorbing light wavelength λ_D is reported for rubidium at constant light power density W_D [2]. The effect is more and more intense as λ_D decreases. In the considered wavelength range no threshold is observed and no resonance effect is detected when λ_D is tuned to either D lines. The LIAD dependence on W_D shows saturation at W_D values (80 mW/cm^2 for Rb [2]) which are orders of magnitude smaller than those reported for all other desorption experiments [16]. From saturation it is possible to derive the total number of adsorbed atoms and hence the surface density σ. The obtained value for rubidium, at room temperature, is $\sigma \cong 10^{10}$ cm^{-2}. The corresponding value of the adsorption energy results (from eq.4) $E_a \cong 0.5$ eV [2]. This value is not negligeable as expected. The explanation, as discussed later, stays in the unvalid hypothesis that the desorbed atoms come only from the outer surface and not also from the coating bulk.

The LIAD dependence on the temperature has been studied for sodium in the range between 273 and 373 K [1,3]. LIAD is observed down to about 273 K and the desorbed atoms appear thermalized to the surface temperature. This is an important issue for possible applications and it deserves more investigations. Similar results have been obtained by us for rubidium and cesium.

By switching on and off the desorbing light and by monitoring continuosly the vapor density, it is possible to study the LIAD dynamics. In fig.4 the rubidium absorption signal as a function of time is reported. When the desorbing light is switched on at time $t = t_1$, vapor density starts to increase with a time constant τ_L inversely proportional, in first approximation, to W_D. When the light is switched off at $t = t_2$, the vapor density goes back to equilibrium with a decay time τ_D, which is in first approximation independent on the previous excitation conditions, but that, at a more deep analysis, shows a more complex behaviour.

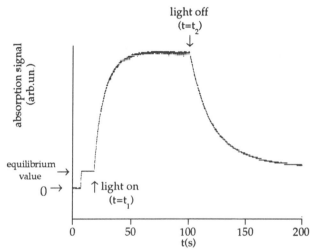

Fig. 4. Rb absorption signal as a function of time in a PDS coated cell at room temperature.

Let us first discuss a simplified model where the atom exchange is only between the gas phase and the outer surface. The following rate equation holds for the total number of atoms in the vapor phase n_V:

$$\frac{dn_V}{dt} = \alpha I_D n_S + \beta n_S - \gamma n_V \tag{5}$$

where $\alpha = \alpha(\lambda_D, S)$ is a factor describing the LIAD efficiency on the desorbing wavelength and on the kind of surface S, β and γ are the desorption and adsorption rates respectively. I_D is the desorbing light intensity and n_S is the number of atoms on the surface. By introducing the total number of atoms $N = n_S + n_V = const$, the solution of the rate equation is:

$$n_V(t) = \frac{N}{\alpha I_D + \beta + \gamma} \left(\alpha I_D + \beta - \frac{\alpha I_D \gamma}{\beta + \gamma} e^{-(\alpha I_D + \beta + \gamma)t} \right) \qquad (6)$$

when light is switched on, and

$$n_V(t) = \left(\bar{n}_V - \frac{\beta N}{\beta + \gamma} \right) e^{-(\beta + \gamma)t} + \frac{\beta N}{\beta + \gamma} \qquad (7)$$

when light is switched back off. $\bar{n}_V$ represents the total number of atoms in the vapor phase at $t = t_2$. When $I_D = 0$ the equilibrium value of n_V is

$$n_{eq} = \frac{\beta N}{\beta + \gamma} \qquad (8)$$

Equations (6),(7) have been used to fit the experimental data on rubidium and to derive the β and γ parameters. The obtained values are $\beta \cong 2.5 \ 10^{-3} s^{-1}$ and $\gamma \cong 5 \ 10^{-2} s^{-1}$ [17].

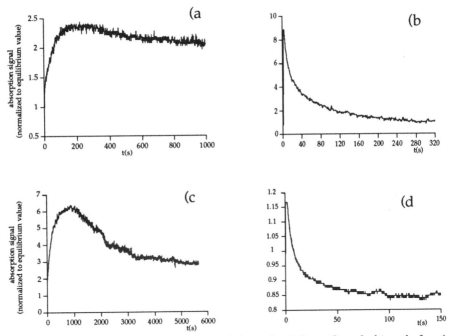

Fig. 5. Absorption signal as a function of time of rubidium (figs. 6a,b) and of cesium (figs. 6c,d) in presence of PDS coating (figs. 6a,c) and OCT (figs. 6b,d).

The decay time of the vapor density in the dark can be derived from the fitting and it results, in this case, $\tau_D = 20$ s. τ_D is much longer than the diffusion time of the vapor in the buffer gas that is of the order of $\tau_{diff} = L^2/D = 10^{-1} + 10^{-2}$ s. This is a further evidence that the atoms do not stick at the surface after the first collision but they make many collisions with the cell walls before being adsorbed.

This preliminary analysis of the LIAD effect is limited to a short time interval, of the order of few minutes. If the time scale is expanded to hours, new features appear which depend on the molecular structure of the coating and on the alkali atom species. In fig.5 the LIAD dynamics for rubidium and cesium with PDS and OCT coatings is shown. The behaviour of the two coatings over a large time scale is quite different. PDS shows a fast density increasing when the desorbing light is switched on ($\tau \cong 50$ s), followed by a slower increasing ($\tau \cong$ few minutes) and then by a much slower decay of the density back to the equilibrium value ($\tau \cong$ few hours). OCT shows an initial fast increasing ($\tau \cong 1$s) immediately followed by a slower increasing ($\tau \cong$ few seconds) and then by a fast density decay to the equilibrium value ($\tau \cong$ minute). Moreover, while PDS does not show fundamental differences between rubidium and cesium, the same does not hold for OCT. OCT, in fact, gives much faster times with cesium and it results that, after an initial density increasing, there is a density decreasing below the equilibrium value (fig.5d). These features can be explained by taking into account both the diffusion of the alkali atoms inside the coatings and their different molecular structures.

4 - Discussion and applications

LIAD, as stated before, is not a thermal effect. Very weak light intensity, like the one generated by standard laboratory illumination, is enough to produce detectable density increasing. Thermal desorption produced by pulsed lasers needs light power densities that are orders of magnitude larger [16]. LIAD wavelength dependence excludes the possibility of a surface-plasmon excitation [18]. A detailed analysis of this effect is still in progress but the basic mechanism, as proposed by Gozzini et al. [1,3], is the following: the atom approaching the surface, collide elastically with the methyl groups. Time by time the atom has enough energy to go through the potential barrier present at the surface and to get inside the coating where it, interacting with the silicon or oxygen atoms and loosing the valence electron, is solved as cation and complexed by a weak bond. When the surface is illuminated, the light breaks this bond and the atom either is released or moves to a neighbour site.

The features shown in fig.5 can be qualitatively explained as follows. PDS coating is formed by many superimposed and entwined polymer layers where atom diffusion is possible. The diffusion velocity has been already evaluated and it is of the order of 1μm/s [17]. When the desorbing light is switched on, the atoms adsorbed on the surface are immediately ejected, while all the others have to diffuse inside the coating before being desorbed. By leaving the light on, the vapor density reaches again the thermal equilibrium value by pushing the atoms very slowly back into the reservoir. The OCT coating, that is not a polymer, has a simpler structure which makes less important the atom diffusion inside it. The atoms are trapped in the centre of the OCT molecule from where are ejected in presence of light. When light is on, they cannot be trapped at all (like for cesium) or for shorter time (like for rubidium). This explains the much faster density decay with respect to PDS or the lower density observed with cesium.

This new effect opens new perspectives both to the study of atom-dielectric surface physics and to applications in many fields. LIAD may be used, for example, to optimize the vapor-cell laser traps, as recently discussed by Stephens et al. [19]. They, in fact, use coated cells to reduce the sticking time and to improve the trap loading. LIAD would give the possibility to control the trap loading by a weak light and to store back the atoms on the cell walls when the laser trap is switched off. Moreover, as this cycle may be repeated many times, this device would be particularly interesting, for example, in the case of rare isotopes.

Acknowledgements

We like to acknowledge the technical help of M.Badalassi, E.Corsi, P.Mannucci and A.Pifferi. This work has been supported by the European Economic Community (contract CHRX-CT93-0366).

5. References

1.   A.Gozzini, F.Mango, J.H.Xu, G.Alzetta, F.Maccarrone, R.A.Bernheim (1993) *Nuovo Cimento* **15D**, 709

2.   M.Meucci,E.Mariotti,P.Bicchi,C.Marinelli,L.Moi (1994)
        Europhys.Lett. **25**, 639-643

3.   J.H.Xu (1995) Photo-ejection of alkali atoms and dimers on polysiloxane
        surfaces, PhD Thesis, Scuola Normale Superiore, Pisa

4.   see for example J.H.de Boer (1968) *The dynamical character of adsorption,*
        Oxford at the Clarendon Press

5.   L.Moi,S.Gozzini,C.Gabbanini,E.Arimondo,F.Strumia eds. (1991) Proceedings
        of the International Workshop on *Light Induced Kinetic Effects on Atoms, Ions
        and Molecules,* ETS, Pisa

6.   F.Kh. Gel'mukhanov and A.M. Shalagin, (1979) *JEPT Lett* **29**, 711

7.   H.G.C.Werij,J.E.M.Haverkort,J.P.Woerdman (1986) *Phys.Rev.***A33**, 3270

8.   S.Gozzini, G.Nienhuis, E.Mariotti, G.Paffuti, C.Gabbanini, L.Moi (1992)
        Opt.Commun. **88**, 341

9.   A.M.Bonch-Bruevich,T.A.Vartanyan,A.V.Gorlanov,Yu.N.Maksimov,
        S.G.Przhibel'skii, V.V. Khromov Sov.Phys.(1990) *JEPT* **70**, 604

10.   M.Stephens, R.Rhodes, C.Wieman (1994) *J.Appl.Phys.* **76**, 3479

11.   M.Oria,M.Chevrollier,D.Bloch,M.Fichet,M.Ducloy (1991)
        *Europhys.Lett.***14**, 527

12.   D.R. Swenson, L.W. Anderson (1988) *Nucl. Instr. Meth.* **B29** 627

13.   M.Allegrini, P.Bicchi, L.Moi, P.Savino (1980) *Opt. Comm.* **32**, 396

14.   J.H.Xu,M.Allegrini,S.Gozzini,E.Mariotti,L.Moi (1987) *Opt.Commun.* **63**, 43

15.   E.Mariotti, M.Meucci, C.Marinelli, P.Bicchi, L.Moi, A.Kopystynska
        in M.Inguscio, M.Allegrini, A.Sasso (eds.), Proceedings of the XII
        International Conference on Laser Spectroscopy, World Scientific (in press)

16.   I.N.Abramova, E.B.Aleksandrov, A.M.Bonch-Bruevich,V.V.Khromov (1988)
        JEPT Lett. **39**, 1372

17.   E.Mariotti,S.Atutov,M.Meucci,P.Bicchi,C.Marinelli,L.Moi (1994)
        *Chem.Phys.***187**, 111

18.   W. Hoheisel, M.Vollmer, F.Trager (1993) *Phys. Rev.* **B48**, 17463

19.   M.Stephens, C.Wieman (1994) *Phys. Rev. Lett.* **72**, 3787

ATOMS IN NANOCAVITIES

S. I. KANORSKY and A. WEIS
Max-Planck-Institut für Quantenoptik
Hans Kopfermann Str. 1, D-85748 Germany

1. Introduction

In this lecture we discuss some of the physical properties of ions and atoms located inside a dielectric cavity of such a small radius, that the ψ-function of the atomic outer electrons "touches" the cavity walls. First of all the question arises whether such small cavities can be realized in practice? At first glance the situation is hopeless. The cavity should have a radius of few Å in order to fulfill the above condition. This is comparable to the inter-atomic separation in condensed matter. Therefore only few atoms would sit on the surface of such a cavity making the very use of the concept of a cavity improper. However, there is one matrix, namely condensed helium, where the large amplitude of the fundamental vibrations effectively smears out the grain structure of the matrix thus enabling to treat it as a continuous medium even on a sub-nanometer scale.

The development, in the past decade, of experimental techniques by which foreign atoms can be implanted into condensed (liquid and solid) helium matrices has opened the road to the investigation of the properties of these defects.

This lecture is organized in the following way. In the second section we review the physical properties of condensed He and discuss the consequences of its quantum nature. The third section addresses positively and negatively charged point defects in condensed helium. We show how the anomalous mobilities of these defects have led to introduce the concepts of *snowball* and *bubble* to describe the structure of their trapping sites. Section 4 then deals with the trapping sites of *atomic* defects in ^{4}He. We introduce the atomic bubble model and show how the structure of the bubble can be related to physically observable spectroscopic properties. The basic assumptions of the bubble model are presented. In section 5 the model is applied to calculate the positions and shapes of the optical excitation lines of atoms trapped in such cavities. These theoretical results are then compared in section 6 to experimental results obtained from a study of the optical resonance lines of Ba atoms in condensed ^{4}He. Section 7 is devoted to further applications of the bubble model. We discuss in particular alterations of the hyperfine structure of implanted atoms and present a tentative explanation for the puzzling quenching

367

M. Ducloy and D. Bloch (eds.), Quantum Optics of Confined Systems, 367-393.
© 1996 *Kluwer Academic Publishers. Printed in the Netherlands.*

of resonance fluorescence of implanted light alkali atoms.

2. Liquid and Solid Helium

Figure 1 shows the p-T phase diagram of ^{4}He. One distinguishes two liquid phases:

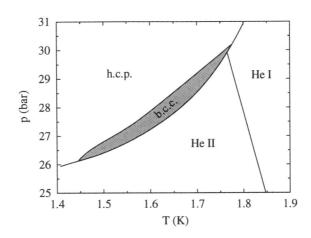

Figure 1: Part of the phase diagram of ^{4}He. Gas phase and *f.c.c.* crystalline phase do not fall into the plotted p-T region

at high temperatures, i.e. above the so called Λ-line[1] the normal fluid phase (He I), and below the lambda line a phase which is usually referred to as superfluid (He II) but which, nevertheless, consists of a temperature dependent mixture of normal and superfluid components. The superfluid content is 100% at T = 0 and 11.5% at T = 1.5 K. Helium can be solidified only under pressure and three crystalline phases are known. At low pressures the structure is mainly hexagonal close packed (*h.c.p.*), with a small island of body-centered cubic (*b.c.c.*) structure. A face-centered cubic (*f.c.c.*) phase exists above 1000 bar; it is however irrelevant for the problems addressed here. Note that all other noble gases crystallize in the *f.c.c.* phase.

2.1. ZERO POINT ENERGY OF CONDENSED ^{4}He

The interaction between rare gas atoms is characterized by a weak attractive potential due to the interaction of mutually induced dipoles (van-der-Waals interaction), and a hard-core type of repulsive potential. Figure 2 shows the pair potentials of

---

[1]The p = 0 point on the λ-line is called λ-point. For ^{4}He it is at T = 2.18 K.

rare gases, i.e. the potential energy of two atoms as a function of their separation. Helium being the lightest of the rare gas atoms has the smallest polarizability

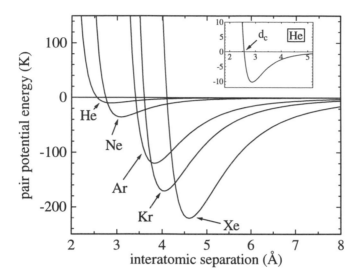

Figure 2: Adiabatic pair potentials for rare gases

and exerts hence a binding force weaker than any other element. The hard core diameter d_c, defined as the zero crossing of the pair potential curve, is 2.6 Å, and the well depth only 10 K (insert of Figure 2). The weak binding is responsible for the low boiling temperatures of the helium isotopes (4.2 K for ^{4}He, and 3.19 K for ^{3}He). Condensed helium is furthermore unique in the sense that even for T $\rightarrow$ 0 it does not solidify under its saturated vapor pressure. A pressure in excess of 25 bar has to be applied to achieve solidification. The reason for this, besides the low binding energy, is the presence of a large zero point energy (ZPE). If one were to construct a one dimensional crystal of He atoms, one would naively expect the separation between two adjacent atoms to be equal to d_{min} - the distance corresponding to the minimum of the pair potential of Fig 2. This is indeed the case for heavier rare gases, but not for helium, the reason being the large ZPE. Note, that in our hypothetical crystal each atom would be localized in a square well potential of dimension $\delta = 2(d_{min} - d_c) = 0.75$ Å. The ZPE of a particle in such a well is

$$E_{ZPE} = \frac{\pi^2 \hbar^2}{2m\delta^2} \quad \text{or} \quad E_{ZPE}[K] \approx \frac{60}{\delta[\text{Å}]^2} \quad \text{for helium.}$$

For $\delta = 0.75$ Å, the ZPE is 107 K and the large positive total energy $E_{tot} =$

$E_{pot} + E_{ZPE} \approx 97$ K makes such a crystal unstable. The inter-atomic separation in condensed helium is indeed much larger then d_{min}. From the density of He II (0.145 g/cm³) it can be estimated as 3.6 Å. If we now construct our one-dimensional helium crystal using this empirical value for the lattice period, the potential energy for a single helium atom will look as shown in Fig. 3. The loss

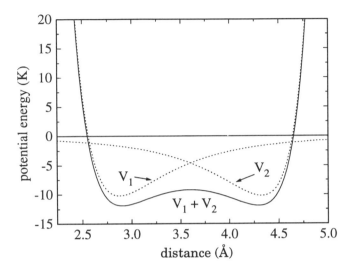

Figure 3: Potential energy of a helium atom in the one-dimensional potential of two neighboring atoms fixed at 0 and 7.2 Å respectively.

of (negative) potential energy due to the imperfect overlap of potential minima is well compensated by the decrease of the kinetic energy due to weaker localization. The total energy of a real *3-dimensional* crystal is indeed slightly negative.

The actual molar volume of condensed helium is approximately three times larger than expected from equilibrium conditions of the inter-atomic potential only. The large molar volume makes helium highly compressible. When increasing the pressure from the Saturated Vapor Pressure (SVP) at 1.5 K (5 mbar) to 25 bar the molar volume changes by 17%. Among all rare gases the contribution of the zero-point kinetic energy to the total internal energy of the condensed phase is particularly large for helium, due to the smallness of the helium mass. In Table 1 we compare the properties of the different rare gases at low temperatures. The last column gives the ratio of the rms value of the zero-point vibration amplitude to the inter-atomic separation R_{A-A}. Due to the large oscillation amplitudes helium is thus also characterized by a strong overlap of the wave functions of adjacent atoms, which explains the large tunneling rate of atoms and vacancies. Examination of the inter-atomic potentials in Fig. 3 also reveals a strong similarity with the interaction potential in an ideal gas of hard spheres of approximately 2.6 Å in diameter. In

Atom	$\lambda = E_{ZPE}/E_{pot}$	$R_{A-A}[\text{Å}]$	$R_{vib}/R_{A-A}[\%]$
^{4}He	1	3.6	24
Ne	0.25	3.2	9
Ar	0.1	3.8	5
Kr	0.057	4.0	4
Xe	0.031	4.3	3

Table 1: Contributions of zero point energy and zero point vibrations in condensed rare gases

combination with the strong overlap of the atomic wave functions this allows to expect that the description of condensed helium as a continuous medium will be a reasonable approximation.

So far we have been talking about "condensed" helium having primarily in mind the *liquid* phase for which the approximation of a continuous medium should be fairly good. The large amplitude of the zero point vibration also affects the properties of the solid phase. First, helium can be solidified only under a pressure of more than 25 bar. Furthermore, one expects the rare gases to crystallize in a closest package structure (*f.c.c., h.c.p.*), which allows a maximal number of atoms to be packed into a given volume. The heavy noble gases do indeed present only a cubic close-packed structure, which is equivalent to a face-centered cubic (*f.c.c.*) lattice. The theoretical justification why the *f.c.c.* structure is prefered over the hexagonal close packed structure (*h.c.p.*) is a delicate problem [1]. At low pressures ^{4}He and ^{3}He favor the body-centered cubic (*b.c.c.*) lattice, which provides more space per atom, and thus allows to minimize the ZPE. For T $\to$ 0 the *b.c.c.* phase in ^{4}He is not stable and the crystalline structure is *h.c.p.* Only for very high pressures in excess of 1000 bar the structure of helium becomes the same as for other rare gases, i.e. *f.c.c.*. Being a crystal with "long-range order" solid helium exhibits similar quantum properties as liquid helium:

- strong delocalization of helium atoms, amplitude of zero point vibrations comparable to the lattice period;

- large overlap of atomic wave functions;

- approximate equality of the potential and kinetic energies.

This quantum nature of the crystal makes it *very soft*, so that already moderate local perturbation will override the weak bonding of the helium atoms and, therefore, determine the local structure of solid helium. From this point of view, we can expect solid helium to be similar to liquid helium and use the model of a continuous medium for its description as well. The above arguments are definitely not a rigorous justification of the model. Only comparison of theoretical predictions obtained within this "continuous model" with experimental results can substantiate it.

3. Excess charges in liquid and solid ^{4}He

The two decades following the first investigation of the mobility of charges in liquid helium by R. L. Williams in 1957 [2] were marked by a wealth of theoretical and experimental papers trying to answer the basic question of the structure of the point defect formed by a charged particle in liquid helium. The methods used to implant the charges were among others the extraction of positive or negative charges from tracks produced by α - particles, β - radiation or the photoelectric effect [3]. One of the key experiments was the study of the mobility of the implanted charges. Let us remind that the mobility μ is the ratio v/E of the asymptotic velocity v that the charged particle reaches in a viscous medium under the effect of an electric field of strength E. For low field strengths the mobility is independent of E and was found to increase with decreasing temperature as $\mu \propto \exp(\Delta'/kT)$, with $\Delta'/k = 8.8$ K for positive and 8.1 K for negative ions respectively. In this regime the ion motion is limited by collisions with rotons whose density n_r varies as $n_r \propto exp(-\Delta_r/kT)$, where $\Delta_r/k = 8.7$ K is the roton creation energy [4]. Two major questions arose from the mobility studies:

- Why are the mobilities so small? The diffusion coefficients of ions in liquid ^{4}He inferred from the mobility measurements were found to be *orders of magnitude* smaller than the diffusion coefficient of ^{3}He in ^{4}He.

- Why are the mobilities of positive ions approximately 50% larger than those of the negative ions?

3.1. POSITIVE IONS: SNOWBALLS

In 1959 Atkins [5] partly explained this behavior by the following analysis of the structure of the defect formed by a positive ion. At velocities which are low compared to the velocity of sound the ion drags with it a large number of helium atoms and therefore has a large effective mass. The binding of the helium atoms to the positive charge is due to electrostriction; a polarizable dielectric medium placed in an electric field tends to be pulled into the region of higher field. One can show that in the Coulomb field of the positive ion the helium density increases as $\rho(r) - \rho_{bulk} \propto r^{-4}$. Inside a sphere of radius 6.3 Å the density becomes so high, that helium becomes solid. This has given the structure the name of "snowball". The effective mass associated with a snowball is approximately 100 m$_{He}$.

The geometrical cross section for collisions of the hard core of a snowball with a roton (radius $\approx$ 4 Å) is approximately $\pi(4 + 6.3)^2$ Å$^2 = 3.3 \cdot 10^{-14}$ cm^2. Remembering that in gas kinetic theory the following relation between mobility and scattering cross section holds

$$\mu = \frac{e}{m} \cdot \frac{\lambda}{v},$$

where $\lambda = 1/n\sigma$ is the mean free path, n being the density of scatterers (here rotons), and v - the average relative velocity of collision partners. Using the above value of the effective mass the measured mobilities yield cross sections $\sigma_+ =$

$5.7 \cdot 10^{-14}$ cm^2, for positive ions. This value is in fair agreement with the value estimated above. However, the snowball model based on electrostriction only can not account for the difference in the measured mobilities of positive and negative ions. By "negative ion" one usually means an excess electron. From the mobility measurements the scattering cross section for the latter could be estimated as $\sigma_- = 9.3 \cdot 10^{-14}$ cm^2.

3.2. NEGATIVE IONS: BUBBLES

In 1965 Woolf and Rayfield [6] have measured the potential barrier ε_b for the penetration of electrons into liquid helium. From the shift of the spectrum of photo-effect in liquid helium they found $\varepsilon_b = 1.02(8)$ eV. The origin for this potential barrier is the so called exchange potential, or "Pauli exclusion force". Helium has two tightly bound electrons forming a closed shell; at distances shorter than the hard core radius $r_c = 1.3$ Å such a closed shell will exhibit a strong repulsion on an approaching third s-wave electron.

In contrast to the much heavier positive ions, a confined electron has a very large positive ZPE leading to a strong delocalization. The structure of the point defect formed by an electron can thus be guessed on the basis of the following simple consideration [7]: the strong "hard core" repulsion along with the delocalization of the electron should lead to the formation of a cavity from which helium is expelled - the so called "electron bubble". The radius R_b of this cavity and the ψ-function of the trapped electron can be found by variational minimization of the total energy of the system:

$$E = \int \psi^*(\mathbf{r}) \left(-\frac{\nabla^2}{2m} + V(r) \right) \psi(\mathbf{r}) \, d^3r + 4\pi\sigma R_b^2 + \frac{4\pi}{3} R_b^3 p,$$

where $V(r)$ is a spherical square well potential, with $V(r) = -1.02$ eV for $r < R_b$ and zero outside the bubble; the third term represents the surface energy of the bubble and the fourth term the pressure-volume work for the cavity formation. This simplest form of the energy functional neglects the long-range polarization potential and quantum mechanical volume kinetic energy due to the helium density gradient on the bubble interface. Nevertheless, already this simplified model turned out to be extremely successful in describing the experimental findings. For the ground state of the system the bubble radius was found to be 18 Å. Further refinements of the model did not change this number significantly. Since the potential for the electron has a central symmetry, the electron states in such a bubble can be classified according to the values of angular momentum, the ground state being a $1s$-state. The model also predicts the existence of excited electron p-states with principal quantum numbers up to $n = 3$. The impressive confirmation of the model was the direct observation of the corresponding optical transitions in the infrared absorption of the electron impurities in liquid helium [8, 9]. The pressure dependence of the resonance frequency of the $1s - 1p$ transition line was in good agreement with the predictions by the bubble model. Moreover, it was shown in [10] that the same bubble model for the structure of the point defect can be successfully extended to the solid phase of the helium matrix. Today the bubble model is recognized as the one which gives an adequate description of the properties of electron impurities in condensed helium. This concept of bubble structure will also be essential for the understanding of the properties of *atomic* impurities in condensed helium.

4. Atoms in Helium Matrix

Next we shall address the question of the structure of the point defect for neutral atoms. Can the bubble model still be used in this case? The problem of defining the structure of the point defect is equivalent to finding the density profile $\rho(\mathbf{R})$ of helium around the impurity atom. However, this density profile cannot be directly measured in experiment. We shall therefore relate it to "observables": shifts and broadenings of atomic spectral lines, shifts of hyperfine structure levels etc.

4.1. STRUCTURE OF THE POINT DEFECT FORMED BY A FOREIGN ATOM IN LIQUID HELIUM

4.1.1. A simple model
For the time being we shall restrict ourselves to alkali and alkaline-earth atoms which were most studied experimentally. A very lucid discussion of the shape of the trapping site for the impurity atom has been given by Dupont-Roc [11]. The helium matrix around the impurity atom is taken as a continuous fluid described by a local equation of state, which relates the local pressure p and the density ρ

$$p = \rho_0 \varepsilon [(\rho/\rho_0)^b - 1],\tag{1}$$

with $\rho_0 = 2.2 \times 10^{22} cm^{-3}$, $b = 5.73$, and $\varepsilon = 4.75K$. Helium atoms in the vicinity of the impurity atom feel an additional potential due to the helium-impurity interaction V_{int}. The relation between the local pressure p and the interaction potential V_{int} is given by [12]

$$-\int_0^{p(R)} \frac{dp'}{\rho(p')} = V_{int}(R),\tag{2}$$

where $\rho(p')$ is the density at pressure p'.

Solving Eq. 1 for $\rho(p)$ and substituting the result into Eq. 2 gives the following expression for the pressure $p(\mathbf{R})$ at position $\mathbf{R}$:

$$p(\mathbf{R}) = \varepsilon\rho_0 \left[\left(1 - \frac{V(\mathbf{R})}{\varepsilon\eta} \right)^\eta - 1 \right],\tag{3}$$

where $\eta = b/(b-1)$.

It is natural to assume that helium will be expelled from a certain volume occupied by the foreign atom. The surface of the thus formed cavity can be determined from the condition that the increase in local pressure due to the presence of the impurity atom is compensated by the pressure under the curved surface of the helium interface:

$$p(\mathbf{R}) = \sigma C(\mathbf{R}),$$

where $C(\mathbf{R})$ is the curvature of helium surface.

Expanding Eq. 2 with respect to the small parameter $V(\mathbf{R})/\varepsilon\eta$ and taking the interaction of the impurity atom A in the state $|n\,l\rangle$ with the He atoms as a sum of pair potentials

$$V_{nl} = \sum_i V_{A-He}^{nl}(\mathbf{R}_i)$$

gives the following generic equation for the surface of the cavity:

$$\langle n\,l|V_{A-He}^{nl}(\mathbf{R}_s)|n\,l\rangle = \sigma C(\mathbf{R}_s)/\rho_0, \qquad (4)$$

where $\mathbf{R}_s$ is the radius-vector of a point on the surface and $C(\mathbf{R}_s)$ is the local curvature of the surface.

Pair potentials are characterized by a strong repulsion at short distances and have a van-der-Waals attractive part for large separations. For typical radii of curvature on the order of a few Å the right hand side term in Eq. 4 is on the order of a few Kelvin and can be neglected compared to the steep repulsive pair potentials, giving the following very simple equation for the shape of the cavity:

$$\langle n\,l|V_{A-He}^{nl}(\mathbf{R}_s)|n\,l\rangle = 0. \qquad (5)$$

In other words, the surface of the cavity formed in the helium matrix by an impurity atom coincides to first approximation with the surface of zero pair potential. Figure 4 shows the results for the ground and the excited singlet states of the Ba atom, using the pair potentials of [13].

Another important consequence of this considerations is that the symmetry of the point defect reflects the symmetry of the ψ-function of the impurity atom. For example, impurity atoms in symmetric S-states will form spherical cavities. The central symmetry of the Hamiltonian is therefore preserved, the total angular momentum is still a good quantum number, and the general selections rules for the optical transition are still valid for such atoms.

4.1.2. A refined model

Basic model assumptions. A rigorous treatment of the problem would require the solution of the Schrödinger equation for a system of N+2 interacting particles (N helium atoms, core and valence electron of foreign atom). For large N such a problem is obviously intractable. Therefore, the following reduction scheme is usually used:

- Born-Oppenheimer approximation: the total quantum system (N ground state helium atoms and the foreign atom A) is segmented into two components - the "fast" optical electron of the foreign atom described by its

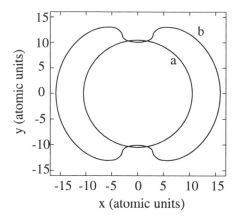

Figure 4: Shapes of the cavities formed in liquid helium by a Ba atom in the ground $6s^2\,{}^1S_0$-state (a) and in the singlet excited state $6s6p\,{}^1P_1$ (b), according to Eq. 5.

radius-vector $\mathbf{r}_e$ and a "slow" subsystem containing the atomic core and helium atoms, described by a manifold of coordinate vectors $\{\mathbf{R}_i\}$:

$$\Psi(\mathbf{r}_e, \mathbf{R}_i) = \psi_a(\mathbf{r}_e, \mathbf{R}_i) \times \phi_a(\mathbf{R}_i). \qquad (6)$$

In the ψ-function of the fast subsystem $\mathbf{r}_e$ is a true dynamic variable, whereas $\mathbf{R}_i$ are parameters, and the state of the slow system depends on the current state $|a\rangle$ of the optical electron.

- Frank-Condon approximation: the electronic transition $|a\rangle \to |b\rangle$ occurs on such a short time scale that the parameters $\mathbf{R}_i$ do not change.

- The dipole matrix elements between electronic states $|a\rangle$ and $|b\rangle$ are practically constant over the range of $\{\mathbf{R}_i\}$ spanned by $\phi_a(\mathbf{R}_i)$. In this case the golden-rule predicts the rate W of the $|a\rangle \to |b\rangle$ transition as:

$$W \propto \left|\langle \psi_a(r, \tilde{R})|r|\psi_b(r, \tilde{R})\rangle\right|^2 \times \left|\langle \phi_a(\tilde{R})|\phi_b(\tilde{R})\rangle\right|^2 \times \delta(E_a + \hbar\omega - E_b), \qquad (7)$$

where $\tilde{R}$ is the "most likely" or equilibrium set of $\mathbf{R}_i$.

The problem is thus reduced to finding the stationary functions $\psi_a(\mathbf{r}_e, \mathbf{R}_i)$, $\phi_a(\mathbf{R}_i)$ via functional minimization of the total energy of the system

$$\langle E \rangle = \frac{\langle \Psi|H|\Psi\rangle}{\langle \Psi|\Psi\rangle}.$$

Neglecting many-body interactions, the approximate Hamiltonian for the complete system can be written as:

$$H_{tot} = H_{He} + H_a + H_{int} \tag{8}$$

where the first term is the Hamiltonian of a system of N helium atoms interacting pairwise:

$$H_{He} = -\frac{1}{2M} \sum_{i=1}^{N} \nabla_i^2 + \sum_i \sum_{k<i} U(R_{ik}),$$

H_a is the Hamiltonian of the free atom, whose explicit form will be considered later, and the last term is the Hamiltonian of the impurity atom – helium interaction, which is again given by the sum over all pair interactions:

$$H_{int} = \sum_i^{N} V_{A-He}(\mathbf{r}, \mathbf{R}_i).$$

In Eq. 8 we neglect again many-body interactions and assume that the core of the impurity atom has infinite mass and is fixed at the origin.

Practical realization. The steps described above are common to all existing treatments of the problem and differ mainly in the adopted forms of trial helium ψ-functions.

One very popular approach to the problem was introduced by Hickman et al. [14] when treating the spectra of metastable helium atoms in He II. The authors used the trial function for the helium in the form introduced earlier by Hiroike et al. [15] for electron bubbles. Helium around the impurity atom is treated as an *incompressible* continuous medium, described by a wave function $\phi(\mathbf{R})$ related to the local density of the liquid $|\phi(\mathbf{R})|^2 = \rho(\mathbf{R})$. The density is zero at the position of the impurity atom $(R = 0)$ and asymptotically approaches the bulk liquid density as $R \to \infty$ (see Fig. 5):

$$\rho(R; R_0, \alpha) = \begin{cases} 0 & , R < R_0 \\ \rho_0\{1 - [1 + \alpha(R - R_0)]\exp(-\alpha(R - R_0))\} & , R > R_0 \end{cases}, \tag{9}$$

where α and R_0 are the parameters to be determined by variational minimization of the energy functional, which now becomes:

$$\langle E \rangle = E_{bubble} + E_{atom}.$$

Here the bubble energy is the sum of the pressure volume work for the cavity formation E_v, the surface energy E_s, and the volume kinetic energy E_{vk}:

$$E_{bubble} = pV_{bubble} + \sigma S_{bubble} + \frac{1}{8M}\int \frac{(\nabla\rho)^2}{\rho}d^3R. \tag{10}$$

As we have seen, atoms with angular momentum $l \neq 0$ form aspherical cavities.

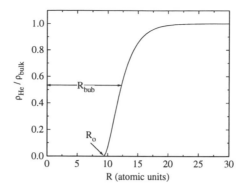

Figure 5: Model profile of helium density around the atomic defect according to Eq 9. The radius of the bubble R_{bub} is usually defined as the center of mass of the interfacial region.

To lowest order in a multipole expansion this deviation from spherical symmetry can be taken into account by adding a quadrupolar angular dependence to the parameter R_0

$$R_0(\vartheta) = R_0 \left(1 + \beta \frac{3\cos^2(\vartheta) - 1}{2}\right) \qquad (11)$$

The atomic energy can be written as the expectation value of the remaining terms in the Hamiltonian (Eq. 8):

$$E_{atom} = \langle \psi(\mathbf{r}_e) | H_a + H_{int} | \psi(\mathbf{r}_e) \rangle . \qquad (12)$$

Further significant simplification can be achieved by eliminating the electronic ψ-function in Eq. 12 using adiabatic atom-helium pair potentials[2] known from quantum chemistry calculations (see e.g. [16]). Eq. 12 then takes the form

$$E_{atom} = E_0(nl) + \int d^3R\, \rho(\mathbf{R}) V_{nl}^{p.p.}(\mathbf{R}) \qquad (13)$$

For symmetric S-states the interaction is isotropic , while for P-states it is anisotropic and depends on the mutual orientation of the angular momentum $\mathbf{L}$ of the atom and the radius-vector $\mathbf{R}$ pointing to the position of the helium atom:

$$V_{nS}^{p.p.}(\mathbf{R}) = V_{\Sigma}^{nS}(R) \qquad (14)$$

$$V_{nP}^{p.p.}(\mathbf{R}) = V_{\Sigma}^{nP}(R) + \left(\frac{\mathbf{L} \cdot \mathbf{R}}{\hbar R}\right)^2 \left(V_{\Pi}^{nP}(R) - V_{\Sigma}^{nP}(R)\right) \qquad (15)$$

---

[2]These quantities are *potential energies* of corresponding atom-helium pairs and should not be confused with potentials in the Schrödinger equation for the atomic electron (Eq. 12)

The total energy of the point defect is now a function of only three bubble shape parameters R_0, α, and β (for spherical cavities $\beta \equiv 0$ and the number of parameters is reduced to two). The problem of defining the equilibrium bubble configuration is thus reduced to the determination of these parameters by a variational energy minimization. The results for the ground and the first excited states of Ba are shown in Fig. 6. These results ought to be compared to those of the simplified model (section 4.1.1). The difference is most pronounced for the excited state. The bubble terms Eq. 10 and in particular the surface tension, are responsible for the smooth shape of the cavity. The very attractive feature of this procedure is

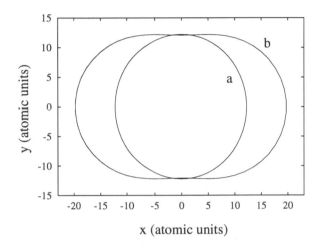

x (atomic units)

Figure 6: Equilibrium shapes of the bubbles formed by a Ba atom in the ground (a) and the first excited (b) states found by minimizing the total energy Eqs. 12–15 using the model helium profile of Eqs. 9–11. The values of the bubble shape parameters are: $R_0 = 8.9$ a.u., $\alpha = 0.65$, $\beta = 0$ for the ground state, and $R_0 = 11.48$ a.u., $\alpha = 0.6$, $\beta = 0.35$ for the first excited state.

its simplicity, however one has to keep in mind its limitations originating from the basic model assumptions. Let us recall them once more:

- The helium matrix is described as *continuous incompressible* fluid.

- The helium contribution to the total energy is taken in the form of Eq. 10, which uses the macroscopic concepts of pressure-volume work and surface tension for cavities as small as several Å in diameter. Although partial justification of this approximation can be found in [15], it is not clear what value of the surface tension should be used.

- The atomic contribution to the total energy is found neglecting many particle interactions by summing energies of all pair atom-helium interactions, which are strictly speaking not additive.

In the following chapters we shall analyze these model limitation in more detail and discuss the possibilities for further refinements of the model.

5. Optical spectra

Once the equilibrium structure of the point defect is determined, the corresponding shift of the optical line can be found from Eqs. 7 and 13 as the difference of the atom-helium interaction energies in the initial and final states

$$\Delta\nu = \frac{1}{\hbar} \int d^3 R \, \rho(\mathbf{R}) \left(V_b^{p \cdot p \cdot}(\mathbf{R}) - V_a^{p \cdot p \cdot}(\mathbf{R}) \right). \tag{16}$$

Note, that only the foreign atom – helium interaction energy contributes to the shift because, due to the Frank-Condon principle, the shape of the bubble does not change during the transition.

In order to be able to make predictions for the line shape, the *dynamics* should be included in the model. To our knowledge this has not been done in a self consistent way so far. One possibility to handle this problem for atomic impurities was introduced by Bauer et al. [17]. In that approach the line broadening comes from oscillations of the bubble radius around its equilibrium value. In order to find the amplitude of these oscillations the bubble shape parameters α and β are fixed to their stationary values and the total energies of the point defect for the impurity atoms in the initial and final states are calculated as functions of the bubble radius R_{bub}, defined as a "center of mass" of the cavity interfacial region (Fig. 5). Energy curves obtained in such a way are shown in Fig. 7. The total energy $E_{tot}(R_{bub})$ is then treated as the potential for an effective "bubble oscillator". By quantizing the motion of this oscillator one gets the probability distribution $|\phi(R_{bub})|^2$ for the bubble radius R_{bub}. The spectrum is obtained from Eq. 7 by projecting $|\phi(R_{bub})|^2$ onto the energy curve of the final state

$$I(\nu) \propto \frac{|\phi(R_{bub})|^2}{\partial E_{tot}(R_{bub})/\partial R_{bub}} \tag{17}$$

This model gives the right qualitative description of the observed spectra - a large blue shift and asymmetry of the excitation lines and a relatively smaller shift and broadening of the emission lines. However, comparison with experiment tells that this procedure usually underestimates the observed broadening. In principle this is not surprising since the model takes into account only the lowest order "breathing" mode of the bubble oscillation.

A next step of refinement has recently been done by Kinoshita et al. [18], who allowed for "quadrupole" oscillations i.e. oscillations in the two-dimensional parameter space ($R_{bub} \times \beta$), where β is the quadrupolar deformation parameter introduced in Eq. 11. The allowance for the distortion of the equilibrium bubble shape lifts the three-fold degeneracy of the excited P-states giving rise to additional line broadening and asymmetry via the Jahn-Teller effect. This improves the agreement with the experiment in the case of the Cs D2 line, but still the theoretical linewidth is smaller than the experimental one by a factor of almost two. In principle, by further increasing the number of oscillatory modes taken into account one could expect to get a correct description of the optical spectra. However, for higher order modes the calculations become very cumbersome. This approach has yet another problem related to the determination of the effective masses of the bubble oscillator for different oscillatory modes. The expressions usually used to estimate these parameters [7], [19] were obtained with the assumption that helium around the atomic point defect is an *incompressible* liquid. At the same time, for typical amplitudes of the bubble shape oscillations of ~ 0.5 Å a considerable deviation from this assumption can not be excluded.

An elegant way around this problem is the use of the standard statistical line broadening theory in its static limit [20]. Within this approach the line broadening originates not from *correlated* motions of helium atoms around the impurity, such as bubble shape oscillations, but rather from *uncorrelated* fluctuations of the helium density. This fluctuations are considered to be of the same type as for an ideal

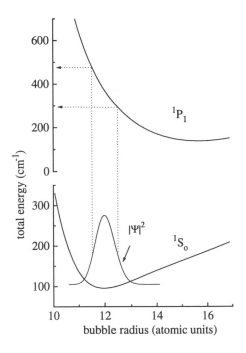

Figure 7: Total energies of Ba atom in a spherical bubble as functions of the bubble radius. The bell-shaped profile is the probability distribution for the bubble radius R_b.

gas. This assumption is justified as we have shown in section 2 that condensed helium can be described as an ideal gas of hard spheres. A further argument for the assumption of uncorrelated fluctuations comes from the consideration of the relevant time scales. Optical lines observed in a helium matrix have typical widths of several nanometers, which corresponds to the optical dipole autocorrelation time τ_{ac}^{opt} given by the reciprocal of the linewidth on the order of 10^{-14} s. This value determines the relevant time scale of the optical line broadening process. On this short time scale one can neglect correlations between helium atoms because the time needed for these correlations to build up, on the order of the reciprocal of the roton frequency, is $\tau_{helium} \sim 10^{-12}$ s. Furthermore, during this short time helium atoms would move by only about 0.1Å, which is much smaller than typical bubble radii and even interfacial widths. We shall therefore completely neglect the motion of helium atoms and treat the problem as *static*.

Within this approach the spectrum is found as the Fourier transform of the optical dipole $d(t)$ autocorrelation function $C(\tau)$:

$$I(\omega) \propto \int_{-\infty}^{\infty} e^{i\omega\tau} C(\tau) d\tau, \tag{18}$$

where

$$C(\tau) = \langle d^*(t)d(t+\tau) \rangle_t \tag{19}$$

Perturbation from helium atoms will shorten the autocorrelation time and thus broaden the line. For the quantitative analysis we consider the impurity atom and N helium atoms to be confined in a box of volume V. Let $\rho(R)$ be the number density of helium:

$$\int_V d^3R\,\rho(R) = N$$

According to the basic model assumption the shifts of atomic energy levels by different perturbers are additive:

$$\Delta\omega(t) = \sum_{j=1}^{N} \delta\omega_j \left(R_j(t)\right). \tag{20}$$

Substituting Eq. 20 into Eq. 19 and using the ergodic hypothesis to replace the time average by an ensemble average:

$$\langle \ldots \rangle_t \to \frac{1}{N} \int d^3R \ldots \rho(R),$$

the autocorrelation function is found in the static limit $R_j(t) = const$:

$$C(\tau) = exp\left\{ - \int \left(1 - e^{-i\Delta\omega(R)\tau}\right) \rho(R) d^3R \right\}. \tag{21}$$

The spectrum is then obtained by substituting in Eq. 18, 21 the density from Eq. 9 with equilibrium parameters R_0, α, and β

6. Comparison with experiment

We have measured the excitation spectra of the singlet $6s^2\,{}^1S_0 \to 6s6p\,{}^1P_1$ transition of Ba atoms in matrices of liquid and solid helium at different pressures [21]. It is instructive to compare the experimental findings with the predictions of the bubble model described in previous chapters.

6.1. LINE SHIFT

Figure 8 demonstrates the measured shift of the excitation line barycenter as a function of helium pressure. The dotted line is the prediction obtained by using the

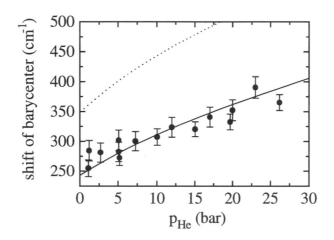

Figure 8: Measured pressure shift of the excitation spectra of the $6s^2\,{}^1S_0 \to 6s6p\,{}^1P_1$ transition of a Ba atom in liquid helium. The lines represent the results from the bubble model discussed in the text.

Ba–He pair potentials of [13]. The theory reproduces the observed pressure shift rate $\partial\lambda/\partial p = -0.13(2)\,nm/bar$ but overestimates the absolute value of the shift by about 3 nm. Can this contradiction be resolved within the bubble model? A closer look at the results of total energy minimization reveals two such possibilities [22].

- First, the two bubble shape parameters - R_0 and α turn out to be strongly correlated: a change of the total energy due to a variation in one of them can be compensated by a corresponding adjustment of the second parameter. In this situation the reliability of the variational method is not high.

- Second, for typical bubble radii of 5 Å the main contribution to the bubble energy comes from the surface tension. Therefore, the procedure is very sensitive to the value of the surface tension coefficient. It is not clear, whether the value $\sigma = 0.34$ erg/cm^2 obtained from the analysis of the infrared spectra of electron bubbles [8] is still valid for the much smaller atomic bubbles.

In order to check these possibilities we have minimized the total energy by varying only R_0 with fixed values of α and σ. The latter parameters were then varied in order to fit the experimentally observed pressure dependence of the excitation line barycenter. The result is shown in Fig. 8 as solid line. The best agreement is achieved for "optimal" values of parameters $\alpha = 0.6$ and $\sigma = 0.32$ erg/cm^2. The latter value is compatible with the value used in the complete model, while the optimal value of α is two times smaller than the one obtained with the total energy minimization procedure. Remarkably, the optimal value of the parameter α corresponds to a width of the bubble interface comparable to mean interatomic separation in liquid helium.

6.2. LINE SHAPE

After this parameter adjustment described above the quasistatic line broadening theory (Eqs. 18 and 21) can be applied to calculate the shape of the absorption line. In Fig 9 we compare the theoretical profiles with the experimental data. Note that the only parameter used in this comparison is a trivial amplitude scaling factor.

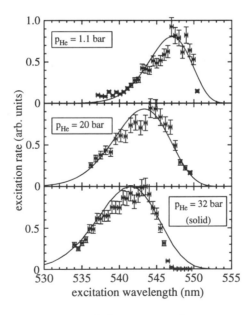

Figure 9: Line shape of the $6s^2\ ^1S_0 \rightarrow 6s6p\ ^1P_1$ transition of Ba atoms implanted in helium matrix at different pressures.

7. Further development of the model

A possible way to further refine the model would be to remove the assumption of helium *incompressibility* and to use a more elaborate energy density functional. One such functional was introduced by Dupont-Roc et al. [23]. The helium energy is written as

$$E = E_q + E_b + E_c \tag{22}$$

where E_q is usual volume kinetic energy term

$$E_q = \int d^3\mathbf{r} \frac{\hbar^2}{2m} |\nabla\sqrt{\rho(\mathbf{r})}|^2, \tag{23}$$

E_b is the sum of all two-body interactions

$$E_b = \frac{1}{2} \int\int d^3\mathbf{r} d^3\mathbf{r}' \rho(\mathbf{r})\rho(\mathbf{r}')V_{LJ}(|\mathbf{r}-\mathbf{r}'|) \tag{24}$$

where V_{LJ} is the standard Lennard-Jones potential screened at short distances:

$$V_{LJ}(r) = 4\varepsilon \left[\left(\frac{r_c}{r}\right)^{12} - \left(\frac{r_c}{r}\right)^6 \right] \quad, r \geq h \tag{25}$$

$$V_{LJ}(r) = V_{LJ}(h) \left(\frac{r}{h}\right)^4 \quad, r < h \tag{26}$$

and, finally, the last term E_c accounts for the increasing contribution of the hard core when the density increases

$$E_c = \int d^3\mathbf{r} \frac{c}{2}\rho(\mathbf{r})\bar{\rho}(\mathbf{r})^{1+\gamma}, \tag{27}$$

where $\bar{\rho}$ is the "coarse grained density", defined by averaging $\rho(\mathbf{r})$ over a sphere with a radius h.

With the following values of parameters: $\varepsilon = 10.22$ K, $r_c = 2.56$ Å, $h = 2.377$ Å, $c = 1.04554 \times 10^7$ K Å$^{3(1+\gamma)}$, and $\gamma = 2.8$ this functional correctly reproduces the equation of state of bulk helium over a wide range of densities. First calculations of the binding energy of alkali atoms to the surface of liquid helium using this energy density functional have been recently performed by F.Anzilotto et al. [24]

7.1. QUENCHING OF THE RESONANCE STATES IN LIGHT ALKALIS

Light alkalis (Li, Na, K) provide a situation where the bubble model in its standard form cannot be applied for the analysis of the emission spectra. Neither the

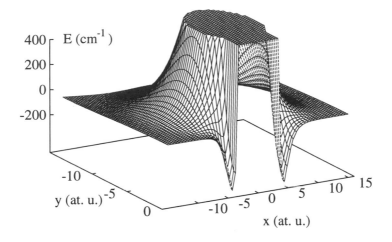

Figure 10: Three dimensional plot of Na*(3P)–He pair potential

assumptions of continuity and incompressibility nor the form of the trial function Eq. 9 are applicable for the excited states of these atoms.

Figure 10 shows a three dimensional plot of the Na*(3P)–He adiabatic pair potential from [16]. This potential has a deep attractive minimum in the nodal plane of the P-state electronic ψ-function. Several helium atoms will be strongly localized in this 650 K deep potential well in a ring of ~4 Å diameter thus forming a molecular complex Na*(3P)He$_n$. This complex, in turn, will reside in a kind of deformed bubble. It is clear that the main perturbation of the atomic energy levels will come from these closely localized helium atoms.

The number of helium atoms in such a complex can be estimated in the following way. The Hamiltonian of the system is written as

$$H = -\frac{1}{2m}\nabla^2 + \sum_j^n U_{He-Na}(\mathbf{r}_j) + \sum_{j,i>j}^n U_{He-He}(|\mathbf{r}_j - \mathbf{r}_i|).$$

Using pair potentials of [16] and [25], and performing the minimization of the total energy on the trial ψ functions for helium atoms in cylindrical coordinates (z, r, φ) of the form

$$\psi(z, r, \varphi) = \phi(z)\chi(\rho) \exp\left(-\frac{1}{2}\left\{\frac{\varphi - 2\pi(i-1)/n}{\Delta\varphi}\right\}^2\right)$$

one finds that a minimum of the total energy $E_{min} = -1810 \ cm^{-1}$ is achieved for n=5. Five helium atoms localized at a distance of 2 Å from the sodium atom will cause the 3S and 3P levels to cross. We believe that this explains the long standing puzzle of the absence of fluorescence from light alkalis in condensed helium. A complementary approach to the problem can be found in [11].

7.2. HYPERFINE STRUCTURE OF Cs ATOM IN HELIUM MATRIX

Until recently the only source of experimental information about atomic impurities in helium matrices were optical spectra (excitation, emission, recombination [26, 27]) of these atoms. Within the accuracy relevant to the problem the pair potential approach is sufficient for the analysis of these spectra. Recently the microwave transition between the $F = 3$ and $F = 4$ hyperfine sublevels of the ground state of Cs atoms in a matrix of liquid [28] and solid [29] helium was detected. Besides being interesting by itself this experiment provides a new valuable test of the model calculations. An important difference here is that one has to find not only the change of atomic energy levels but also the modification of the electronic ψ-function of the impurity atom in the helium matrix. To do this one has to solve the Schrödinger equation for the Cs valence electron.

For atoms in an s-state the hyperfine constant is proportional to the electron density at the nucleus:

$$A_{hfs} \propto |\psi(0)|^2 \ .$$

To find this value for a heavy multielectron atom is a heavy task. Our problem however is somewhat simpler - we are looking only for a change of this quantity due to the presence of surrounding helium atoms. Let us first consider a free Cs atom. At short distances (within the core of the Cs^+ ion) the wave function is given by the solution of a system of Hartree-Fock equations. Let the radial part of the solution be

$$N \times \phi_{short}(r) \tag{28}$$

where N is a normalization factor and the explicit form of ϕ is irrelevant for the present discussion. The normalization constant N can be estimated by using the fact that the main contribution to the normalization integral comes from large distances. In this region the motion of the outer electron is nonrelativistic and the wave function $\phi_{long}(r)$ can be found by the solving the Schrödinger equation with the effective Hamiltonian (see e.g. [16])

$$H_0 = -\frac{1}{2m}\nabla^2 + \frac{1}{r} + V_{pol} + V_{s.r.} . \tag{29}$$

Here the term

$$V_{pol}(r) = -\frac{1}{2}\frac{\alpha_d}{(r^2 + h^2)^2} + -\frac{1}{2}\frac{\alpha_q}{(r^2 + h^2)^3} , \tag{30}$$

(α_d, α_q are dipole and quadrupole core polarizabilities, and h is a cut-off parameter) accounts for the Cs core polarizability and the last short range term

$$V_{s.r.}(r) = B\exp(-Cr^2) \tag{31}$$

called pseudopotential simulates the repulsion exerted on the electron by the core at short distances. Values of these parameters for Cs can be found in [30]: $\alpha_d = 15.0$, $\alpha_q = 230$, $h = 2$, $B = 14.76732$, $C = 0.541614$ in atomic units. The normalization factor for $\phi_{short}(r)$ can be found by matching the two solutions in the vicinity of the classical turning point r_x:

$$N\phi_{short}(r_x) = \phi_{long}(r_x). \tag{32}$$

When a Cs atom is placed in a helium matrix, the *shape* of the wave function at short distances will not be significantly changed, and the change in the amplitude can be found from Eq. 32, giving

$$\frac{A_{hfs}^{free}}{A_{hfs}^{He}} = \frac{|\psi_{free}(0)|^2}{|\psi_{He}(0)|^2} \approx \frac{|\phi_{free}(r_x)|^2}{|\phi_{He}(r_x)|^2} \tag{33}$$

The presence of the helium matrix is taken into account by adding additional terms to the H_0 Eq. 29:

$$H = H_0 + \int d^3\mathbf{R}\, \rho(\mathbf{R})V_{e-He}(\mathbf{R} - \mathbf{r}) \tag{34}$$

The electron-helium interaction potential consists of two parts: a *local* polarization part $V_{loc}(|\mathbf{R} - \mathbf{r}|)$ and a *non-local* pseudopotential part $V_{ps}(\mathbf{R} - \mathbf{r})$.

The local part of the potential contains usual polarization terms of the form given by Eq. 30 with $\alpha_d^{He} = 2.3265$, $\alpha_q^{He} = 0.8573$, $h^{He} = 1.0$, three-body (electron-Cs core-helium) cross-term

$$V_{CT}(\mathbf{R}, \mathbf{r}') = -\frac{\alpha_d^{He}\xi}{(R^2 + h^2)(r'^2 + h^2)} + \frac{1}{2}\frac{\alpha_q'(3\xi - 1)}{(R^2 + h^2)^{3/2}(r'^2 + h^2)^{3/2}}, \tag{35}$$

where $\mathbf{r}' = \mathbf{R} - \mathbf{r}$, $\alpha_q' = 5.0933$, and $\xi = \hat{\mathbf{r}}' \cdot \hat{\mathbf{R}}$.

The helium atom has a closed s-shell. Therefore, an electron whose wave function is an s-wave when projected onto the helium core will feel an additional repulsion, the so called Pauli force. For electrons with $l \neq 0$ the interaction will be attractive due to incomplete screening of the nucleus and the polarization of the helium core. This effect is taken into account by the non-local part in V_{e-He}.

$$V_{ps}(\mathbf{r}') = \sum_{l=0}^{\infty}\sum_{m=-l}^{m=l} B_{He}^l \exp(-C_{He}^l r'^2) |Y_m^l(\hat{\mathbf{r}}')\rangle\langle Y_m^l(\hat{\mathbf{r}}')|, \tag{36}$$

where $Y_m^l(\hat{\mathbf{r}}')$ is a spherical harmonic centered on the helium atom, $B_{He}^{l=0} = 2.03$, $C_{He}^{l=0} = 0.463$, and $B_{He}^{l>0} = -1.0$, $C_{He}^{l>0} = 1.0$.

Fortunately, one does not have to perform an infinite summation over l. Using the property of completeness of the system of spherical harmonics Eq. 36 can be reduced to

$$V_{ps}(\mathbf{r}') = \left(B_{He}^{l=0}\exp(-C_{He}^{l=0}r'^2) - B_{He}^{l>0}\exp(-C_{He}^{l>0}r'^2)\right)|Y_0^0(\hat{\mathbf{r}}')\rangle\langle Y_0^0(\hat{\mathbf{r}}')| + B_{He}^{l>0}\exp(-C_{He}^{l>0}\, r'^2). \tag{37}$$

The last term contains no projection operators and can be included into the local part of the potential. The remaining non-local part of the form $V(\mathbf{r}')|Y_0^0(\hat{\mathbf{r}'})\rangle\langle Y_0^0(\hat{\mathbf{r}'})|$ does not allow to write down the Schrödinger equation in closed form, since this part of Hamiltonian depends on the shape of the ψ-function. We therefore have to use an iterative procedure:

- We start with a numerical integration of the radial Schrödinger equation for the free Cs atom. This gives us the zero-order approximation $\phi^{(0)}(r)$.

- Using this $\phi^{(0)}(r)$ and a helium density profile $\rho_0(R)$ in the form (9) with some "reasonable" starting values for the parameters R_0 and α we get an expression for the He-interaction part of the Hamiltonian.

- We can now get the next approximation for $\phi(r)$ by integrating the Schrödinger equation with the Hamiltonian obtained in the previous step. This also gives us the atomic energy E_{atom}

- The "outer" loop of the iteration is the variational minimization of the total energy $E_{atom} + E_{bubble}$ with respect to the bubble shape parameters R_0 and α.

After convergence this procedure yields the radial part of the electron wave function at large distances $\phi_{He}(r)$, the atomic energy and the equilibrium bubble shape. The shift of the hyperfine transition frequency is then found from Eq. 33. The projection integrals, which arise in the second step of the integration procedure are easy to calculate for the Cs ground state wave function:

$$\langle\phi(r)Y_0^0(\hat{\mathbf{r}})|Y_0^0(\hat{\mathbf{r}'})\rangle\langle Y_0^0(\hat{\mathbf{r}'})|\phi(r)Y_0^0(\hat{\mathbf{r}})\rangle = \frac{1}{4\pi}\left|\int d\Omega_{\hat{\mathbf{r}'}}\,\phi(|\mathbf{R}+\mathbf{r}'|)\right|^2$$

The results of the calculations for the shift of the hyperfine transition frequency

$$\frac{\nu_{hf}^{He} - \nu_{hf}^{free}}{\nu_{hf}^{free}}$$

are given in Table 2.

Experimental conditions	hfs shift theory	hfs shift experiment
He II, SVP at 1.5 K	0.54%	0.6% [28]
Solid He, 27 bar	1.53%	2.2% [29]

Table 2: Comparison between theory and experiment for the shift of the relative hf-transition frequency of Cs atoms in condensed helium

8. Summary and outlook

In this lecture we have discussed some physical properties of atomic point defects in condensed helium matrices. We have shown that for atoms of the first two groups of the periodic table these properties can be explained in most cases within the simplest, so called "bubble model", which treats helium around the atomic defect as a continuous medium. The experimentally observed optical and microwave spectra are in reasonable agreement with the results of model calculations. It is the quantum nature of the helium matrix which makes this "jelly" model applicable on a sub-nanometer scale. Remarkably, the model is successful even in the case of *solid* helium. Nevertheless, one should be careful while applying it to other elements. The example of Na atoms in the excited $3P$ state shows, that this model is only valid when the depth of the attractive part of the foreign atom — helium potential does not significantly exceed that of the helium — helium potential. If this is not the case, several helium atoms can be strongly localized in the vicinity of this deep attractive minimum and one has to use a molecular-like model for the description of the formed complex.

One very important consequence of the softness of helium the matrix is that the symmetry of the trapping site reflects the symmetry of the implanted atom. In particular atoms in spherically symmetric s-states form spherical cavities. The central symmetry of the Hamiltonian of an atom in such a cavity will be preserved. As a consequence, the total angular momentum of the atom remains a good quantum number, and the selection rules for the optical transitions based on the conservation of the angular momentum remain valid [3]. This opens a very promising root of investigations not covered in the present lecture - the study of spin properties of implanted paramagnetic atoms. Optical pumping, magnetic level crossings and optical - r.f. double resonance experiments have been recently demonstrated with alkali atoms in superfluid [32] as well as in solid [33] He matrices. Extremely long electronic spin relaxation times exceeding 1 second have been observed [34]. It was actually this perspective, which has been at the basis of our interest in helium matrix isolated atoms, as it may eventually lead to a novel experimental approach to search for permanent atomic electric dipole moments forbidden by parity conservation and time reversal invariance [35].

9. Acknowledgments

The authors gratefully acknowledge the contributions of M. Arndt, S. Lang, R. Dziewior, and S. B. Ross to the experimental results presented in this paper as well as the constant encouragement and support of this work by Prof. T. W. Hänsch. One of us (S.I.K.) thanks the Max-Planck-Institut für Quantenoptik for hospitality and financial support. This work was supported in parts by the EC program "Human Capital and Mobility" (Grant # ERBCHRX CT 930366).

---

[3]Note that this may not be the case for atoms with higher angular momentum. A possible indication for that in the case of metastable magnesium atoms was reported in [31]

10. References

1. Niebel, K.F. and Venables, J.A. (1976), The crystal structure problemin Klein, M.L. and Venables, J.A. (eds.), *Rare Gas Solids*, Academic Press, London, pp. 558–589.
2. Williams, R.L. (1957), Ionic Mobilities in Argon and Helium Liquids, *Canadian J. Phys.* **35** 134–146.
3. Meyer, L. and Reif, F. (1958), Mobilities of He Ions in Liquid Helium, *Phys. Rev.* **110**, 279–280.
4. Landau, L.D. and Lifshitz, E.M. (1980) *Course of Theoretical Physics* **9**, Statistical Physics, Part 2, p.88, Pergamon Press, Oxford.
5. Atkins, K.R. (1959), Ions in Liquid Helium, *Phys. Rev.* **116**, 1339–1343.
6. Woolf, M.A. and Rayfeld, G.W. (1965), Energy of Negative Ions in Liquid Helium by Photoelectric Injection, *Phys. Rev. Lett.* **15**, 235-237.
7. Fowler, W.B. and Dexter, D.L. (1968), Electronic Bubble States in Liquid Helium, *Phys. Rev.* **176**, 337-343.
8. Grimes, C.C. and Adams, A. (1990), Infrared Spectrum of the Electron Bubble in Liquid Helium, *Phys. Rev.* B41, 6366–6371; (1990), Infrared Spectrum of the Electron Bubble in Liquid Helium, *Phys. Rev.* B45, 2305–2310.
9. Parshin, A.Ya, and Pereversev, S.V. (1992), Spectroscopic Study of Excess Electrons in Liquid Helium,*Sov. Phys. JETP* **74**, 68–76.
10. Golov, A.I. and Mezhov-Deglin, L.P. (1992), Measurement of the Infrared Absorption Spectrum by Negative Charges in Solid Helium, *Sov. Phys. JETP Lett.* bf 56, 514–516.
11. Dupont-Roc, J. (1995), Excited p-states of Alkali Atoms in Liquid Helium, to be published in *Proceedings of Symposium on Ions and Atoms in Superfluid Helium, Heidelberg, Z. Phys.* **B**.
12. Cole, M.W. and Bachman, R.A. (1977), Structure of Positive Impurity Ions in Liquid Helium, *Phys. Rev.* **B15**, 1388–1394.
13. Czuchai, E. unpublished.
14. Hickman, A.P., Steets, W. and Lane, N.F. (1975), Nature of Excited Helium Atoms in Liquid Helium: A Theoretical Model, *Phys. Rev.* **B12**, 3705–3717.
15. Hiroike, K., Kestner, N.R., Rice, S.A. and Jortner, J. (1965), Study of Properties of an Excess Electron in Liquid Helium. II. A Refined Description of Configuration Changes in the Liquid, *J. Chem. Phys.* **43**, 2625–2632.
16. Pascale, J. (1983), Use of l-dependent Pseudopotentials in the Study of Alkali-Metal-Atom – He Systems, *Phys. Rev.* **A28**, 632–644.
17. Bauer, H., Beau, M., Friedl, B., Marchand, C., Miltner, K. and Reitner, H.J. (1990), Laser Spectroscopy of Alkaline Earth Atoms in HeII, *Phys. Lett.* **A146**, 134–140.
18. Kinoshita, T., Fukuda, K., Takahashi, Y. and Yabuzaki, T. (1995), Optical Properties of Impurity Atoms in Pressurized Superfluid Helium, to be published in Günther, H., zu Putlitz, G., Tabbert, B. (eds.) *Proceedings of the Symposium on Ions and Atoms in Superfluid Helium, Heidelberg, Z. Phys.* **B**.

19. Lerner, P.B., Chadwick, M.B., and Sokolov, I.M. (1993), Inhomogeneous Broadening of Electronic Transitions in liquid Helium Bubble: The Role of Shape Fluctuations, *J. Low Temp. Phys.* **90**, 319–328.

20. Anderson, P.W., (1952), *Phys. Rev.* **86**, 809.

21. Kanorsky, S.I., Arndt, M., Dziewior, R., Weis, A., and Hänsch, T.W. (1994), Pressure Shift and Broadening of the Resonance Line of Barium Atoms in liquid Helium, *Phys. Rev.* **B50**, 6296–6302.

22. Kanorsky, S.I., Weis, A., Arndt, M., and Hänsch, T.W. (1994), Pressure Dependence of Atomic Resonance Lines in Liquid and Solid Helium, to be published in Günther, H., zu Putlitz, G., Tabbert, B. (eds.) *Proceedings of the Symposium on Ions and Atoms in Superfluid Helium, Heidelberg, Z. Phys.* **B**.

23. Dupont-Roc, J., Himbert, M., Pavloff, N., and Treiner, J. (1990), Inhomogeneous Liquid Helium ^{4}He: A Density Functional Approach with a Finite-Range Interaction, *J. Low Temp. Phys.* **81**, 31–44.

24. Ancilotto, F., Cheng, E., Cole, M.W., and Toigo, F. (1995) The Binding of Alkali Atoms to the Surface of Liquid Helium and Hydrogen, to be published in Günther, H., zu Putlitz, G., Tabbert, B. (eds.) *Proceedings of the Symposium on Ions and Atoms in Superfluid Helium, Heidelberg, Z. Phys.* **B**.

25. Aziz, R.A., and Slaman, M.J. (1991) An Examination of *ab initio* Results for the Helium potential Energy Curve, *J. Chem. Phys.* **94**, 8047–8053.

26. zu Putlitz, G. and Beau, M. (1992), Dye Laser Spectroscopy of Isolated Atoms and Ions in Liquid Helium, in Struke, M. (ed.) *Dye Lasers 25 Years*, Topics in Applied Physics **70** Springer, Berlin.

27. See Günther, H., zu Putlitz, G., Tabbert, B. (eds.) *Proceedings of the Symposium on Ions and Atoms in Superfluid Helium, Heidelberg 1995, Special Issue Z. Phys.* **B**.

28. Takahashi, Y., Fukuda, K., Kinoshita, T., and Yabuzaki, T. (1995) Sublevel Spectroscopy of Alkali Atoms in Superfluid Helium, to be published in Günther, H., zu Putlitz, G., Tabbert, B. (eds.) *Proceedings of the Symposium on Ions and Atoms in Superfluid Helium, Heidelberg, Z. Phys.* **B**.

29. Lang, S., Kanorsky, S.I., Arndt, M., Ross, S.B., Hänsch, T.W., and Weis, A. (1995) The Hyperfine Structure of Cs Atoms in the b.c.c. Phase of Solid ^{4}He, *Europhys. Lett.* **30**, 233–237.

30. Bradsley, J.N. (1974) Pseudopotentials in Atomic and Molecular Physics, *Case Studies in Atomic Physics* 4(5), 299–368.

31. Günther, H., Foerste, M., Hönninger, C., zu Putlitz, G., Tabbert, B. (1995) Lifetime of Metastable Magnesium Atoms in Superfluid Helium, to be published in Günther, H., zu Putlitz, G., Tabbert, B. (eds.) *Proceedings of the Symposium on Ions and Atoms in Superfluid Helium, Heidelberg, Z. Phys.* **B**.

32. Yabuzaki, T., Kinoshita, T., Fukuda, K., and Takahashi, Y. (1995) Laser Spectroscopy and optical Pumping of Alkali Atoms in Superfluid Helium, to be published in Günther, H., zu Putlitz, G., Tabbert, B. (eds.) *Proceedings of the Symposium on Ions and Atoms in Superfluid Helium, Heidelberg, Z. Phys.* **B**.

33. Weis, A., Kanorsky, S.I., Arndt, M., and Hänsch T.W (1995) Spin Physics in Solid Helium: Experimental Results and Applications, to be published in Günther, H., zu Putlitz, G., Tabbert, B. (eds.) *Proceedings of the Symposium on Ions and Atoms in Superfluid Helium, Heidelberg, Z. Phys.* **B**.

34. Arndt, M., Kanorsky, S.I., Weis, A., and Hänsch, T.W. (1995) Long Electronic Spin Relaxation Times of Cs Atoms in Solid ^{4}He, *Phys. Rev. Lett.* **74**, 1359–1362.

35. Arndt, M., Kanorsky, S.I., Weis, A., and Hänsch, T.W. (1993) Can Paramagnetic Atoms in Superfluid Helium Be Used to Search for Permanent Electric Dipole Moments?, *Phys. Lett.* **A174**, 298–303.